SELECTIONS FROM THE CHRONICLE

The Fascinating World of Early Tools and Trades

This Great Seal of the Early American Industries Association was designed by Harvey Jeacock. There are thirteen different tools illustrated, one for each of the thirteen original colonies, and each representing an early trade or occupation: the farmer, silversmith, blacksmith, tinsmith, wheelwright, shipwright, housewright, weaver, cooper, shoemaker, printer, currier, and cabinetmaker.

The CHRONICLE
Founded in 1933

SELECTIONS FROM THE CHRONICLE
The Fascinating World of Early Tools and Trades

Introduction by
Charles Hummel
Deputy Director for the Museum and
Library Department, Winterthur

Preface by
Elliot Sayward
Editor of *The Chronicle*

Edited by
Emil Pollak and Martyl Pollak

THE ASTRAGAL PRESS
Mendham, New Jersey

Library of Congress Catalogue Card Number:
91-72503
International Standard Book Number:
1-879335-00-X

Jacket design by Nina Gaelen

Manufactured in the United States of America

Published by
The Astragal Press
P.O. Box 239
Mendham, New Jersey 07945-0239

Table of Contents

PREFACE

ELLIOT M. SAYWARD
Editor of *The Chronicle*
March 1991

In the introductory chapter to his innovative work, *The Colonial Craftsman*, (1950) Carl Bridenbaugh said, "Crafts—and the artisans who practice them—played a most important part in early American life, *even more than historians have generally supposed.*" While the emphasis on his last seven words is mine, I do not doubt that Bridenbaugh would find it appropriate. He had already said in his Preface that he had long wanted "to read a work about colonial craftsmen, but no such treatment existed."

In order to fully understand the life of the early craftsman, it is necessary to resort to documentation beyond the written or printed word, documentation that lies in the objects which the worker used to perform his or her work.

To state it another way, in order to understand the facts of the early workers' existence, one most have reference to the implements they used in the pursuit of their livelihoods. These implements are the documents of everyday history. Only through their collection, identification, and preservation can we continue to understand our past.

In his work on the Dominy craftsmen of East Hampton, New York, *With Hammer In Hand,* Charles Hummel dedicates one hundred and seventy pages to an illustrated catalog of the tools used by this important family of wood and metal workers. In his Preface, Hummel says, "... discovery and preservation of tangible evidence documenting our American cultures have been left to antiquarians, collectors of antiques, and museum personnel. Objects that these people have preserved provide flesh for the dry bones of history by adding dimension to the written work."

The lesson that history must be concerned not only with famous people and great events, but also with the everyday activities of average men and women, has been a long time sinking into the consciousness of historians and laymen alike. However, it is a seldom noted truth that the dedicated amateur has often led the professional in the early advancement of many disciplines.

It is often left to lay people, unfettered by what the trained expert already "knows," to make the departures from accepted doctrine that allow new ground to be opened.

Schliemann, a retired business man, found Homer's City of Troy beneath a mound called Hissarlik in Turkey near the mouth of the Dardenelles in spite of the opinions of the archaeological fraternity that he was wasting his time. The professionals, once shown the way, then advanced Schliemann's discoveries.

The Schliemann of early tools and trades and the man who first pointed out to us the value of the nation's junk piles of old implements was Henry Chapman Mercer whose book, *Ancient Carpenters' Tools*, first published in 1929, opened the way—at least in the United States—to the study of a subject that had hitherto been beneath notice. University trained, Dr. Mercer became a professional anthropologist whose credentials in the sister science, archaeology, are attested by his early curatorship of American and prehistoric archaeology at the University of Pennsylvania. The field he was about to enter in 1897 was in no way part of either the anthropology or the archaeology of the times, although both disciplines contributed much to the manner in which he was to handle his new interest. Mercer went out one day to find a pair of antique fire tongs. Visiting a dealer in "Penny Lots" acquired at country auctions he found himself, as his long-time disciple, Joseph Sandford tells us, ". . . in the midst of a hoard of obsolete utensils ranging from wagons to tin dinner horns." Mercer was enchanted. In the next seventeen years he acquired an enormous collection of these things and virtually founded a new field of study, eventually to be called "historical archaeology." Mercer himself called it "archaeology turned upside down."

In 1913, the building of the Mercer Museum was begun in Doylestown, Pennsylvania. It was to house what would become the largest collection of the implements of everyday life ever assembled. Mercer soon was not alone. In 1927, as a result of the work of Dr. W. A. R. Goodwin and the financial support of John D. Rockefeller, Jr., both amateurs in the cause of history, the restoration of the colonial town of Williamsburg in Virginia was begun. Colonial Williamsburg presents, as never before, a merging of great historical events with daily life as it was lived in 18th century Virginia. These and other developments heralded the burgeoning interest in a broader view of the past that is typified by the words of Carl Bridenbaugh quoted in the sentence that began this Preface.

It can be seen that new approaches to history were brewing when, on August 31, 1933, sixteen men and women, collectors of antique artifacts and amateurs all, sat down to luncheon in the Wiggins Tavern in Northampton, Massachusetts. They were there for a purpose that reached far beyond the enjoyment of a midday meal. When they rose, they were founding members of the Early American Industries Association, an organization that was soon to state its reason for being as: ". . . to encourage the study and better understanding of early American industry, in the home, in the shop, on the farm, and on the sea, and especially to discover, identify, classify, preserve and exhibit obsolete tools, implements, utensils, instruments, vehicles, appliances and mechanical devices used by American craftsmen, farmers, housewives, mariners, professional men and other workers."

Since that day, the Association has created an enviable record. It has published, either as principal or co-publisher, more than forty books and significant booklets. It has awarded dozens of Grants-in-Aid to people pursuing research and publication in subjects within the context of the Association. It has established a Library of nearly 2,000 volumes plus films, tapes and ephemera, all supporting the Association's purpose. It has taken part in scores of "events"—exhibits, seminars, programs. It has served as a focus and a forum for the fraternity of implement collectors and students of trades and crafts. It has established *The Directory of American Tool Makers*, certainly the most ambitious project of its 58 years. Perhaps most important, from a date only two-and-a-half months after its founding, the Association has published a quarterly journal, *The Chronicle* of the Early American Industries Association.

Under the *Chronicle*'s first editor, Steven C. Wolcott, standards were established which have maintained the high quality of information published for 58 years. Over this period, *The Chronicle* has been guided by eleven editorships. These were: Steven C Wolcott; William B. Sprague; John Davis Hatch; Josephine H. Peirce; Minor Wine Thomas, Jr. & William D. Geiger; William D. Geiger & Raymond R. Townsend; Raymond R. Townsend; Daniel B. Reibel; John S. Kebabian; Elliot M. Sayward & Harvey F. Jeacock; Elliot M. Sayward. There have also been various sub-editors and editorial assistants. I regret that there is insufficient room to list all of these, especially since some of them were never carried on the masthead.

Rich as it has been in editors, *The Chronicle* has been vastly richer in contributors. Almost all of the non-editorial content of every *Chronicle* since the first issue has been produced by members of the association on a volunteer basis. The number of contributors and the variety of subject matter they have produced has been extraordinary. Evaluated against the level of knowledge at the time each contribution was submitted, the quality of material has been outstanding.

Of course, the EAIA did not long remain an organization exclusively composed of amateurs. Very soon its ranks were expanded and enriched by academic and museum people and the institutions to which they belong. And very soon *The Chronicle* became an important source of information not just to collectors and early technology enthusiasts, but to historical societies, museums, restorations, and a host of professionals, teachers, writers and the like, who had previously enjoyed little access to this type of information.

This new information covered a tremendous variety of subjects. There have been studies of activities across the whole spectrum of trades and crafts and reaching beyond, including farming, hunting, fishing, seamanship, household affairs, power and machinery, food preservation, distillation and brewing and many more. Conservation, preservation, display, record keeping, and all the other procedures related to the collection of artifacts have been discussed. The pre-American roots of our occupations have been examined, and their evolution over some 300 years of our presence in America has been charted. One of the most challenging and entertaining categories has been the identification of objects whose functions have been forgotten. Called "What-sits" today, such mysteries were first offered for solution in *The Chronicle* in the year following EAIA's founding and have reappeared regularly ever since.

The importance of *The Chronicle* to the developing interest in the tools and technology of a bygone day was enormous, but its influence was slow in generating a literature outside its own pages. At the time of the Twenty-fifth Anniversary of EAIA in 1958, it was still possible for Loring McMillen, who outlined the Association's history in a special Anniversary issue, to deplore that as yet "the sole scholarly writings concerning the tools of our Early American Industries" were the articles from *The Chronicle* and Dr. Mercer's book, *Ancient Carpenters' Tools*. The ensuing twenty-five years was to see a major change, but one which was led and inspired by the men and women who constituted the Association and wrote its *Chronicle*.

Indeed, this volume is the brainchild of one of those men and women, Emil Pollak, whose interest in the long-term aims of the Association has led to the founding of Astragal Press, one of the still relatively few publishing endeavors concentrating on the documentation of history by the study of the tools and other artifacts of the past. The selection of *Chronicle* articles reprinted here gives a fascinating introduction to the concerns of the material historian and collector, both amateur and professional. *Selections From The Chronicle* provides both great reading and a wealth of information to all who have an urge to understand how life was lived and work was done in the days before their oldest living relatives were born. For some, they will be the inspiration for a new avocation. For all, they will be an enriching experience.

INTRODUCTION

CHARLES F. HUMMEL
Deputy Director for the Museum and Library Department, Winterthur
March 1991

In the Fall of 1959, I traveled to Colonial Williamsburg to attend my first meeting as a member of the Early American Industries Association. The impetus for joining the organization was its quarterly publication, *The Chronicle*.

A year earlier, I had volunteered to undertake a task that no one else at Winterthur was interested in doing—the sorting, identification, cataloguing, and installation of a large collection of handcraft tools ranging in date from 1750 to 1860. Much of the information that I was able to discover about various tools owned by three generations of craftsmen in the Dominy family in East Hampton, Long Island, could only be found in the pages of *The Chronicle*.

Reference books then available were quite few: Henry Chapman Mercer's *Ancient Carpenters' Tools*, booklets about tool collections at the Shelburne Museum by H. R. Bradley Smith and Frank Wildung, pamphlets by the research staff at Colonial Williamsburg, and William L. Goodman's *The History of Woodworking Tools*. Salaman's excellent dictionaries of tools in various crafts and Paul Kebabian's wonderful *American Woodworking Tools* were still years away from completion.

Then, as now, (peruse the Table of Contents of this volume), one turned to the pages of *The Chronicle* for information about tools used by blacksmiths, shipwrights, coppersmiths, tinsmiths, brass and iron founders, tanners, coopers, wheelwrights, cordwainers, pumpmakers, housewrights, lumbermen, and colliers. As I quickly discovered at my first and subsequent meetings, the forms, functions, and names of tools and equipment no longer in universal use in the United States were preserved in the collective memories of members and friends of the Early American Industries Association and in organizations that were founded more recently than 1933.

The very democratic nature of this organization encouraged contributions from carpenters, moldmakers, machinists, bankers, doctors, librarians, curators, entrepreneurs, teachers, and people from all walks of life. Male and female!

A desire to share information has characterized membership of EAIA during its history with which I am familiar. As current editorial policy of *The Chronicle* indicates, "most people write a lot better than they think they do." To encourage the preservation of knowledge in print, members are asked not to worry about how to write, but, instead, to get their information on paper, then in print, preferably in the pages of *The Chronicle*.

Not that it has always been easy to obtain articles. Bless you, "J. Didsbury," the pseudonym under which Raymond Townsend, long-time editor of *The Chronicle*, often wrote on a variety of craft-related subjects because a quarterly issue was due and not enough articles had been received to complete a particular issue.

The Chronicle, therefore, is a reflection of a collectors' organization with an ongoing commitment to research and publication. Its members' joys include aesthetic considerations (tools are beautiful) as well as obtaining information about obsolete tools, equipment, machines, and processes. "Whatsits" was coined by EAIA as a regular feature for *The Chronicle* and as a regular feature of its meeting.

A Grants-in-Aid Program encourages researchers to finish their work and make it available through publication. Linda LeMieux's article on "The Salem Waterworks," reprinted in this volume, is an example of research undertaken and completed with the help of an EAIA grant-in-aid.

The selection of articles from *The Chronicle* does reflect the fact that almost all of the material that appears in it is of a practical nature. Few articles of a conceptual or philosophical nature have been submitted to it for publication. That is not a factor of editorial policy, but rather a reflection of a broad scope of its members' interest and the strong "pull" of a journal published by a collectors' organization to be concerned about identification and use of tools.

If a slight criticism can be tolerated, it is that this selection of articles from *The Chronicle* does not reflect the fact that some EAIA members have always had an interest in machinery—big machinery. Agricultural machinery, mechanical lathes—both metalworking and woodworking—sewing machines, early jig saws, etc. The December 1990 issue, for example, contains a brief note about Koehring concrete mixers.

Of necessity, too, a number of articles in the last decade have begun to chronicle the use of hand tools in late-nineteenth and early twentieth-century occupations. Material relating to electricians', plumbers', and automobile mechanics' tools appear with some frequency. But these are absent from this selection.

This introduction began with a brief reference to attending my first EAIA meeting in 1959. It was not uncommon, then, for critics of EAIA to refer to it as a "chowder and marching" society. Admittedly, there were some truly dedicated, fun-loving, high-living spirits who regularly attended meetings. My second edition, unabridged, *Random House Dictionary of the English Language,* however, defines chowder as a "thick soup or stew made of clams, fish, or vegetables, with potatoes, onions, and other ingredients and seasonings." The membership of the Early American Industries Association is certainly made up of many "ingredients and seasonings." *The Chronicle,* while not always representing a perfect image of its membership, has reflected the broad range of knowledge and information represented by its members and their collections.

It has been, and is, the "marching" vehicle for transmitting that knowledge and information. Since 1933 we are all better for its existence.

Hoofbeats of Destiny

By ANNA K. CUNNINGHAM

July 1951

At the time she wrote this article, Ms. Cunningham was Curator of History at the University of the State of New York.

This is the story of Henry Burden, a young Scotch-immigrant engineer of Troy, New York, who was the first man in America to devise a mass production method for a transportation necessity. He invented a machine which could make a horseshoe from a bar of iron in four seconds.

Across the Forth River from Scotland's ancient capital of Stirling, and somewhat west of the road that runs north from Glasgow to Perth, lies the little town of Dunblane. This was the home of the Burden family—the father was a yeoman farmer—and here on the family homestead Henry Burden was born April 22, 1791.

He was an inventive genius even as a boy working around the farm. He figured out a better method of threshing; he tinkered and mended farm equipment, the neighbors' as well as the Burden's. As he grew older, he wanted to become a trained engineer; so evenings he went to the home of a local scholar, William Hawley, whom the countryside had long respected as an accomplished arithmetician. In the daytime, young Burden was busy erecting grist mills and making various farm implements. The pupil soon outstripped the master and went on to Edinburgh to study engineering, mathematics and drawing.

Equipped with letters of introduction to United States Senators Thomas Benton and John Calhoun and to Stephen Van Rensselaer of the powerful patroon family, Henry Burden landed in America in 1819 and went to work for the Townsend and Corning Iron Shops of Albany, New York. Twenty-eight years of age at this time, a trained and seasoned engineer, it was said of him later, "An accomplished mechanic, he could make a better piece of work than any man he could find in his shops; he could deal a heavier blow with the sledge than any of his strikers at the forge. Thus physically favored, his business forecast was only equalled by his genius for invention, and his productions and triumphs are acknowledged as among the most important and revolutionary known to the history of iron manufacture."

Across the river from Albany, in the early 1800's, a little mill, known as the "Troy Iron and Nail Factory," had been started by a local stock company. Three years after he came to work in Albany, the stockholders of the Troy mill asked young Burden, who meanwhile had been winning prizes at county fairs with improved agricultural equipment, to take over the superintendency there. And from this small beginning was to grow the giant Burden Iron Works so vital to the Federal Government during the Civil War.

So important was this single plant in the turbulent 1860's that at one time the United States government contemplated a step regarding it which was eventually to be taken with the railroads during World War I—government management. So coveted were the secrets of the Burden Iron Works that the Confederacy in its last dark days was to hatch a plot to infiltrate spies at the Troy plant to learn them.

But in 1822 it was just a little nail factory with insufficient waterpower. And a new superintendent who had a thousand ideas. He had been mulling many of them over for a long time and was still to work long years on some of them. One of his pet theories was that horseshoes could be made mechanically.

Up until the invention of the Burden machine, horseshoes were pounded out by the blacksmith by hand on his anvil. But it was a slow way to make a much-needed article. And so the young superintendent of the little Troy Iron and Nail Factory set to work figuring, drawing and re-figuring. In 1835, the United States Patent Office issued to "Henry Burden, of Troy, New York, Letters Patent," covering a "Machine for Making Horseshoes." America, and the world, would move faster now that old Dobbin could be more quickly shod.

Henry Burden perfected his patent as the years passed. He wanted to construct a machine which could take a bar of iron from the roll-train and finish a shoe without re-heating. So, in 1843 he added improvements to his original design, which reduced

BURDEN'S NEW PATENT HORSE-SHOE MACHINE.

the operations to two movements; and again in 1857, whereby, after receiving the heated bar, the machine cut, bent and forged it into a perfectly shaped shoe in one movement. Indeed, his last of many patents, in 1862, was that for final improvement on the horseshoe machine.

In an age of radar, armored tanks and atomic power, to mention the political importance of manufacturing horseshoes by machinery may seem like a bit of local boasting, but if Henry Burden had not discovered this mechanical method by which hundreds of thousands of sets of shoes could be made for the great number of horses and mules used by the Northern armies, it is very likely that some of the most important cavalry movements of the Civil War, and they were advantageous to Federal strategy, could not have been made. Nor could the wagon trains with the Union provisions have moved so readily, for, except during the first few months of war, the Federal Blue was an invading army. It might be said that in a large measure the success of the Union army rode upon the famous Burden horseshoe.

The Burden Iron Company sent a replica of this first horseshoe machine to the Centennial Exposition in Philadelphia in 1876. It was badly placed and to this circumstance we are indebted for a particularly clear, non-technical, description of the machine which an irate *New York Times* reporter wrote and his newspaper published. He said:

"I do not propose to let this perfect piece of machinery remain hidden away, and in case it be too late to have it installed in its proper position in Machinery Hall, which ought surely not to be the case, I will tell the readers of *The New York Times* all about it and its wonderful doings. In the first place the ordinary working machine is neither large nor complicated. On the contrary its simplicity is one of its leading features. The two things about it which first strike the eye are some wheels and a long reheating furnace adjoining them, the latter looking very much like a horizontally built chimney. Bars of iron forty feet in length are brought hot from the rolling mill and are placed in the reheating furnace for the purpose of annealing the iron. As soon as this process is completed, the bar of iron is passed into the machine through two rollers, which act as automatic feeders, the rollers being kept in constant pressure on the bar by an attachment to the cutting lever, which, working on a cam, cuts off the bar just the necessary length of iron to make one horseshoe. At the moment of cutting off, the bending tongue catches the piece cut off and carries it into the first die, which gives the shape and form to the shoe. It then passes on to another roll and shaft, called the creasing shaft, on which is a die to which the creasers are attached, and by means of which the creasing of the shoe is produced, while, at the same time, the holes for the nails are pierced. Here is then a finished shoe for horse or mule, which has only to be heated and fitted by the local blacksmith before being used. One great advantage of this machine is that any shoe of any shape, size or pattern can be made upon it with the utmost ease. A buyer may send a shoe of peculiar pattern, and one which would puzzle a blacksmith to make, with a secret private mark upon it, accompanied by an order. His order, no matter if it amounts to 1,000 tons, will be delivered to him exactly to his pattern, any possible variation in a single shoe in the whole consignment requiring a microscope to discover it."

When the Burden horseshoe machine was first put into operation and word spread abroad of the new invention, the countryside flocked around to see it work and marvel at it. Production at the mill soared. A capacity of 51,000,000 shoes annually was to be achieved as the years passed. Two storage warehouses with space of 250,000 kegs of shoes were built. Nine machines were in operation, each ultimately making sixty shoes a minute. An army of more than twelve million horses could be shod every year with the product of the Burden works. Where were these shoes sold? Everywhere throughout the United States and Canada. On a loading platform of the warehouse one day might be seen hundreds of kegs tagged "San Francisco, California;" on another, a shipment consigned to "Portland, Oregon." Western desperadoes probably "went thattaway" riding hell for leather on Burden horseshoes.

Today, on Mill Street in Troy, not too far from the old Wynantskill that long ago furnished power for the little Troy Iron and Nail Factory, there is a bronze tablet on which is this inscription:

Near This Place
in
1835, 1843, 1857
HENRY BURDEN
(1791-1871)
First operated machines which he had invented for the manufacture of
horseshoes and which became an important factor in the
Civil and Military development of the Country.

Blacksmiths' Hammer Signals

By DAVID G. PERCH

March 1977

Editor: Noise, smoke and movement were characteristics of a blacksmith's shop.. Communication between a blacksmith and his helper could at times be difficult. This article describes one solution.

While leafing through a copy of *The Scientific Canadian* for the month of April, 1880, I came upon the following note of blacksmiths' hammer signals and thought that it might be of interst to the members of the EAIA so I am sending it as it was printed:

When the blacksmith gives the anvil quick light blows, it is a signal to the helper to use the sledge, or to strike quicker.

The force of the blows given by the blacksmith's hammer indicates the force of the blow required for the sledge.

The blacksmith's helper is supposed to strike the work in the middle of the width of the anvil, and when this requires to be varied the blacksmith indicates where the sledge blows are to fall by touching the required spot with his hand-hammer.

If the sledge is required to have a lateral motion while descending, the blacksmith indicates the same to the helper by delivering hand-hammer blows in which the hand-hammer moves in the direction required for the sledge to move.

If the blacksmith delivers a heavy blow upon the work and an intermediate light blow on the anvil, it denotes that heavy sledge blows are required.

If there are two or more helpers, the blacksmith strikes a blow between each helper's sledge hammer blow, the object being to merely denote where the sledge blows are to fall.

When the blacksmith desires the sledge blows to cease, he lets the hand-hammer fall upon the anvil, and continues its rebound upon the same until it ceases.

Thus the movements of the hand-hammer constitutes signals to the helper, and what appears desultory blows to the common observer constitutes the method of communication between the blacksmith and his helper.

Shot Tower

By G.A.R. GOYLE

June 1940

We do not know who G.A.R. Goyle was but we feel quite sure that this was not his real name. Ed.

After the Revolutionary War, pioneer living along the eastern seaboard of the United States became more and more secure. The Indians had been subdued and wild animals decimated. From about 1800 on, we find this comparative security reflected in the changing of rifles into the smoothbore shotguns. No longer was it a matter of life and death to be ever alert to shoot and kill for protection from lurking Indians or the onslaught of wild animals. The fire-arm of the frontiersman became a hunting weapon.

But shot-guns required shot, shot which could be produced cheaply, not by the tedious method of individual casting in a mold, or of cutting from sheet lead small cubes to be roughly rounded in a tumbling barrel. A simpler method of casting shot is said to have been invented by Watts, of Bristol, England, about 1782. Watts dreamed one night that he was out in the rain and every drop of it was a round pellet of lead. Pondering over this curious dream, he became interested in the question whether dropping melted lead would form round pellets. To solve the question, he melted some lead and poured it from the top of St. Mary Redcliffe Church into the water below. The test was successful, and he sold the invention for a handsome sum of money.

The new process of making shot was soon introduced in the United States. At the third annual exhibition of the Columbian Society of Artists, held in Philadelphia in 1813, was an exhibit which indicates the building of a shot tower in Richmond at that time. It was the exhibit of the architect Robert Mill, a "Design of the shot tower now building in the city of Richmond, Virginia, diameter 30 feet; elevation to the top of the battlement 160 feet."

There also was a shot tower in the City of New York. Bernhard, Duke of Saxe-Weimar Eisenach, saw it on his travels through North America, during 1825 and 1826, and he described it as follows:

"A high square pyramidal tower attracted our attention; it is a shot tower, one hundred and sixty-six feet high. The melted lead which is thrown through a tin box, whose apertures are suited to the size of shot wanting, falls from the whole height into water; while falling it forms itself into shot and becomes cold as it falls in the water. The different numbers of shot are intermixed; in order to separate the perfect from the imperfect shot, they put them in a flat basin, and by a certain motion in an oblique direction, the perfectly round ones roll down into a receiver, whilst the imperfect remain in the basin. After this they throw the good shot into a box of the shape of a bureau, with rockers like a cradle; the drawers have perforated tin bottoms, the upper drawer has the largest holes, and the lower the smallest; when the upper drawer is filled with shot, it is locked, and then the whole box is rocked for some minutes. Through this the shot is separated according to the size, and I believe there are fourteen different numbers. In order to give the shot a perfectly smooth surface, they throw it into a box which is attached to a wheel turned by water, and in this manner they are rolled for some time. They are then packed according to their number, in bags, and carried into the warehouse."

A plan of Philadelphia in *The American Traveller or Guide Through the United States* (second edition) by H.S.Tanner (Philadelphia, 1836), shows the location of two shot towers, one near the Delaware and the other near the Schuylkill River. Sparks' Shot Tower, in the Southwark district, was in a block south of Christian Street, between Front and Second Streets, and Beck's Shot Tower is marked in the block bounded by Cherry, Arch, Twenty-first and Twenty-second Streets. At that time (1836) the shot towers must have been noteworthy sights of the town, to be specially pointed out on a small-scale plan of the city of Philadelphia.

A lengthy description of a shot tower in Knight's *American Mechanical Dictionary* (New York, 1877) proves that it was then still the recognized means of casting shot. While originally the molten lead had been poured from a colander or sieve, by that time a trough-like ladle with serrated edge had been found more suitable to pour even streams of molten lead down the tower into reservoirs of water.

The inventive spirit did not rest, and the direction in which to look for improvement was, of course, to do away with a cumbersome tower from 150 to 180 feet high, on top of which the furnace had to be kept going to melt the lead. In 1849, a process was patented by one Smith, which directed the dropping of the metal through a tube up which a strong current of air was driven. Glasgow and Wood, of St. Louis, in 1868, patented their process of dropping molten lead through a column of glycerine or oil.

Finally a method was evolved which through its simplicity and efficiency spelled the end of the shot tower. The molten lead was dropped upon a speedily revolving horizontal disk which was fenced in with side walls of perforated sheet brass. The centrifugal force threw the molten lead in uniform drops through the perforations against a surrounding canvas screen. The highly agitated air around the revolving disk cooled the drops sufficiently so that they kept their spherical shape and dropped down as round shot from the canvas.

Subsequent letters to the editor are printed on the following page.

September 1940

Apropos of the article in our last issue, entitled "Shot Tower," Mr. Newton C. Brainard sends us the following, clipped from *The* (London) *Field* of June 22nd, 1940:

BRISTOL'S SHOT TOWER

SIR,—There are only three of the old shot towers left in England: Bristol's (the original, 150 years old); the London one, seen from the Embankment, due for demolition; and the other in Chester.

The shot is formed by dropping the molten lead, containing arsenic, through perforations in the piece of metal shown. This forms into beads underneath, which drop down for 120 ft., solidifying on the way.

The invention dates to 1782, when the method came to one William Watts, in a dream. Once a mansion, he cut the floors of the tower away to allow a clear space down through the wine cellars to a well below. Watts sold the patent for £10,000.

Yours faithfully,

F.R. WINSTONE

Bristol.

September 1940

We are indebted to Mr. J.D. Hatch, Jr., for the following description of Paul Beck's shot tower at Philadelphia, from an entry of May 12th, 1814, in the manuscript diary of Erastus Skinner [the spelling is Skinner's]:

"____Went of Paul Becks Shot Tower; went into it and all over viewed it. The Shot manufactory or Tower is square, 38 feet at bottom and 15 at top. Wall 5 feet thick at bottom and 14 inches at top. 166½ feet high, ascended by 262 steps winding round and inside. At the top is 2 small furnaces, where the lead is melted. And then turns into a copper seive at the top; is shook all the time, and falling so far separates the lead into different sizes. After which the shot are sifted to get the dust and pieces of lead which are not in shape out and then put into a barrel with Blklead where they are rolled or churned to make them smooth and glossey. Afterwards put into a set of seives made of parchment perhaps 12 or 14, one above the other. The top one large enough for Buck Shot, the next smaller, and so down to the lower which is very small. They are all shook at once and the shot go through untill they get to their proper size where they remain and are numbered accordingly. The lead and wood is carried to the top by a windless or tackle drawn by a horse. Have a fine view from the top of the tower—see all the city and country."

A Double Twist Auger

By ROBERT H. CARLSON

March 1969

By 1875, the methods of producing the typical *Spiral Auger* and *Auger Bit* were fairly well standardized. The auger had evolved from a seemingly crude, but nonetheless effective, boring tool, almost completely handmade by local blacksmiths operating under one roof (the factory), to a sophisticated machine-made instrument. Nevertheless, many of the operations depended on, and still do, the personal skill and knowledge of the individual workman. Many complicated dies and processes had been patented by this time, most of them by Connecticut manufacturers, who were the pioneers and leaders in the auger bit business throughout the world. The process, typified by the early Collins Axe Company, of the same workman carrying the tool from the cast steel billet through to the finished article, was now superseded by the forerunner of our modern production line, where each worker is a specialist.

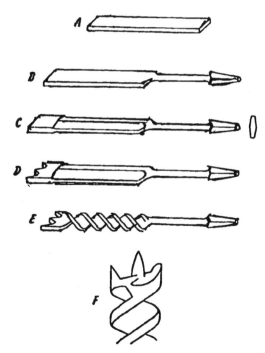

Figure 1. Steps in making a Double Twist Auger. a - Blank. b - Roughing. c - Plating. d - Rough Heading. e - Twisting. f - Final Heading.

Obviously, there were many variations on the norm here described. Each manufacturer had secret or patented processes that he jealously and zealously guarded from his competitors, which he mentioned in glowing but carefully vague terms in his advertisements. Thus, his bit was always "the best." As one catalog put it, "No chocking, no breaking." "No swearing, no tearing." "A bit without bending. A joy without ending."[1]

Nevertheless, there was a basic method. The one we describe is based on information supplied by Martin W. Wright, Jr., retired President (1956-1961) of the Connecticut Valley Manufacturing Company, Centerbrook, Connecticut, who spent fifty-two years in the business, and who also remembers the factory he knew as a boy in the late 1800's.

The raw material for a typical $14/16$ inch bit was a piece of hot-rolled cast steel $1/4$ by $7/8$ by $5 1/2$ inches, *Figure 1-a*. It was cut from mill lengths of 18 inches on a Bradley Helve Hammer. This blank was heated in a forge fire which had been started with charcoal and then fed with chestnut coal, the size favored by the Hammerman. The forges were under forced draft without a hood or stovepipe; consequently, at the start of the work day, the forge room was dense with smoke and full of sparks, creating a tenfold version of Longfellow's *Village Blacksmith, "*And children coming home from school/Look in at the open door/They love to see the flaming forge/And hear the bellows roar/And catch the burning sparks that fly/Like chaff from the threshing floor." But by the time the Hammerman had oiled the bearings on his Bradley Hammer, set out his tongs, and adjusted his iron seat from a discarded mowing machine suspended by the chain from the rafters, the atmosphere had cleared somewhat because of the open doors and the vented cupola on the roof, and the blanks in the forge were a good cherry-red.

The first operation was the forging of the round shank, leaving a long section for the twist, and a short section for the tapered tang, which was formed in the same heat and stamped with the size in sixteenths. This was done to eliminate any possible future errors as the various sized bits went through subsequent operations. Remember, these bits were calibrated in sixteenths; and in the rough form, it would be only too easy to misjudge size. At this point, the bit had the form of *Figure 1-b,* and had been stretched to $9 1/2$ inches. Next the Plating Dies were set, and the Hammerman, on the next heat, trimmed and drew out the Flat, which became the twist, to the shape shown in *Figure 1-c.* The reason for this shape is to make the finished bit wider in the center, and thus stronger and stiffer. If this description of the process seems long, the efficiency and speed of a good experienced Hammerman seems short; he could, and did, produce some 600 bits a day. (Besides using his 80 pound hammer to crack his lunch egg with a whisker blow at noon!)

The next operation was Heading, where the "business end" of the bit was cold drop forged to shape the center screw and spurs, *Figure 1-d.*

The trickiest operation, where the personal skill of the worker was most important, was forming the spiral. This was a hand operation and required great accuracy in relating the heat of the steel, the size of the bit, the speed of the rotation, and other factors less definable, in order to come up with a true and clean spiral. At Convalco, they used a sort of lathe, with a fixed tail stock, that had a broached slot to receive the rectangular end of the bit, while the head stock was a hand crank, similar to one on a grindstone, with a square hole to receive the tapered tang. The worker quickly grasped a properly heated bit from the forge with special tongs, placed it between the centers, and twisted it to a 45 degree spiral. In some factories the process was reversed, i.e., the head stock of the lathe was powered and received the head of the bit, while the workman held the tang stationary.[2] In either case, it took a lot of skill and experience, *Figure 1-e.*

In spite of the skill of the Twister, the spiral was never quite perfect. The imperfections were corrected in a Forming Die, which can best be described by comparing it to the interlocking knuckles of your right and left hands, which crimped the irregular twist into a perfect spiral.

Again the bit was heated, then placed vertically in a Heading Press, to reform the rough shape of the Floor Lips, the Spurs, and the Center Screw, Figure *1-f*. After all this torture to the original cast steel blank, it was once more heated and then annealed slowly in a bed of lime and bone meal for two or three days, then pickled in sulphuric acid and descaled.

Even now the bit was not absolutely straight. Again, the sensitive touch of the human hand and the accuracy of the human eye came into play; it was straightened by eye and hand on a level anvil with a hammer.

Up to this point, nearly all the work had been "blacksmithing" operations: forging, upset, shaping, etc.; very little stock had been removed from the original blank. In fact, Mr. Wright pointed out that this was one of the "bonuses" of bit-making, that there was almost no waste by machining of the raw steel stock that came into the plant. But now a few shavings and some steel dust were removed. The bit went into the Fitting Room where it was turned to a diameter 1/64 inch larger than the final size; this allowed for the final grinding and polishing that would bring it to exact size. The bit was not turned with parallel sides. The diameter at the cutting end was left about 1/32 inch larger than the diameter at the shank end. This slight taper served to eliminate binding and facilitate chip removal. The center point and spurs were trimmed, and the cutting edges of the floor lips were formed on a rasping machine. The next operation was a tricky one, cutting the threads on the center lead screw. This was especially difficult on a Double-Twist Bit, where it was necessary to cut two threads, one "inside" the other, so that the bit would balance and draw evenly into the wood. The business end of the bit was "rough filed" and the whole bit carefully heat-treated and tempered, another process that required long experience and care control. Finally, the bit was polished, in four runs, using as many different grades of abrasives, and "fine-filed" by hand. In the shipping room each bit was inspected and hand stamped by a member of the Wright family, a point of pride, as well as very effective advertising, oiled, wrapped, and packaged.[3]

So, will you believe that in 1900 a set of these bits, thirteen bits, from ¼ to 1 inch, by sixteenths, packed in three section wooden box with a groove for each bit, retailed for $3.00.[4]

References:
1. Irwin Bit Company, Catalog 1904.
2. Connecticut Valley Manufacturing Company Catalogs, and personal information from Martin W. Wright, Jr.
3. Personal information from workers at the Russell Jennings Manufacturing Company.
4. Montgomery Ward Catalog, 1895; et al.

The Nail Making Industry in Early Virginia

By JOHN T. KEENE, JR.

March 1972

At the time this article was written, John T. Keene was Historic Sites Historian for the Illinois State Historical Library, Springfield. He was educated at William and Mary, where he did the research for this paper which formed a portion of his Master's thesis.

When European colonists first approached American shores in the early years of the seventeenth century, their motivations were many and varied. Yet Champlain at Port Royal off the Bay of Fundy and the Virginia Company settlers on Jamestown Island in Virginia faced the same basic problems. The chief obstacles to be overcome were environmental; the Europeans had to learn to adapt themselves to a new climate and to a world in which the only existing civilization was radically different from that which they had known in Europe. Survival was a real test, to be met by developing an adequate food supply and constructing suitable shelter.

As soon as they were able, most settlers constructed homes similar to those they had left in England or on the Continent. This was the pattern followed by settlers in Virginia. Very few years passed before heavy oak framed buildings or houses of native burned brick were being erected. This meant that there was a heavy importation of building materials which could not be, or were not being, produced locally, including glass, lead, ironware and stone. Perhaps most important of these was ironware—locks, casement window frames, hinges, tools and, of course, nails. A misapprehension of the type of architecture employed by early Virginians led several early students of Virginia economic history to conclude that few, if any, nails were needed by the early colonists.[1]

Evidence to the contrary is actually quite abundant. For example, the Virginia Company, in addition to supplying tools and weapons to tenants sent to cultivate the public lands, also provided nails of many sizes.[2] Robert Evelyn, a mid-seventeenth-century Virginian, compiled a list of necessities every immigrant should bring. It included merchandise to sell for cattle, provisions, seed, tools, weapons, ammunition, and nails.[3] Edward Williams, author of *Virginia Richly Valued* (London:1650), a promotion pamphlet, included a similar list.[4]

The practice of proprietors providing nails to help their tenants get established was a tradition which did not die with the Virginia Company in 1624. French Huguenot refugees being settled on the Northern Neck were offered enough nails and hardware to build a home twenty-six feet to twenty-eight feet by fourteen feet to sixteen feet in 1687,[5] while William Fitzhugh also provided his new tenants with nails in 1690.[6] An advertisement in the *Virginia Gazette* in 1736 offered prospective freeholders willing to remove to Georgia enough nails to build a house sixteen feet by twenty-four feet.

That proprietors should offer to supply nails to tenants clearly indicates that they considered nails a necessity for establishing a homestead and farm. But for the freeholder who did not have his necessities provided him, procuring nails was from the first a very expensive operation.

An act passed in the General Assembly which met in February 1644 indicates that the Virginians were most resourceful in procuring nails for their personal use, although this did not always serve to promote the general welfare:

> *It shall not be lawful for any person so deserting his plantation as foresaid to burne any necessary houseing that are scituated thereupon, but shall receive so many nailes as may be computed by 2 indifferent men were expended in the building thereof for full satisfaction. . .*[7]

Thereafter, the burning of buildings for nails, in theory at least, required specific authorization. Such permission was granted to the person who would occupy the site on the James River opposite the unused Fort Charles in 1646. He was to be allowed to enjoy the housing of the fort for timber, or burning them for nails or otherwise.[8] Nor was this practice restricted to Virginia, for when a new courthouse was authorized for Kent County, Delaware, in 1691, it was also resolved that ". . .wee does alsoe Concent that ye old Courthouse may be burnt to gett the nailes."

With nails in such short supply and so expensive when imported, it should not be surprising that Virginians should attempt to produce this relatively simple article for themselves. Iron, it should be noted, was one of Jamestown's first exports to England during the period when the colony was seeking a profitable export to sustain the venture. The first full-scale ironworks in America was constructed on Falling Creek approximately sixty miles above Jamestown in 1619, and was expanded in 1622. Unfortunately, the Indians in that year destroyed the works and massacred the workers.[9] Because of the capital required, no other substantial ironworks was built in the colony until the eighteenth century. Nevertheless, some ore was refined by the bloomery process throughout the rest of the century.[10]

Whether native iron was used to produce nails in the young colony is at this time an unanswerable question, for imported nail rods could also have been employed. Few records exist of the production of nails themselves. One of the earliest producers thus far discovered was Thomas Wyatt of Northampton County on the Eastern Shore, who was producing nails in his shop in 1635.[11] Two years later the Northampton court ordered John Parramore to pay Wyatt for 2,000 four-penny nails, 1,700 six-penny nails, 140 three-penny nails, and 820 two-penny nails,[12] so apparently Wyatt ran a fairly sizable operation.

The operations used by Wyatt in his nail production, although not specifically recorded, are not difficult to determine. Albert Sonn reports that a nail found in the Roman Forum was indistinguishable from one taken from a 1724 Wethersfield, Connecticut house,[13] reflecting that the craft remained basically unchanged from ancient times until the first introduction of nail cutting and nail making machinery in the closing years of the eighteenth century. The necessities were few and simple—a small anvil with a tapered swage hole on the horn, a hammer with a cross peen for cutting the nail rod, and a small fire which did not even require a forced draft. "Few blacksmith shops were so specialized as not to make carpenter's nails."[14]

Adam Smith, however, points out that by far the most efficient nail making came from shops devoted exclusively to the craft. He estimated that a common smith unused to the trade could produce but 200-300 nails per day—and very poor ones at that. A smith accustomed to making nails as a side line fared better—his daily production would average between 800-1000 nails. However, boys under twenty, trained exclusively in the trade, could by exertion produce up to 2,300 nails per day.[15]

The basic raw material was nail rod, strips of soft and malleable forged iron, varying from one-eighth of an inch to one-fourth of an inch square, depending on the size of the nails being made, and approximately five feet long. Except for the smallest nails, the rod had first to be heated. Then the heated end would be pointed either on four sides, to produce a fine point, or on two to produce a flat point. The rod would then be partially cut by a hammer blow, the pointed end put in the swage hole, and the rest of the rod broken off and placed back in the fire. The projecting end of the nail was then hammered down and spread around the hole, forming a head. The cooling nail was then knocked from the swage hole. Rivets were made in a similar manner, except the swage hole was round instead of square.[16] Depending upon the use for which the nail was intended, rose heads (four or more facets), T-heads, or L-heads could be formed.

One improvement which came into common use by the late 1700's was the hand-held nail header. Grasped in the left hand, headers replaced the swage hole in the anvil and could be simply inverted and tapped to release the headed nail. Another form of nail header could be held in a bench vise.[17] These are simply the most rudimentary forms of the craft. A wide variety of tools was actually available to the nail maker, particularly by the late eighteenth century, yet all were actually quite simple variations on the devices here described. Diderot's *Encyclopedia* illustrates some slightly more sophisticated tools employed by French nail makers of the period, but few similar examples have thus far been discovered in America.[18]

For the general smith, nail rod proved a handy item to have in his shop for jobs in addition to nail making. Coopers' cressets or fire baskets, gridirons, trivets, spits for reflector ovens, meat hooks, grappling hooks, and quarry tools are among the products a versatile smith could fashion from nail rods.[19]

While everywhere the nail making process was essentially the same, there still were regional variations. In the northern colonies nail making was a household industry in which the whole family participated during the long winter months. Each fall a supply of nail rod would be obtained, and, with the kitchen hearth serving as a forge, the evenings would pass to the tune of hammer blows on a small anvil. If resources were short, scrap iron would be employed in place of the nail rod. The product of the winter's work could be used for spring repairs and summer construction on the farm and the surplus sold for profit. Although the domestic manufacture of nails was common throughout New Jersey, Plymouth and Bristol Counties in Massachusetts and Litchfield County in Connecticut were early centers of the trade. The nail making industry is the only metal working occupation which was so centralized during the handicraft period.[20] When nail making became mechanized, eastern Massachusetts, not surprisingly, was a leading center of the industry.

Harry B. and Grace M. Weiss, in their study *Trades and Tradesmen of Colonial New Jersey*, point out one of the basic problems in studying the early nail making industry: "Nailers during colonial days seldom received public attention unless they were runaways or got into trouble of some sort."[21] In addition to this general shortage of information on the trade, the Iron Act of 1750 induced colonial nailers to avoid publicizing illegal activities. As a result material on early Virginia nail making is fragmentary at best.

The record of Thomas Wyatt's shop of 1635 on the Eastern Shore, previously mentioned, seems to indicate that nail making was his principal occupation. However, it could be that his was a blacksmith's shop which happened to produce nails for local customers. In any case, records of such specialized shops do not appear again in Virginia until the mid-eighteenth century. Thus, unlike the English pattern, nail making was not a geographically centered or distinct trade in Virginia.

Apparently, however, in Virginia the household industry of nail making by farm families was small or non-existent. What nail making was done in Virginia was primarily the product of the local or plantation blacksmith. In the early years of the colony, men of many trades were brought over in an attempt to provide the community with all necessary services. Nailers were not listed among them, but blacksmiths were. However, with only a few smiths to repair all the ironwork in the colony and fabricate some of it, there probably was little enough time to produce nails on a large scale. Harold Gill, in his recent study of the blacksmith in Colonial Virginia, reaches this conclusion: "The work of most seventeenth-century smiths was evidently confined to repairing broken articles and making small tools. Iron work for special construction was usually imported."[22] David and William Geddy illustrate the multiple talents of colonial nail makers, for they are best remembered as gunsmiths and as armourers for the colony.

Not all Virginia smiths, however, were as well established as the Geddys. Lack of towns meant that most smiths in the colony worked on plantations by the eighteenth century, and since plantations sought cheap labor, these were most often indentured servants, or, by the early eighteenth century, slaves.

Because of lack of money, as well as other handicaps to the success of craftsmen, most blacksmiths who followed their trade in Virginia, especially during the seventeenth and first half of the eighteenth century, were indentured servants who operated plantation smithies.[23]

Merely because the majority of smithies were on plantations during the colonial period, it should not be assumed that this meant a general lack of commercial smithies. Only the largest of plantations could afford to maintain its own blacksmith, and even then it was often difficult to find enough to keep this man fully occupied. A planter thus would not be at all reticent to allow his smith to perform work for his neighbors on a commission basis. If even this proved to be insufficient to keep the smith busy, nail making was a perfect answer, for nails were always needed, and surplus could be sold with no trouble. A slave blacksmith who could produce nails had much to recommend him as a purchase, and could command top prices, as the following advertisement illustrates:

> *For sale, for ready money or 12 months credit, an exceeding good plantation BLACKSMITH, who is very well acquainted with many other kinds of work in that branch of business, such as nail making, tiring wheels &c. . .*[24]

Robert Carter of Nomini Hall was one of Virginia's more prominent planters who maintained a nailery. He was also one of the more progressive planters of the colony, being one of the first to shift from tobacco growing to the production of foodstuffs for the export trade. In addition, Carter was interested in America's infant manufactures, and held a significant interest in the Baltimore Iron Works, from which he probably obtained his nail rod. His nailery must have had a fairly substantial output, for it is recorded that he shipped nails to London in quantity when the price was good.[25] This was most certainly a reversal of normal trade patterns.

One additional source for nails was Virginia's infant ironworks which were beginning to reappear in the eighteenth century. The Accokeek Iron Works, operated by Nathanial Chapman in Stafford County, was established in response to 1727 and 1729 Acts for Encouraging Adventurers in Iron Works.[26] In 1731, Chapman sold John Mercer of Marlborough, Virginia, several hundred nails of various sizes and types.[27] Chapman produced iron utensils and farm implements in addition to nails from native ores at this same site.

Virginia always faced a shortage of blacksmiths because of competition from other pursuits. Many young European smiths would accept an indenture to obtain passage to America. Yet, when their terms expired, most found the lure of the land too strong—

they abandoned the trade in favor of establishing a homestead and becoming a farmer. Jefferson noted this allurement of the land even as late as the post-Revolutionary period:

> *Such is our attachment to agriculture, and such our preference for foreign manufactures, that it be wise or unwise, our people will certainly return as soon as they can, to the raising of raw materials, and exchanging them for finer manufactures than they are able to execute for themselves.*[28]

Because of this combination of circumstances, Virginia during the colonial period could produce but a small percentage of the nails she needed. The remainder were imported largely, but not exclusively, from England. The Netherlands, France, Pennsylvania, and the New England colonies were all direct or indirect suppliers of Virginia planters.

Large shipments were not uncommon even as early as the mid-seventeenth century. Witness a 1644 invoice of goods shipped abroad the *Sivilla* and consigned to Adam Bland, Peter Burr, and John Jewell.[29]

A barrell with 23,000 of 6d. nails at 3s.10d. £ 4 28s.4d.
A firkin with 30,000 of 4d. nails at 3s.17d. £ 2 13s.9d.
A firkin with 32,000 of 3d. nails at 3s.16d. £ 2 5s.4d.
A firkin with   6,000 of 10d. nails at 3s. 4d. £ 1 18s.0d.
A firkin with   4,000 of 20d. nails at 10s. 3d. £ 2 1s.0d.

———————

£13 16s.5d.

Williams Fitzhugh was one planter who seldom sent an order without including nails among the desired commodities, nor were they always of English manufacture. In a letter to Jonathan Cooper dated March 30, 1684, he notes that he received all the goods ordered except the Dutch nails and tacks, which he presumed had been omitted in packing.[30]

In 1695, he wrote John Taylor, a London merchant, informing that he had sent thirteen hogsheads of tobacco consigned to him

> *. . .which I hope will come to good Market, this come now to desire you to serve me in goods for what I shall have in your hands after the sail of the said Tob°; let it be in Nails, hoes, axes, kerseys, cotton, and other coarse Goods.*[31]

By his selection of "coarse Goods" one might presume that he had an eye towards the possible resale of these standard staples.

London merchants still did some direct business with large and small customers alike as records of John Norton & Sons indicate. In 1770 John Robinson of York County ordered 147,000 nails and brads of varying sizes,[32] while the following year George Wythe of Williamsburg ordered a single cask of nails.

Nails were apparently an important item in the coastwise trade from New England, which in turn bought much of its nail rod from Philadelphia. Old ironware would be bought for scrap from Tidewater plantations and refined and recast into new implements for sale in the South and West Indies. Nails and furniture were among the staples in this trade.

The nails Virginians used during the colonial period, then, came from a wide variety of sources. Planters like Robert Carter, however, who had a separate nail store, well filled from his own nailery, were certainly the exception. It seems from the fragmentary evidence available that most of Virginia's nails before the Revolution were imported.

Acting Governor John Blair affirmed this in his report to the Lords of Trade in September 1768:

> *Our pig-iron and some bar iron is chiefly shipped to Britain. We do not make a saw, auger, gimlet, files, or nails, nor steel; and most tools in this country are imported from Britain. . .*[33]

This might be considered suspect because of its being dated after the passage of the Iron Act of 1750, but Lieutenant Governor Gooch, whose term ended in 1749, made much the same answer in response to the query concerning what British manufactures were imported and in what quantities: ". . .all manner of Iron Ware such as Locks, Hinges, Nails, Carpenters, Joyners, and Smiths Tools, Axes, Fire Armes, and Small Anchors."[34]

It was the onset of the Revolution which spurred Virginians to take an objective look at their largely neglected manufactures. Only the iron industry had made significant progress before the war. However, in March of 1775, a convention was held in Richmond to elect delegates to the upcoming Continental Congress. They appointed a committee to prepare plans for the encouraging of arts and manufactures. That committee in its report recommended unanimously that the manufacture of iron into nails, wire, and other necessary articles be undertaken.[35] Indicative of the importance placed upon nail manufacture was its location between resolutions calling for the manufacture of gunpowder and steel.

The impending breach with Britain inspired nationalistic feeling of many kinds, including hostility to imperial trade restrictions, as evidenced by this letter to Alexander Purdie, editor of the *Virginia Gazette*, early in 1776:

Much has been said concerning the independence of the colonies, and some people have been made to believe that such a state is not desirable, and that we should wish for no more liberty than we enjoyed in 1763; but let any man consider that, at that time, we were restrained from making nails and hats, and might with equal justice have been hindered from building houses or making stockings; that we were cruelly and wantonly restricted in our trade, in some instances, as it were merely to show we were the slaves of Britain.

AN AMERICAN[16]

William Pitt, the colonists' champion in Parliament, also recognized their plight, for he is said to have stated, "If the horse flung a shoe, an American could not, of right, so much as make the nails required to set it."[37]

Effective measures were taken immediately by the county Committees of Correspondence throughout Virginia. One of their most important duties was to see that non-importation agreements were complied with, and apparently the agreements did operate with a good deal of effectiveness. As early as February 1775 an announcement was published informing the public that sundry goods, including nails and smith's tools, were to be sold by order of the Elizabeth City County Committee.[38]

Committees also exercised control over prices in order to insure that no excessive profiteering resulted from the shortage of imported goods. In Port Royal, Virginia, one Robert Johnston, newly moved from Philadelphia, had been retailing nails at higher prices than the local merchants. Accordingly he was approached by the Committee and reprimanded, after which he voluntarily signed the Association.[39]

Virginia resorted to other sources. Many nails, as mentioned previously, were imported from the Continent via the West Indies. Many *Virginia Gazette* advertisers of the late 1770's noted that their nails came from such points as Holland, Nantes, the West Indies, Martinique, St. Croix, and France.

This might seem to indicate that under the pressure of war, Virginians merely shifted the source of their imports to meet their demand for nails. However, the early resolutions to establish and encourage domestic production were not empty promises. Advertisements for nailers from areas as widely separated as Alexandria and Portsmouth appeared for the first time. The Shenandoah Valley was the great center of the iron industry. Most of the forges were located there because of the proximity to furnaces.[40] It was to here also, in Augusta County, that William Freal moved from Pennsylvania and established one of the oldest nail factories in the country. This factory was said to be in operation as early as 1777.[41]

Wanted at Hunter's iron works, to whom suitable encouragement will be given, a masterly hand in a wire mill; also nailers...[42]

This advertisement, which appeared in the *Virginia Gazette* in December 1779, shows that nails were being produced at one of America's largest ironworks at the time. Hunter's Works, which were located on the Rappahannock River slightly more than two miles above Fredericksburg, were described by Ebenezer Hazard in 1774.[43] The operation was so extensive as to actually resemble a small village, Hazard noted. There were houses for the workmen, a grist mill, saw mill, cooper's shop, saddler's shop, brass founder's shop, wheelwright's shop, steel works, and, of course, a large blacksmith shop. A slitting and planing mill and furnace were under construction, and a steel works was nearby. A three-quarter mile canal supplied the necessary water power. Nails and nail rod, after the slitting mill had been completed, were among the most important products of the works.

Virginia did not depend solely on private enterprise to supply nails to the Commonwealth. Under the sponsorship of the War Office, Virginia actively engaged in promoting industry by having James Anderson, a Williamsburg blacksmith, operate a nail factory for the Commonwealth. Anderson's greatest problem seemed to be getting enough qualified workers to keep up with the business which poured into his shop:

> *I will pay extraordinary wages to a good BLACKSMITH and*
> *NAILER that is capable of acting as foreman in my shops.*
> *—James Anderson*
> *Great wages will be given by the subscriber to journeymen*
> *Gunsmiths, Blacksmiths, and Nailers, that are good workmen. None*
> *others need apply. Six or eight boys are wanted as apprentices.*
> *—James Anderson*[44]

Colonel Muter of the War Office was most enthusiastic about the venture. When the contract with Anderson expired in 1780, Muter uged that it be renewed because of the demand for nails at that time.

Nail rod was supplied from Fredericksburg, in all probability the product of Hunter's Iron Works. Early in 1781 Anderson held out for better contract terms, so the state entered into direct negotiation with the nailers in order to keep supplying that necessary article. Their terms seem strikingly contemporary: "The tradesmen are willing to work for the State, provided they are protected from militia duty and drafts," as well as being assured of an adequate supply of provisions.[45] Eventually, however, a new contract was signed with Anderson.

Anderson was again contracted to manage the public shops in 1782, but by that time there were difficulties in securing nail rod which nearly stopped production. Davies reported to Governor Harrison that, "Mr. Anderson is exceeding uneasy at not being able to be of use, either to the public or himself, in his department."[46] Later that year

Anderson retired from public service and moved back to Williamsburg from Richmond, where he continued his blacksmithing business until his death.

The Virginia nail making industry did not disappear after the war, but Britain again captured the lion's share of the market. Nail rod often had to be imported from as far away as Philadelphia, and it was difficult to obtain and retain nailers. British nails were cheap and could be obtained with little effort. In the year ending September 30, 1790, Virginia imported 913,000 pounds of nails and spikes, almost entirely from England.

It was during this same period, however, that the nail industry itself was undergoing profound changes. The age-old method of laboriously shaping each nail from nail rod was both tedious and time consuming. Few really significant changes had been made in the process since Roman times. The late eighteenth century was the dawning of the industrial ages, and on both sides of the Atlantic men sought and first developed mechanical means to produce nails during this period.

The first advance came from Cumberland, Rhode Island, where Jeremiah Wilkinson developed a method of cutting tacks from sheet metal plates about 1775.[47] The process could not be applied immediately because the plates required to make nails were much thicker than the thin sheets required for tacks. The invention of a machine for cutting nails from plates is credited to another New Englander, Ezekiel Reed (Rud) of Bridgewater, Massachusetts, in 1786. Two years later he got a patent for cutting and heading nails in one operation, although like most early mechanical headers, it was not particularly successful.[48] Early cut nails continued to be headed by hand for many years.

In 1790, Jacob Perkins of Newburyport, Massachusetts, developed what was to prove a more successful design of machinery to cut nails from iron plates. Five years later he patented another machine which would cut and head nails in one operation.[49] Later that year J.G. Pierson of New York invented another nail making machine. In all, eleven patents were registered during the decade of 1795-1806 for nail, tack, and brad making machines.[50]

The first mass production nail factory in the country was opened in Amesbury, Massachusetts, in 1798. It was incorporated in 1805 with a capital of $450,000. By 1810 there were 410 nail factories producing 15,000,000 nails annually in the country.[51] As early as 1791 Hamilton had written that private enterprise fostered by protective state duties had already brought prosperity to some industries, among them nail making.[52] In 1789 the new Federal Government had placed a one cent per pound duty on all imported nails and spikes, over the objections of the South which had few naileries at that time. Significantly, tacks and brads were duty free, for it was acknowledged that the British could produce superior goods more cheaply in the smaller sizes.

The British were mechanizing their industry at about the same time. Thomas Clifford developed a method of forming nails in a groove between rollers in 1790. The nails emerged with heads and tails joined, and had to be separated later. This advance was quickly followed by nearly forty other patents in the next few years. America, however, still retained credit for the first machine to cut and head nails in one operation. The cut nails produced by the new machines had several advantages over the older wrought nails, for they could be driven into more solid timber without drilling, were less liable to split the board, had better holding qualities, required no flattening of the head, and more significantly, sold for less per pound and numbered more per pound than wrought nails. Thus the new cut nails almost universally superseded wrought nails as soon as they reached the market, usually about 1800.[53] Wrought nails continued to be produced for special applications, especially where clinching the nail was necessary, for cut nails were quite brittle and could not be bent over without breaking.

The operation of the early cutting machines was quite simple. Nail plates about a foot long and with width to suit the size of the nail being made would be heated to a deep red, then held with a wooden handle with a clamp at one end. A double crankshaft with a band wheel at one end and flywheel at the other powered the heavy wrought iron cutter blade which rose and fell in a heavy frame. At each stroke a tapering strip was cut from the plate to form a nail. Then the operator turned the plate over to obtain a uniform taper on the nails. Larger sizes were slower to produce—8d nails could be cut at eight times the speed of 20d nails.[54] Earlier machines for smaller sized nails were foot powered, but by the early nineteenth century most were belt driven.

Fortunately, very complete records exist for one of Virginia's first mass production naileries, that of Thomas Jefferson at Monticello. Jefferson's nailery was only one of several operations which he developed along Mulberry Row at Monticello, including spinning, weaving, and blacksmithing. The operation was begun in April 1794 when forty bundles of nail rod arrived. For labor he used slave boys aged ten to sixteen (the girls were employed in spinning). Apparently, however, he experienced some difficulty taming their youthful spirits, for at one point he resolved to make Cary, possibly a runaway, an example to the others ". . .in order to maintain the police so rigorously necessary among the nail boys."[55]

Jefferson threw himself enthusiastically into the new activity. He wrote Henry Knox in 1795 that what time he did not spend on his farming was spent on his nailery, so that he rarely had time to read or write. It was with obvious pride, however, that he undertook this activity, as reflected in one of his letters; "I am myself, a nail-maker. . . my new trade of nail-making is to me in this country what an additional title of nobility or the ensigns of a new order are in Europe."[56]

Originally only wrought iron nails were produced, but on 22 February 1796 a new nail making machine from England was introduced. It differed from most American machines in that it still used nail rods rather than plates. Rods were heated to a black heat, then inserted in the machines, which cut off blanks of the required length, which were then struck by a moving die to form a head. It is not clear if the pointing was accomplished in the cutting operation or had to be done by hand.

Nail rod had to be imported from Philadelphia, since after Hunter's Works had closed, there were no slitting mills in Virginia. Jefferson's operation was carried out in combination with his blacksmithing in a wooden shop eighteen by thirty-seven feet. An addition was planned in 1796, but it is not known if this was completed. At any rate a new nailery of stone was built in 1801.

He began by using only one fire, but by 1796 was using three fires and producing 10,000 or more nails per day.

At first, fearing the wrath of the importing houses, merchants refused to take his nails, but Jefferson realized the inexorable laws of economics were on his side. Early in 1796 he wrote Archibald Stuart that

> . . .*it is tolerably certain that the moment my deposit opens there will be an entire stoppage to the sale of all imported nails, for nobody can retail them in the upper country at Richmond wholesale prices, advanced only 5. or 10. percent (for commission and transportation).*

Not long after he wrote:

> *I retail at 5. per cent on the Richmond wholesale prices. This I presume will soon give me a clear field, and defeat in this particular this effort of the general system of Scotch policy to suppress every attempt at domestic manufacture.*[57]

When Jefferson was at home to manage the nailery it was quite successful. As early as 1795 he wrote, "The best resource of quarterly paiment in my power is nails, of which I make enough every fortnight to pay a quarter's bill (for groceries, $100-$125)."[58]

Although Jefferson's unavoidable absences from Monticello, especially during his terms as President and Vice-President, complicated the problems of running the nailery, he still took a direct interest, writing to Howell and Jones, suppliers of nail rod in Philadelphia, and his observer, Gabriel Lilly, concerning the nailery. He sought to make improvements when possible, such as writing Benjamin Perkins in an attempt to secure a set of drawings of his brother's nail making machine.

The most serious problem caused by his absences was that although production continued, no one collected debts while Jefferson was away. This caused acute shortages of income, and attempts at collection immediately upon his return were more often met with evasion than success. Nevertheless, the nailery was in continuous operation through 1812. It then had to be closed for the next three years because of the war; nail rod could not be shipped from Philadelphia. During the period from 1815 to 1823 it operated only intermittently.

During the height of its operation, Jefferson's nailery supplied much of the Piedmont area under Charlottesville and most of the Shenandoah Valley with nails. It was his practice to grant a monopoly on the distribution of his nails to but one retailer in each town. In addition he supplied individuals on occasion. For example, Monroe purchased the nails to construct Ashlawn from his close friend's nailery.[59]

In short, Jefferson's venture was one of the most promising attempts to establish the nail industry on a sound basis in Virginia. The demand was large, and the product of good quality. It failed principally because of mismanagement. Jefferson's governmental responsibilities kept him away from Monticello for so much of the time that he was not able to give his nailery the degree of attention it required. Had Jefferson been able to continue to devote the time to it he spent in the first years of operation the outcome might have been entirely different.

Jefferson was not the only Virginia nailor of the period, however, for in Richmond he had at least two competitors, of whom unfortunately little is known. In the closing years of the century, the state undertook the construction of a "Penitentiary House" at Richmond under the direction of the architect Benjamin Henry Latrobe, which, of course, required nails. Robert Paul, who operated a nail factory in the city, offered to supply the state with nails "of best quality" in 1797 for one shilling per pound.[60]

Thomas and William Callis were apparently among Paul's competitors, for in May 1799 they wrote the Governor that they would execute the grates for the penitentiary at 4d. per pound if the state would furnish the coal, iron, and tools, and that they would furnish the nails at 10½d. per pound and provide the iron.[61]

After the penitentiary was completed, the state once again got into the nail making business. Unwilling to receive no compensation at all for maintaining the inmates, and willing to test Jefferson's enlightened ideas on penology, which involved both solitary confinement and work in common, the state sought to have prisoners produce some goods which would be salable. Nail making was one of the activities chosen. The use of convicts as nailors was not, however, an innovation, for a 1774 issue of the *Virginia Gazette* had offered a ten-dollar reward for the return of Daniel Dunn, a convict servant and nailor by trade.[62] It was Latrobe who urged that a nail cutting machine be installed, in a letter to the Governor in 1796: "...I have drawings of all the Philadelphia Machines for making nails. We ought to think of making some, I am willing to take any trouble and give any assistance."[63]

Apparently the instruction program was well under way in July of 1800, for Martin Mims, Keeper of the Penitentiary, reported that he had employed George Anderson Stile at $36.00 per month to instruct and work with the prisoners in nail making.[64] In 1806 it was reported by a committee appointed by the Governor to inspect the penitentiary that thirteen hands were normally engaged in the manufacture of cut nails, most in heading the nails.

By the mid-nineteenth century the nail industry in Virginia was well established. Wheeling, then in Virginia, had a nail factory in 1811 and in time became one of the four or five largest centers of commercial nail manufacture in the country.[65] The Tredegar Iron Works in Richmond also became a large producer. The Belle Island Works which became the Old Dominion Nail Works in 1856 had 48 machines in operation, and produced over 1,000 tons of nails annually.[66] The Old Dominion Nail and Iron Works, established at Lynchburg in 1884, expanded its plant and added a blast furnace just three years later.[67]

The history of the nail making industry in early Virginia is a surprising one in that it developed so slowly and haltingly. Nail making operations had begun as early as 1635 on the Eastern Shore, yet throughout the remainder of the seventeenth and early eighteenth century, very few nails seem to have been produced even by local blacksmiths. Similarly, the household industry of nail making by individual families, as it was practiced throughout the Northeast, appears to have been minimal, if existent at all. Virginia relied on England chiefly, and on her sister colonies, principally Massachusetts and Pennsylvania, for nails.

It was the onset of the revolution that first shocked many Virginians with the realization that by depending on imports for almost all their nails, their supply of that vital commodity was very vulnerable. Blockade running with European nails from the West Indies proved a somewhat less than satisfactory method of supply. Unfortunately, Virginia's response to the critical shortage of nails was mostly in the form of stopgap measures rather than an attempt to establish the industry as a permanent part of the state's economy.

The real precursors of Virginia's nineteenth-century nail making industry were such operations as Jefferson's nailery at Monticello and the nailery at Richmond Penitentiary. Both of these operations were progressive, incorporating the latest machinery in an attempt to mass produce nails for the domestic market. These were probably the first shops devoted exclusively to nail production, rather than operations turning out nails as a sideline. Their success proved that Virginia naileries could compete favorably with foreign producers, and that such operations, if well managed, could be profitable.

Notes

1. For example see, Philip A. Bruce, *Economic History of Virginia in the Seventeenth Century* (New York, 1907), II, 147.

2. *Ibid.*, I, 233. Bruce contradicts his own thesis.

3. *Ibid.*, I, 339.

4. Cited *Ibid.*

5. *Tyler's Quarterly Historical and Genealogical Magazine*, V.3 (Jan.,1924), 168.

6. *Virginia Magazine of History and Biography*, III, 1 (1895), 8.

7. William W. Henning, ed., *The Statues at Large; . . .of Virginia* (Richmond, 1810), I, 291.

HAND WROUGHT NAILS 17th 18th and early 19th centuries

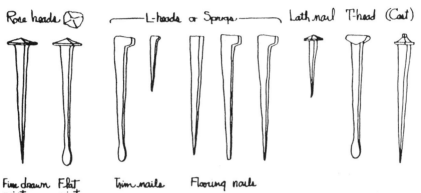

Rose heads

L-heads or Sprigs

Lath nail T-head (Cut)

Fine drawn point   Flat point

Trim nails

Flooring nails

Other characteristics: shanks taper on both faces, iron fibers run lengthwise, lack of uniformity (especially heads)

MACHINE CUT NAILS 1790s to Present

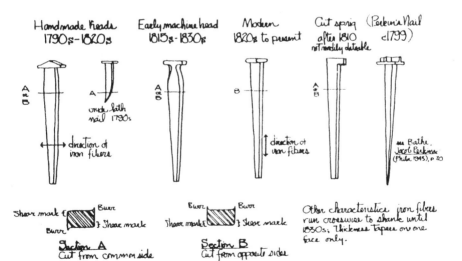

Handmade heads 1790s-1820s

Early machine head 1815s-1830s

Modern 1820s to present

Cut spring (Perkin's Nail after 1810 c1799) not readily datable

crude lath nail 1790s

direction of iron fibers

direction of iron fibers

see Bathe, Jacob Perkins (Phila. 1943), p. 20

Shear mark   Burr

Burr   Shear mark

Section A Cut from common side

Burr   Burr

Shear mark   Shear mark

Section B Cut from opposite sides

Other characteristics: iron fibers run crosswise to shank until 1830s. Thickness tapers on one face only.

You can see by the examples here that the changing technology changed the shape of the nails. This appeared in Technical Leaflet 15, American Association for State and Local History and is printed with their permission and the permission of the author, Lee H. Nelson.

8. *Ibid.*, I, 327.

9. J. Leander Bishop, *A History of American Manufacture from 1608 to 1860,* 3d.ed. (Phila., 1868), I, 28-29.

10. Harold G. Gill, "The Blacksmith in Colonial Virginia" (unpublished), *Colonial Williamsburg Research Report* (1965), 6-8.

11. Northampton County records, I, 48, cited in Susie M. Ames, *Studies in the Virginia Eastern Shore in the Seventeenth Century* (Richmond, 1940), 127.

12. *Ibid.*, I, 67; Ames, 127.

13. Albert H. Sonn, *Early American Wrought Iron* (New York,1928), I, 4.

14. Edward Tunis, *Colonial Craftsmen and the Beginnings of American Industry* (Cleveland, 1965), 58.

15. Adam Smith, *The Wealth of Nations.*

16. Harry B. and Grace M. Weiss, *Forgotten Mills of Early New Jersey* (Trenton,1960),63.

17. For example see illustrations in Tunis, above.

18. J. Didsbury, "The French Method of Nail-Making," *The Chronicle,* XII, 4 (Dec.,1959), 47.

19. Lawrence S. Cooke, "Nail Rod and Some of its By-Products," *ibid.,* XIV, 1 (March,1961), 6-7.

20. Victor S. Clark, *History of Manufactures in the United States* (New York, 1949), I, 516.

21. Harry B. and Grace M. Weiss, *Trades and Tradesmen of Colonial New Jersey* (Trenton, 1965), 88.

22. Gill. 19.

23. *William and Mary Quarterly* (Series I), XXIII, 4 (1915), 273.

24. Purdie, *Virginia Gazette* (Williamsburg), June, 1777, 3. Hereafter *V.G.*

25. Louis Morton, *Robert Carter of Nomini Hall* (Williamsburg, 1945), 171.

26. Henning, IV, 228-31, 296-300.

27. C. Malcolm Watkins, *The Cultural History of Marlborough, Virginia* (Washington, D.C., 1968), 25.

28. Thomas Jefferson, *Notes on the State of Virginia,* William Peden, ed. (Chapel Hill, 1955), 164.

29. *Virginia Magazine,* LXXII, 1 (1964), 34. Hereafter *V.M.*

30. *Ibid.,* I, 3 (1894), 268.

31. *Ibid.,* V, 1 (1897), 32.

32. Frances N. Mason, ed., *John Norton and Sons; Merchants of London and Virginia* (Richmond, 1937), 119, 169.

33. Acting Governor to Lords of Trade, 20 Sept. 1768 in Alonzo T. Dill, "A Documentary History of the Governor's House at New Bern, N.C.," unpublished, 77.

34. *VM,* III, 2 (1895-6), 117.

35. Pinkney, *VG,* 30 March 1775, 3.

36. Purdie, *VG,* 29 March 1776, 1.

37. William Pitt, cited by Lorenzo Savine. *Biographical Sketch of the Loyalists of the American Revolution* (Boston, 1864), I, 62.

38. Dixon and Hunter, *VG,* 4 Feb. 1775, 3.

39. Purdie, *VG,* 21 April 1775 (Supplement), 4.

40. Emory Q. Hawk, *Economic History of the South* (New York, 1934), 302.

41. Kathleen Bruce, *Virginia Iron Manufacture in the Slave Era* (New York, 1931), 22.

42. Dixon and Nicolson, *VG,* 4 Dec. 1779, 2.

43. "The Journal of Ebenezer Hazard in Virginia, 1777," *VM,* LXII, 4 (1954), 417.

44. Dixon and Nicolson, *VG,* 16 April 1779, 2; and Purdie, *VG,* 3 Oct. 1777, 2.

45. *Calendar of Virginia State Papers and Other Manuscripts,* William Palmer et al., ed. (Richmond, 1875), I, 402.

46. *Ibid.,* III, 206.

47. Bishop, I, 388.

48. Weiss and Weiss, *Mills,* 67.

49. Bishop, I, 492.

50. Weiss and Weiss, *Mills,* 69.

51. *Loc. cit.*

52. Alexander Hamilton, *Report on Manufactures* (Dec.,1792), cited in John C. Miller, *The Federalist Era* (New York, 1960), 63.

53. Henry C. Mercer, "Dating Old Houses," *A Collection of Papers Read Before The Bucks County Historical Society* (1926), V. 537.

54. Weiss and Weiss, *Mills*, 74-75.

55. Edwin M .Betts, ed., *Thomas Jefferson's Farm Book* (Princeton, 1953), 19.

56. *Ibid.*, 426.

57. *Ibid.*, 432, 433.

58. *Ibid.*, 429.

59. Hamilton W. Pierson, *Jefferson at Monticello*, James A. Bear, Jr., ed. (Charlottesville, Va., 1967), 64.

60. *Calendar of State Papers*, VIII, 450.

61. *Ibid.*, IX, 22.

62. Purdie and Dixon, *VG*, 10 Nov. 1774, 4.

63. *Calendar of State Papers*, VIII, 498-99.

64. *Ibid.*, IX, 120.

65. Clark, I, 516.

66. Bruce, 293.

67. W. Ashbury Christian, *Lynchburg and its People* (Lynchburg, 1900), 349.

Metal Alloys

November 1937, January 1938, March 1938, June 1938, December 1938

NOTE: This article is for historical reference only. There are serious hazards that we now know in handling and combining the various recipes and processes that are described and they should not be attempted.

[The following are extracts from Mackenzie's *Five Thousands Receipts in All the Useful and Domestic Arts*, Kay's edition, Phila., 1829. While presumably there will be few, if any, who will attempt to memorize these formulae, or even to read them all at once, many who do not own the book will be glad to have them in their files for reference purposes.—*Chronicle* Ed.]

Ormolu—Mosaic gold. Melt together equal parts of copper and zinc, at the lowest temperature that will fuse the former, stir them well to produce an intimate mixture of the metals, and add by degrees small quantities of zinc; the alloy first assumes a yellow colour like brass, on adding a little more zinc it becomes purple, and lastly perfectly white, which is the proper appearance of the desired product, when fused. The quantity of zinc to be used altogether, should be from fifty-two to fifty-five parts out of a hundred.

Queens' metal. Melt together 4½ pounds of tin, ½ pound of bismuth, ½ pound of antimony, and ½ pound of lead. A very excellent alloy will be formed by using these proportions; it is used for making tea-pots and other vessels which are required to imitate silver. They retain their brilliancy to the last.

Tombac. Melt together 16 pounds of copper, 1 pound of tin, and 1 pound of zinc.

Red tombac. Put into a crucible 5½ pounds of copper: when fused, add ½ pound of zinc: these metals will combine, forming an alloy of a reddish colour, but possessing more lustre than copper, and also greater durability.

White tombac. When copper is combined with arsenic, by melting them together in a close crucible, and covering the surface with muriate of soda, to prevent oxidation, a white brittle alloy is formed.

Common pewter. Melt in a crucible 7 pounds of tin, and when fused throw in 1 pound of lead, 6 ounces of copper, and 2 ounces of zinc. This combination of metals will form an alloy of great durability and tenacity; also of considerable lustre.

Best pewter. The best sort of pewter consists of 100 parts of tin, and 17 of regulus of antimony.

Hard pewter. Melt together 12 pounds of tin, 1 pound of regulus of antimony, and 4 ounces of copper.

Tutania or *Britannia metal.* Melt together 4 ounces of plate brass, and 4 ounces of tin. When in fusion, add 4 ounces of bismuth, and 4 ounces of regulus of antimony. This is the hardening, which is to be added at discretion to melted tin, until it has the requisite colour and hardness.

Another.—Melt together 2 pounds of plate brass, 2 pounds of a mixture of copper and arsenic, either by cementing or melting, 2 pounds of tin, 2 pounds of bismuth, and 2 pounds regulus of antimony. This is to be added at discretion, to melted tin.

German tutania. Melt together 2 drachms of copper, 1 ounce of regulus of antimony, and 12 ounces of tin.

Spanish tutania. To 8 ounces of scrap iron or steel, at a white heat, add 1 pound of antimony in small portions, with 3 ounces of nitre. Melt and harden 1 pound of tin with 2 ounces of this compound.

Common solder. Put into a crucible 2 pounds of lead, and when melted, throw in 1 pound of tin. This alloy is that generally known as solder. When heated by a hot iron, and applied to tinned iron with powdered rosin, it acts as a cement or solder; it is also used to join leaden pipes.

To plate looking-glasses. This art is erroneously termed silvering, for, as will be presently seen, there is not a particle of silver present in the whole composition.

On tin-foil, fitly disposed on a flat table, mercury is to be poured, and gently rubbed with a hare's foot: it soon unites itself with the tin, which then becomes very splendid, or, as the workmen say, is quickened. A plate of glass is then cautiously to be slid upon the tin-leaf, in such a manner as to sweep off the redundant mercury, which is not incorporated with the tin; leaden weights are then to be placed on the glass, and in a little time the quicksilvered tin-foil adheres so firmly to the glass, that the weights may be removed without any danger of its falling off. The glass thus coated is a common looking-glass. About 2 ounces of mercury are sufficient for covering three square feet of glass.

The success of this operation depends much on the clearness of the glass; and the least dirt or dust on its surface will prevent the adhesion of the amalgam or alloy.

Liquid foil for silvering glass globes. Melt together 1 ounce of clean lead, and 1 ounce of fine tin, in a clean iron ladle; then immediately add 1 ounce of bismuth. Skim off the dross, remove the ladle from the fire, and before it sets, add 10 ounces of quicksilver. Now stir the whole carefully together, taking care not to breathe over it, as the fumes of the mercury are very pernicious. Pour this through an earthen pipe into the glass globe, which turn repeatedly round.

Brass. Put 4½ pounds of copper into a crucible, expose it to heat in a furnace, and when perfectly fused, add 1½ pounds of zinc. The metals will combine, forming that generally used alloy, called brass.

For brass which is to be cast into plates, from which pans and kettles are to be made, and wire is to be drawn, braziers use calamine of the finest sort, instead of pure zinc, and in a greater proportion than when common brass is made; general 56 pounds of calamine to 34 pounds of copper. Old brass, which has been frequently exposed to the action of fire, when mixed with the copper and calamine, renders the brass far more ductile, and fitter for the making of fine wire, than it would be without it; but the German brass, particularly that of Nuremburg, is, when drawn into wire, said to be preferable to any made in England, for the strings of musical instruments.

Pinchbeck. Put into a crucible 5 ounces of pure copper; when it is in a state of fusion, add 1 ounce of zinc. These metals combine, forming an alloy not unlike jeweller's gold; pour it into a mold of any shape. This alloy is used for inferior jewellery.

Some use only half this quantity of zinc, in which proportion the alloy is more easily worked especially in the making of jewellery.

Prince's metal. Melt together 3 ounces of copper, and 1 ounce of zinc: or 8 ounces of brass, and 1 ounce of zinc.

Another.—Melt in a crucible 4 ounces of copper, and when fused, add 2 ounces of zinc; they will combine and form a very beautiful and useful alloy, called Prince Rupert's metal.

Bronze. Melt in a clean crucible 7 pounds of pure copper: when fused, throw into it 3 pounds of zinc, and 2 pounds of tin. These metals will combine, forming bronze, which, from the exactness of the impression which it takes from a mould, has, in ancient and modern times, been generally used in the formation of busts, medals, and statues.

Bell metal. Melt together 6 parts of copper, and 2 of tin: These proportions are the most approved for bells throughout Europe, and in China. In the union of the two metals above mentioned, the combination is so complete, that the specific gravity of the alloy is greater than that of the two metals uncombined.

Brass solder for iron. Thin plates of brass are to be melted between the pieces that are to be joined. If the work be very fine, as when two leaves of a broken saw are to be joined together, cover it with pulverized borax, melted with water; that it may incorporate the brass powder which is added to it: the piece must be then exposed to the fire without touching the coals, and heated till the brass is seen to run.

To gild leather. In order to impress gilt figures, letters, and other marks upon leather, as on the covers of books, edgings for doors, etc., the leather must first be dusted over with very finely powdered yellow resin, or mastich gum. The iron tools or stamps are now arranged on a rack before a clear fire, so as to be well heated, without becoming red hot. If the tools are letters, they have an alphabetical arrangement on the rack. Each letter or stamp must be tried as to its heat, by imprinting its mark on the raw side of a piece of waste leather. A little practice will enable the workman to judge of the heat. The tool is now to be pressed downwards on the gold leaf; which will of course be indented, and show the figure imprinted on it. The next letter or stamp is now to be taken and stamped in like manner, and so on with the other; taking care to keep the letters in an even line with each other, like those in a book. By this operation, the resin is melted; consequently the gold adheres to the leather; the superfluous gold may then be rubbed off by a cloth, the gilded impressions remaining on the leather. In this, as in every other operation, adroitness is acquired by practice.

The cloth alluded to should be slightly greasy, to retain the gold wiped off (otherwise there will be great waste in a few months); the cloth will thus be soon saturated or loaded with the gold. When this is the case, these cloths are generally sold to the refiners, who burn them and recover the gold. Some of these afford so much gold by burnings, as to be worth from a guinea to a guinea and a half.

To tin copper or brass. Boil six pounds of cream of tartar, four gallons of water, and eight pounds of grain tin, or tin shavings. After the materials have boiled a sufficient time, the substance to be tinned is put therein, and the boiling continued, when the tin is precipitated in its metallic form.

To tin iron or copper vessels. Iron which is to be tinned, must be previously steeped in acid materials, such as sour whey, distiller's wash, etc.; then scoured, and dipped in melted tin, having been first rubbed over with a solution of sal ammoniac. The surface of the tin is prevented from calcining, by covering it with a coat of fat. Copper vessels must be well cleansed; and then a sufficient quantity of tin with sal ammoniac is put therein, and brought into fusion, and the copper vessel moved about. A little resin is sometimes added. The sal ammoniac prevents the copper from scaling, and causes the tin to be fixed wherever it touches. Lately, zinc has been proposed for lining vessels instead of tin, to avoid the ill consequences which have been unjustly apprehended.

Metallic watering, or *blanc moire.* This article, of Parisian invention, which is much employed to cover ornamental cabinet work, dressing boxes, telescopes, opera glasses, etc., etc., is prepared in the following manner:

Sulphuric acid is to be diluted with from seven to nine parts of water: then dip a sponge or rag into it, and wash with it the surface of a sheet of tin. This will speedily exhibit an appearance of crystallization, which is the moire.

This effect, however, cannot be easily produced upon every sort of sheet tin, for if the sheet has been much hardened by hammering or rolling, then the moire cannot be effected until the sheet has been heated so as to produce an incipient fusion on the surface, after which the acid will act upon it, and produce the moire. Almost any acid will do as well as the sulphuric, and it is said that the citric acid dissolved in a sufficient quantity of water, answers better than any other.

The moire may be much improved by employing the blow pipe, to form small and beautiful specks on the surface of the tin, previous to the application of the acid.

When the moire has been formed, the plate is to be varnished and polished, the varnish being tinted with any glazing colour, and thus the red, green, yellow, and pearl coloured moires are manufactured.

To gild by burnishing. This operation is chiefly performed on picture frames, mouldings, beadings, and fine stucco work. The surface to be gilt must be carefully covered with a strong size, made by boiling down pieces of white leather, or clippings of parchment, till they are reduced to a stiff jelly; this coating being dried, eight or ten more must be applied, consisting of the same size, mixed with fine Paris plaster or washed chalk; when a sufficient number of layers have been put on, varying according to the nature of the work, and the whole is become quite dry, a moderately thick layer must be applied, composed of size and Armenian bole, or yellow oxide of lead: while this last is yet moist, the gold leaf is to be put on.

For this purpose a leaf of gold is spread on a cushion (formed by a few folds of flannel secured on a piece of wood, about eight inches square, by a tight covering of leather,) and is cut into strips of a proper size by a blunt pallet knife; each strip being then taken on the point of a fine brush, is applied to the part intended to be gilded, and is then gently pressed down by a ball of soft cotton.

The gold will immediately adhere on being pressed by the cotton ball, and before the size is become perfectly dry, those parts which are intended to be the most brilliant are to be carefully burnished by an agate or a dog's tooth fixed in a handle.

In order to save the labour of burnishing, it is a common, but bad practice, slightly to burnish the brilliant parts, and to deaden the rest by drawing a brush over them dipped in size; the required contrast between the polished and the unpolished gold is indeed thus obtained; but the general effect is much inferior to that produced in the regular way, and the smallest drop of water falling on the sized part occasions a stain. This kind of gilding can only be applied on indoor work; as rain, and even a considerable degree of dampness, will occasion the gold to peel off. When dirty, it may be cleaned by a soft brush, with hot spirit of wine, or oil of turpentine.

The Old Babbit Ax Factory
Had a Quality Product

By W. J. STARKS

June 1979

At the time of writing, Mr. Starks was custodian of Deer Hill Reservation.

On a recent stormy day I thought I would make me an Ax helve. After blocking it out with an ax, I took down the old draw shave to smooth it, and I noticed it was getting rather rusty. It was one my grandfather had and one with which my father earned money to get married, shaving shingle and hoop poles. I think a whole lot of it and intend to keep it shiny, so I took some sandpaper and began scouring it.

Soon the name A. J. Babbit showed up, and this reminded me of the old Babbit ax factory of Windsor, Massachusetts, where axes and draw shaves and butcher knives were made 75 years ago. I had a cousin who was a grinder and polisher there, and I used to visit him sometimes and watch the work go on. One day while there I caught my first and only otter, so I remember about the ax factory very well indeed.

It was said that Mr. Babbit learned his trade in England. Anyway, he was an expert at the line, and when one got a Babbit ax, draw shave or knife, he got one as near perfect as could be made. The steel with which these tools were made came in short bars in boxes with the name of a firm in England stamped on them.

To make axes, one of these bars of steel was heated red hot at the forge, and blocks the right size to make the head or pole of an ax were cut off on an anvil. These blocks were then hammered into the right shape, heated to a white heat, set on one end on the anvil and split lengthwise through the center of the other end. Then a piece of steel which had already been shaped for the bit or edge of the ax was fitted into this split and welded. Then a hole was swedged out of the ax for the helve or ax handle to fit into. After shaping the ax up as well as possible on the anvil, it was tempered. This tempering process was a very particular job and required the highest kind of skill. For a long time Mr. Babbit himself was the only man in the factory to do this, but later William Reid was taught the secret and, after Mr. Babbit's death, bought and carried on the business.

After tempering, the axes were placed in small boxes on wheels, and when a certain number were ready, these wheeled boxes were drawn into the grinder room. The grinding was done in a peculiar way. The grindstones were hung just as near the floor as they could be. Beneath each stone was a long, narrow trough, full of water running in and then out constantly, and the grindstone revolved so the lower edge passed through the water. This served two purposes. It kept the grinding surface of the stone wet and cool so the ax would not get overheated by friction and have the temper drawn, and it also kept all grit from the ax and stone washed off out of the way.

About six feet to one side of the grindstones, set firmly in the floor, were blocks or posts about six inches square and two feet high. To the top of each of these posts was fastened, by screws, an ash board about an inch thick and six inches wide. On the upper side of the other end of the board was a sort of cushion. The boards were fastened so the other end was about an inch above the whirling grindstones. A man seated on the cushion with his feet one on each side of the grindstone could with his weight press the board down onto the stone. By the side of each cushion was an ash stick made so that it would fit snugly into the eye of an ax and project on each side of the ax about six inches.

The grinder would put an ax onto his stock, grasp each end of the stick and place the ax under the end of the board so the edge of the ax would be just right on the revolving stone. Then with a sort of teetering motion, putting his weight down on the ax for a moment and then raising it, he would grind the ax. This was a long, tedious job, for although the ax had been hammered as near to a cutting edge as possible, yet to get it in shape to really cut took a long time.

After grinding came the polishing. This was done on revolving stones of different sizes and of different degrees of hardness and coarseness. After the polishing came the stamping of the head with Babbit's name, and the weight of the ax. When the axes were stamped, they were ready to be painted and then boxed ready for delivery.

A new Babbit ax was a thing of beauty and a joy forever. I think one great reason these axes were so satisfactory under all the conditions and tests was the rigid inspection Mr. Babbit gave them at all stages of their manufacture. This very perfection in time brought Mr. Babbit a new branch of the business. In those days, much wood was cut to make charcoal for the Richmond Iron Company in Cheshire and the Follet Lime Company in Adams, and, of course, the axes got dulled and had to be ground sharp again.

Grinding on stones turned by hand soon caused the ax to become thick at the edge and chisel shaped, so it did not enter the wood easily. Finally a man asked Mr. Babbit if he could not draw out the bit of the ax and make it thin again by heating in the forge and pounding on the anvil, as it seemed too bad to throw away such finely-tempered axes because they had become thick.

The steel in the axes was so good that the ax stood this redrawing well and, after retempering and grinding, was good for a lot more chopping. This one job brought in more, and soon Mr. Babbit was doing quite a business drawing out used axes that he had made.

What was true in the way of perfection in axes was just as true of his draw shaves and butcher knives. Each was good till all worn out. I have seen a Babbit butcher knife that was perhaps an inch-and-a-half wide when new worn away until it was not more than half an inch wide, with just the back left, and it would still take a better cutting edge than any knife that could be bought today.

The Spinning Wheel
A Neglected Tool

By WILLIAM RALPH

June 1972

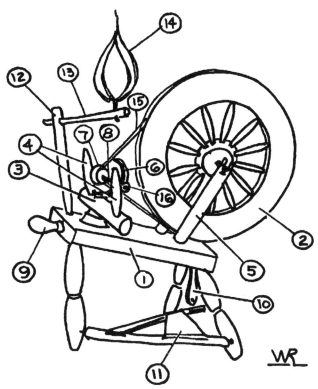

*Figure 1. 1) bed or table; 2) flywheel; 3) mother-of-all; 4) maidens; 5) wheel uprights;
6) flyer; 7) whorl; 8) spool; 9) tension screw; 10) footman; 11) treadle; 12) distaff
upright; 13) distaff crossarm; 14) birdcage-type distaff; 15) drive band; 16) spindle (eye).
All drawings and photographs by author.*

Does the spinning wheel qualify as a tool? If we think of a tool as a device manipulated by man to help him cut, shape, or rearrange natural materials into more useful objects, then the spinning wheel is indeed a tool and worthy of consideration by those who take an interest in tools and early technology. Regrettably, this most important tool of everyday life in times past has been largely neglected. What little interest there is in spinning wheels seems to be confined mostly to people who buy them as decorative accessories to effect an Early American look. A few others spin on them for a hobby, but very few consider the spinning wheel as a tool of serious historical significance. When William Langdon wrote his book *Everyday Things in American Life, 1607-1776,* he barely mentioned the spinning wheel— only in a single sentence— but devoted almost an entire chapter to Chippendale! Yet the spinning wheel ranks in importance with the ax and plough in early American life in helping to provide the three essentials of life: food, clothing and shelter.

In this article I am not going to review the history of the spinning wheel or describe its operation in great detail. These subjects have been adequately covered by previous articles in *The Chronicle*. I am going to describe my personal observations, theories, and conclusions formed over a period of approximately eight years, during which time I closely examined and restored several hundred wheels. For the sake of simplicity and because the subject is so broad, I am going to confine all my remarks here to the three-legged Saxony-type flax wheel *(Figure 1)* which was sometimes also used for wool and its variations.

For those totally unacquainted with spinning and spinning wheels, I hastily describe both in the briefest possible terms. Handspinning is the continuous drawing-out of a few relatively short fibres from a much larger mass of fibres previously combed so that the fibres lie parallel to each other, and twisting them between the fingers into a continuous thread which is then spooled or stored on a bobbin to prevent kinking and to set the twist. The spinning wheel provides only the twisting motion and the spooling.

The spinning wheel is not a machine in the sense that it produces a product automatically. As other basic tools, it requires skillful manipulation. An untrained or unskilled person cannot operate it. Here is an interesting fact that every spinner knows, but few others realize: the spinning wheel does not spin; the spinner spins. The thread is produced outside the wheel and at varying distances from it by the fingers of the spinner. By the time the thread reaches the wheel it is already completely formed. The wheel merely provides the twisting motion which travels up the already completed thread from the eye of the spindle to the fingers of the spinner, where it twists and combines the drawn-out loose fibres into more thread. The finished thread is fed to the spool or bobbin by the flyer, that horseshoe shaped object with the hooks on it. The flyer is firmly attached to the spindle and rotates at the same speed. The bobbin, however, which is separate from but concentric to the spindle, revolves at a slightly faster rate by virtue of its smaller pulley. The bobbin therefore pulls in and spools the finished thread upon itself.

In another sense, the spinning wheel is truly a machine. It employs several mechanical principles including the wheel, the crank, the screw, and the relatively sophisticated principle of differential motion for the spooling mechanism. In its construction, alignment and balance must be carefully observed; otherwise the machine will not work. The craftsman who makes it must have a good understanding of the mechanical principles involved. What is even more astonishing is that the machine is made almost entirely of wood. The early wheels used no nails, no screws, and no glue in their construction. It is possible to take an early spinning wheel completely apart, including the spokes of the flywheel, and place all the parts in a box only slightly larger than a shoe box. The only tools necessary for this operation are a wooden mallet and drive pins to remove the wooden pegs. Conversely, it is possible to reassemble the wheel using the same tools, and with but a few minor adjustments, sit down and spin on it. We are told that itinerant merchants carried the spinning wheels disassembled, with the exception of the flywheel, in their wagons to conserve valuable space, and assembled them on the spot as they sold them. Since this was in the days before interchangeable parts, the merchant had to be careful to keep the parts for each wheel separate and apart from the parts for the other wheels. Perhaps this explains the numbers occasionally written with pen and ink on spinning wheel parts. They may have served to identify all the parts which belonged together.

The spinning wheel has five essential parts, all of which must obviously be present for the wheel to be complete; they are: 1. The Flywheel, which serves to drive the flyer assembly at a much higher rate of speed than its own rotation; 2. The Treadle with which the spinner's foot operates the wheel; 3. The Flyer Assembly consisting of the spindle, flyer whorl (pulley), and bobbin (spool); 4. Tension Adjusting Device which controls the slippage between the flyer and flywheel necessary for good spinning; and 5. The Frame consisting of the table (bed), legs, wheel posts, mother-of-all, maidens, etc. to hold the whole thing together. Of course, it also needs a drive belt, a stout cord looped around the flywheel and the flyer assembly twice, once for the spindle pulley and once for the bobbin pulley. If the wheel is designed for spinning flax, it will also have a distaff assembly to hold the flax in a convenient position for spinning. This usually consists of three parts, the distaff post, the crossarm, and the actual distaff.

Those of us who collect tools know that one of the most common tools from bygone days today is the wooden plane. Few tool collectors would suspect, however, that an equally common tool was the spinning wheel. Marion Channing, in her book *The Magic of Spinning,* states that in the year 1800 there were approximately three and one-half million spinning wheels in the United States alone! It is quite reasonable to assume that until about 1840 every rural family had at least one spinning wheel. Wooden planes are small, sturdy, and practically indestructible, and often lie hidden in the nooks and crannies of old buildings where they were placed many years ago by some long-forgotten carpenter. Thus they have survived the ravages of time. Spinning wheels are bulky and fragile and irresistible to small children who, when they find them, twist and turn them into oblivion. Being useless to modern society except as decorations or mementos, they were converted to garden ornaments and planters, had chairs or lamps made from them, or simply were discarded. Today they are relatively scarce, but at one time they were a great multitude.

Who made this vast horde of spinning wheels and how were they made? Something of a mystery surrounds this question and few reliable answers are to be found. We know who made the early clocks, the guns, the chairs, the tools, the pottery, etc. In many cases we have documented evidence of the size of the shops, the number of workmen, and even the production figures, but similar information does not exist about spinning wheel makers to the best of my knowledge. Many historians, professional or amateur, believe that spinning wheels were made by wheelwrights in their spare time (a popular concept), by a single craftsman making them one at a time, at home by a handy pioneer, or some equally impractical or unlikely suggestion. Only rarely have I heard the opinion that spinning wheels were made in a large specialty shop or factory, by specialized craftsmen, as were most of the other common items of early America. As will be seen, the making of spinning wheels requires special tools and special skills not likely possessed by moonlighting wheelwrights or do-it-yourself farmers.

First and foremost, spinning wheels were made by woodturners, their principal tool being the lathe. This immediately disqualifies the wheelwright, about whom more will be said later on. The average spinning wheel has 42 parts, not counting flyer hooks, wooden pins, etc. Of these, four are made of iron, two of leather, five of wood but not turned, and the remaining 31 parts of wood made entirely or in part on the lathe. The iron parts such as the spindle and the wheel crank were most likely purchased ready-made, probably from England, although we know that some were made here in the United States (Henry J. Kauffman, *Early American Ironware*). Perhaps flyers and bobbins were also available and were purchased ready-made by spinning wheel makers with limited facilities. It is fairly certain that they were sold by stores and itinerant merchants for replacement parts. The flyer is the most fragile part of the spinning wheel and at a time when everything was mended and nothing discarded, many a wheel must have gone through several flyers. These replacement flyers can often be recognized by observing the evenly spaced lines inscribed at the tips of the flyer. I believe these were put there to facilitate cutting the tips to an equal length when the flyer was installed. The flyers were probably made to fit the largest bobbins and cut to size as required. Sometimes the available replacement flyer was too wide and a portion of the "mother-of-all" had to be cut away to clear the tips of the flyer. Incidentally, American and English flyers were always made in one piece; some continental flyers were made with separate "wings" glued or keyed to a center piece.

Getting back to the wheelwright myth, I find this association totally unacceptable. Why not spinning wheels made by coopers, coachmakers, or any other unrelated craft? There is no similarity between a wagon or carriage wheel and the flywheel of a spinning wheel except for the name. Their function is entirely different; their construction is entirely different. Wagon wheels are constructed to endure compression stresses imposed by the weight and the load of the wagon. They are usually "dished." Their rate of rotation is relatively slow. They have only one turned part, the hub, into which the mortises for the spokes are cut with a chisel. The spokes are fashioned with a draw knife and spoke shave. The rim sections (felloes) are cut with a saw. The wheel rotates about a fixed axle.

The flywheel of a spinning wheel carries no load. Its stresses are centrifugal, giving the wheel a tendency to fly apart. The wheel rotates at a relatively high speed. The wheel is not "dished." All parts of the flywheel except the iron crank are turned on a lathe. Running "true" is the most important requirement and even a slight wobble, which could be tolerated in a wagon wheel, will throw the drive belt of a spinning wheel.

The "wheelwright theory" is so prevalent, however, that it cannot be dismissed entirely. Was it originally quoted by some eminent authority and thereafter copied by all who followed, or could it be that in times past the name "wheelwright" was also used for a spinning wheel maker? Alice Morse Earle thinks so and mentions this in her book *Home Life in Colonial Days*.

Let's take a look at some of the construction details of a typical spinning wheel of about 1800. Looking first at the table or bed, we see that it was split from a short oak log. For this reason it is slightly wedge-shaped, being wider at one side than at the other. A piece of wood split from a log will not warp. In constructing the wheel it was customary to place the narrow side towards the spinner. After seasoning for two years or more, the top and sides were planed smooth; the bottom was left rough. Sometimes a bevel was planed on the underside of the side facing the spinner, probably to prevent the spinner's clothing from catching in the rough wood. Occasionally a "bead" molding was planed along the top edges, most likely for the same purpose. Sometimes a meticulous worker would plane and square the table on all sides, but examination of the end grain will betray its split-log origin. Incidentally, this is one way to tell the twentieth century and reproduction wheels from the old ones. Sometimes we find decorations applied in the form of scalloped ends or the top decorated with a design made by using a small circle punch. Occasionally the maker stamped his initials or his name into the end grain, or even in the table top.

Next, the mortise for the tension screw was cut with a chisel and the holes for the legs, wheel posts, and distaff bored, probably with a hand brace. None of the holes were at right angles to the table since the table's normal position is slanted. If we were doing the job today, we would insist upon a drilling jig to guide our drill so that we could drill our holes precisely at the angle specified on the drawing.

The early craftsman was not overly concerned with numerical measurements. His was a time when the yardstick and the blueprint did not yet occupy a position of dominance. Dimensioned drawings for common items were unheard of. Measurements were taken from marks made on tool handles, along the edges of work benches, from templates, from prototypes, or simply from experience, without regard for their numerical value. The early craftsman knew when it was important to be precise and when he could approximate. Take the legs, for example. With a three-legged spinning wheel it doesn't much matter if one of the front legs differs by five or even ten degrees from the position taken by the other, or if one leg is slightly longer than the other. The wheel may list slightly, but it will spin just as well. Three legs always stand fast and steady on the most uneven ground, whereas four-legged devices will wobble unless the floor is perfectly level. This is why most of the early candle stands had three legs.

The spinning wheel maker made much use of a construction method whose principle is almost forgotten in woodworking today. This is the tapered hole and its counterpart, the tapered post. It operates on the principle of the wedge and was used wherever a post had to be securely held but not permanently fastened. When pressed down into its mating hole, the post can neither wobble, turn, or shift position. Pulling or tapping the post out of the hole quickly releases the grip. The machinist will recognize the principle in the Morse Taper used to hold a chuck in the lathe.

The wheel posts supporting the flywheel were almost always tapered. An important advantage of this arrangement was that by inserting small shims around the perimeter of the hole, the posts could be made to lean slightly in a desired direction, thereby facilitating alignment of the flywheel. Also, should the post shrink in diameter and become loose, a slight tap would immediately reset it. The "maidens," the two small posts supporting the flyer assembly, were often tapered. Thus they held the flyer assembly in rigid alignment, but could easily be removed by the spinner to exchange a full bobbin, although other arrangements were also used to accomplish this. The distaff was usually a tapered post so that it would not shift position when in use, but could easily be swung aside when desired.

When we consider the construction of the flywheel, it becomes immediately apparent that a special lathe was needed to turn the rim *(Figures 2 and 3)*. Let us try to reconstruct the procedure probably used by the early wheel maker. The blank from which the rim was to be turned was made by taking four pieces of seasoned oak, split from a log, and planed to size and joined together into a frame, 18 to 24 inches square, depending on the size of the wheel to be made. The parts were held together by a mortise and tenon construction using draw pins and almost always without benefit of glue *(Figure 4)*. The frame was then fastened to a wooden faceplate, most likely with bolts and wing nuts. The faceplate diameter was smaller than the diameter of the finished rim so that the turner had access to both sides of the rim. The faceplate and frame were then affixed to the outboard arbor of what must have been a special lathe, for it had to have a very slow but powerful rotation in a continuous direction (as opposed to the oscillating motion of a pole lathe), to take advantage of and not be hindered by the momentum of the heavy frame. The turning speed must have been around 100 rpm or less. Perhaps the lathe was driven by a hand crank attached directly to the other end of the arbor and equipped with a large flywheel to smooth the motion. Perhaps water power was used to drive the lathe or maybe it had a foot treadle, although I doubt that one man's foot power would be sufficient to do the job. In any case, it must have been a specialized tool.

Another part requiring special attention was the hub. Turning the hub was no problem, but drilling the holes for the crank and the spokes had to be precisely done, otherwise the wheel would wobble. It is reasonable to assume that some sort of jig or drilling machine was used to drill these holes. Interestingly, there is no evidence of indexing the holes for the spokes on most wheels I have examined, for the angle between spokes on the same wheel can often vary by as much as five degrees. Another hole requiring precise alignment was the tension screw bearing drilled into the end grain of the table. This was further complicated by the fact that the outside bearing was larger in diameter than the inside bearing. Perhaps I am not giving the early craftsman enough credit. Perhaps he was able to drill the required holes exactly by "eye" and did not require any special guides or jigs.

Figure 2. A conjectural drawing of a spinning wheel lathe. In addition to the hand power, the wheel could be driven by treadle or water power. The crank-driven flywheel would furnish a turning speed of about 100 rpm and a fairly high surface speed at the rim.

PLAN VIEW

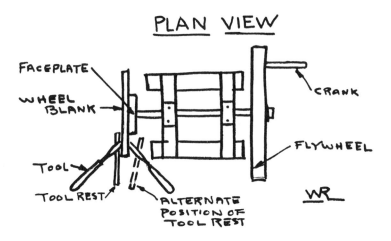

FACEPLATE

WHEEL BLANK

CRANK

TOOL

TOOL REST

FLYWHEEL

ALTERNATE POSITION OF TOOL REST

WR

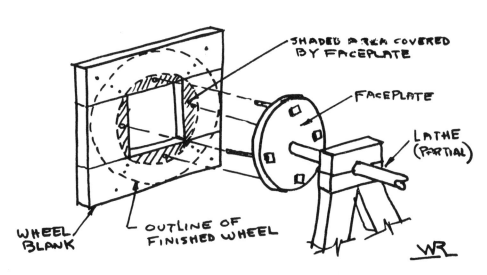

SHADED AREA COVERED BY FACEPLATE

FACEPLATE

LATHE (PARTIAL)

WHEEL BLANK

OUTLINE OF FINISHED WHEEL

WR

Figure 3. Details of a conjectural spinning wheel lathe. A lathe similar to this would be the type a highly specialized shop could use to turn out flywheels on a large scale. A slightly different face plate would be needed to turn the inside of the rim.

Figure 4. This plainly shows the grain of the wood at right angles at the joint, betraying the original frame construction of the blank. The pegs holding the felloes together can be plainly seen, as well as the peg holding the spoke.

Constructing the rest of the wheel presented no special problems. The tension screw and its mating thread was cut with a screwbox and matching tap, a common tool of its day. The remaining parts were mostly simple spindle turnings.

The spool or bobbin is interesting. It was made in three parts: two end pieces turned of fruitwood and a middle part, about four inches long and ⅝ inch in diameter, made of oak with tapered ends to fit corresponding tapered holes in the end pieces, the whole thing being assembled with a press fit. Through the center piece a ⅜ inch hole was drilled longitudinally, leaving the walls of the resulting tube only about ⅛ inch thick. I suppose that this operation must have required a special tool. Spindled pieces of leather were inserted into the bobbin ends to act as bearings and to reduce the diameter of the holes at each end to fit the spindle shaft, which was tapered and therefore of a different diameter at one end than at the other. Incidentally, to remove the bobbin from the spindle, the whorl or pulley must first be removed. Sometimes the whorl is simply a friction fit, but more often it is secured to the spindle with a left-hand thread. To remove the whorl, therefore, it will be necessary to turn the whorl in a *clockwise* direction.

Based on the foregoing, let us try to visualize the shop of an early spinning wheel maker. I believe the shop was a factory, not the kind of factory we think of today, with clanking machinery and rows of bored and tired workers, but a specialty shop making only spinning wheels in quantity, perhaps a dozen at a time, with specialized tools, and employing perhaps a half dozen or more craftsmen and apprentices. The craftsmen probably had their specialized duties, one man doing only the rim turning, another

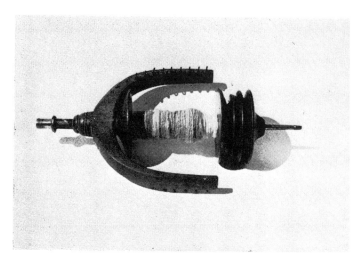

Figure 5. The assembled flyer. One can clearly see the two different sizes of the pulleys. The belt passes over both, driving the bobbin or spool faster than the flyer furnishing the tension.

Figure 6. The flyer disassembled. The pulley fastens onto the shaft of the spindle by a left-handed thread so that it will not come off in operation. The spool spins freely on the shaft so that it can be turned at a faster speed.

Figure 7. Detail of the eye of the spindle and the flyer. One can see the grooves cut by the flax passing from the eye to the spool.

making spindle turnings, a man to do the final assembly and fittings and yet another to select, cut and split the wood for the tables and billets for turning, and so forth. How about the tools and supplies in such a shop? Besides the special rim lathe and perhaps the hub drilling apparatus, we would find a spindle lathe, turning tools, froe, adze, axe, jack plane, molding plane, tenon saw, box saw, brace and bits, taper augers, bow drill, screw box and tap, draw knife, spoke shave, mallet, mortise chisel, gouge, pliers, dividers, calipers, square, bevel gauge (probably with fixed bevel), marker gauge templates, prototype parts, spindles, wheel cranks, flyer hooks, and leather.

I believe a large shop of this kind existed on Long Island. Here we often find a distinctive, lightly built, medium sized wheel with the name "I. PARISH" boldly stamped on the top of the table. The frequency with which these wheels turn up indicates to me a fairly extensive operation. Measurements of these wheels vary slightly, but the characteristics are constant: the stubby "maidens," the thin rim without inside flanges, the simple urn-shaped spokes, the distinctive design of the turnings, even the notch at the center of the flyer. For several years I have tried to discover where and when these wheels were made, but without success. Some say it was in Oyster Bay, others believe it was Glen Cove. Was it in 1870 or in 1830? No one seems to know. Information from informed readers in this matter will be appreciated.

Does anyone know of a documented spinning wheel factory? I have seen the tools and reconstructed shops of wheelwrights, coopers, carriage makers, harness makers, etc., but I have not yet seen, nor does there exist to the best of my knowledge, a collection of tools from a spinning wheel maker's shop. Has the special wheel lathe ever been found? Perhaps it was found and not recognized.

For 350 years, from the date of its inception to the end of the handspinning era, the Saxony wheel did the job. Now handspinning as a way of life is gone, as are many of the early crafts. We who are dedicated to preserving the knowledge and the artifacts of these early crafts should consider the spinning wheel. Its unique and important place in the daily lives of our ancestors deserves this consideration.

The Many Uses of Horn

By MARY EARLE GOULD

December 1962

Horns of animals have been used for many centuries. Prehistoric nations made daggers, handles, pick-axes, dart heads, batons and implements. These were made mostly from reindeer and the red deer and they have been found carved and etched.

In ancient China, the horns of the rhinoceros were used in making a cup which was supposed to have medicinal value. In Biblical times, horns were made into receptacles and we read in First Samuel XVI 13th verse, "Samuel took the horn of oil and annointed him [David]."

Horns were used in their natural shape for cups, handles of knives and forks, the whetstone holder called the mower's horn, powder horns and horns for blowing. Horns for blowing were used in the fox hunt, calling from the fields, for a fire, and for Sunday meeting.

For cups and powder horns, the ends were sealed with a wooden block, the cups having the ends cut off the desired length with a saw. Most commonly in the olden times, a horn cup was used as a holder for drinks, such as mead and ale. The cups were artistically banded with silver and were ornamental at any fireside where drinks were served. From this came the expression "to take a horn" meaning to take a drink of intoxicating liquor.

Making spoons from horn was a common thing. Scotland made them until the 19th century. Many sayings came from them such as: *Solemnly swear by the great horn spoon.*

Again:

The more he thought on it, the madder he grew,
Until he vowed by the great horn spoon,
Unless they did the thing that was right,
He'd give them a l'cking and that right soon.
Unknown author

Horns of the ox, sheep and goats can be split into sheets or plates after they have been soaked and boiled in oil and roasted over a flame. "Horns were to the hornsmith as hides to a tanner." Long soaking broke the horns into layers, each with a pith and kernel. It took 15 days in Summer and a month in Winter. Then they were put into boiling water, then sawed lengthwise with a thin-edged chisel. The horn was separated into layers, again boiled and then placed in a cavity of wood with a sheet of hot iron between. Finally, they were hardened in water. Two pieces of horn could be welded together at the edges by steeping them in hot water.

Thin sheets at one time were used for window panes, before glass was available. And the sides of lanterns were filled with horn, bringing about the early name of lanthorn, shortened eventually to lantern. History tells at great length about the horn book, the primer for learning, at the time of George II in England. The alphabet was put into one side of a frame with a single piece of horn over it. The opposite side had a form of exercise and the Lord's Prayer, with another sheet of horn. The frame was made with a handle by which the book was held. Or else it could be hung at the belt by a cord run through a hole at the end of the handle.

Hoofs of oxen went into combs and buttons. After the horns were steamed and cut, pressed under weight, boiled in oil, the sheets were sawed into the proper width for combs. Then they were rubbed smooth with brick dust, ashes or rotten stone. Instead of steaming the sheets, they were sometimes held over a flame, in long tongs. When in the vise, the combs were sawed into teeth, with a thin saw. The Ancients buried toilet articles and combs in tombs of their deceased.

Knives and fork handles, umbrella and cane handles, drawer knobs, boxes and rims of boxes and buttons were other things made of horn. Nature provided many materials for man by which he existed.

Traditional Soap-Making on the Frontier

By PER E. GULDBECK

June 1963

(Editor's note: In preparation for the 1962 Seminar at the Farmers' Museum, Cooperstown, New York, Mr. Per Guldbeck, Research Associate, and Mr. George Campbell, Curator, experimented with the art of soapmaking as a demonstration for the Seminar. This article is the results of that experiment. In the words of Mr. Guldbeck: "the proof of the pudding is that we even took our annual spring bath with our own soap!")

Soap-making display, Farmers' Museum, Cooperstown, New York.
Leach-tub made from old log.

Highly scented and coloured toilet soaps from France, England and Holland were available in 18th and 19th century America, as were plainer soaps. The better grades were not only attractively made up, but were efficient soaps, made from barilla or soda and olive oil; while the plainer soaps were made from a base of hardwood ash and lime, plus animal fat or cheaper vegetable oils such as linseed. But the various monopolies, taxes, and cost of materials made the better soaps too expensive for the average farm family, while the common grades of soap could be duplicated inexpensively in the farmer's own backyard. To this end the farm wife saved her fireplace ashes and left-over fats until there was enough accumulated to warrant a session of soap-boiling.

STEPS IN PREPARING SOAP

Making the leach-tub: In order to obtain the needed lye for soap, farmers most commonly constructed a wooden tub from something like a two-foot section of an old log that had started to rot in the center. This section of log was propped up on stones, a few inches off the earth, and a small fire built under it. Gradually the rotten side of the interior caught fire and was slowly consumed in a steady even glow of flames that ate its way evenly through the wood. By using a series of long-handled chisels, slicks, or sharpened bark spuds the farmer could then start to enlarge the burnt hole until the log was sufficiently hollowed out, leaving walls perhaps about two inches thick.

At the bottom edge of the tub were carved a series of half-round openings, about an inch in diameter, to allow the lye water to run out when the tub was operating. In order for the log to contain the ashes, a retaining support was set in the bottom of the tub, consisting of small sticks and/or interlaced twigs, over which was laid several inches of grass or hay.

A large flat flagstone approximately a foot larger in diameter than the tub, was laid on a flat stump, which was a foot or foot-and-a-half high. This leaching stone had a series of grooves chiseled on its face which would direct the flow of lye water into an iron kettle set directly below it.

Lye-running: The tub was set atop the stone, filled almost to the top with finely sifted hardwood ash, and a bucket or two of rainwater was poured in. Hardwood ash was considered to make much better lye than ash from coniferous trees; and rainwater, which was free from dissolved minerals, was also preferred over spring water or well water.

As the water gradually trickled through the ash, it dissolved or leached out the potassium salts, and the resulting solution drained off into the iron kettle. Although European soap boilers mixed chalk or quicklime with the ash to produce a stronger lye, American housewives usually used plain hardwood ash and depended on re-filtering or boiling down a solution that was not judged to be of proper strength.

Lacking a proper stone or tub, a person could still make lye by simply boiling up a mess of hardwood ash in water. The brew was set aside to cool, and the ash gradually settled to the bottom in a heavy sludge. The clear liquid was decanted and tested for strength, and boiled again if necessary.

Testing for strength: A fresh egg was dropped into the lye water, and if the water's specific gravity was just a little less than that of the egg, that is, if the egg v e r y slowly sank to the bottom, it was considered of proper strength.

Preparing the fats: If a family had enough extra deer or sheep tallow or bayberry wax, it might be used in soap making, although these ingredients were most commonly saved for candles. The usual fats for soap-boiling at home would be whale oil, fish oil (for those who were near the coast), vegetable oils, left-over cooking fats, and grease and oils left over from skinning of game or domestic animals. If there was any lard, tallow, or suet, it would first be tried out over a slow fire, and then squeezed or crushed to extract all the oil. Then all the fats and oils would be boiled up in a kettle of water and skimmed off and strained through a fine cloth. This helped to eliminate any old spices, bits of hair, flesh, char, or any other foreign material which might contaminate the fats or give the final product a bad smell.

Soap-boiling: The fat mixture was sometimes tempered by mixing a proportion of weak lye-water with it and boiling it for a few minutes before adding the main lye solution to it. This was thought to insure the soap mixture against curdling. Others simply added the lye water directly into the boiling fat and stirred constantly as the lye was added. The resulting brew might be boiled anywhere from 10 minutes to an hour until it was thought to be done. This depended in some measure on whether the housewife wanted soft soap or hard soap. The best clear yellow soft soap was made by a quick furious rolling boil for about 10 to 15 minutes.

The proportion of lye to fat varied according to the maker, who judged the correct proportions either by rubbing some between his thumb and finger to see whether it had the correct feel, or by tasting it to find out whether it had a sweet taste or a harsh bite. Before the days of scientific controls and qualitative measuring, there were probably as many "secret" techniques and rules-of-thumb for ensuring good soap as there were for making butter "come" in the churn.

Additives: Musk, floral essences, oil of rhodium, lemon, cloves, and other essential oils were often added to enhance the product, or to disguise the odor when blubber or rancid fats were used. Various combinations of salt, sal soda, or bayberry wax were added when hard soap was wanted. When alcohol was added a clear, gelatinous soap resulted; and for clearing up skin ailments, compounds with sulphur, birch- or pine-tar were mixed in.

Shaping the soap: Soft soap was poured into a bucket to be kept handy until the housewife wanted to wash the floors, or do the family wash, at which time the soap could be easily poured out and dissolved. Hard soap could be taken when still warm and moulded by hand into balls, or poured into wooden break-away-sided individual cake moulds. If a large batch was run, it would be poured in a large single mould, and when cool could be sliced with a knife, or a taut wire, and lifted out with a spatula.

Although the plain unscented soft soap did not saponify as well as the hard soda-based soaps, it worked well enough and was efficient for bathing, imparting a softness to the skin; and with its own homely fragrance, long remained a favorite with country people.

Irons

By CHARLOTTE WOODHULL

December 1961

Judging from accounts of the home life of our early ancestors, they were not particularly fastidious about laundering. Their garments, furthermore, were not usually made of washable materials. Considering the scarcity of implements for the job and the crudeness of those available, it is small wonder that laundering was not the important weekly household chore it is today. When washing was necessary, it was done in the streams and the clothing was beaten on boards or stones.

Records of the early 16th century mention "Pressing Machines." One of the earliest contrivances was the "Box Mangle." It consisted, literally, of a box filled with stones and worked by a cranked handle. Articles to be pressed were wound round loose wooden rollers, and over these the box ran backwards and forwards. This sort of pressing was sufficient for sheets or blankets only.

"Mangling Boards" were in use prior to any mechanical contrivance. As with the "Box Mangle," clothes were wound round a single small roller, like a rolling pin, and smoothed with a piece of wood shaped like a wide spatula with a handle at one end, Fig. 1. We find many Dutch examples of these "Smoothing Boards." Some are beautifully carved with hearts and arrows and were given as love tokens to prospective brides. It is said that at Ephrata, Pennsylvania, in the Old Bissell Community of Seventh Day Baptists, the fresh and precious linen used in "Love Feasts" and other religious services was always "smoothed" but never "ironed."

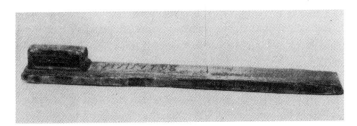

Fig. 1. Smoothing Board. 28" long, 2½" wide, and 1" thick.

With the introduction of starch into England in the middle of the 16th century, all sorts of irons came into use. In the early 17th century a special implement was used for the ruffles, flounces and frills which were the fashion of the times. It was called a "Goffering Iron" or a "Tally Iron," Fig. 2, "Tally" being a corruption of Italian, after the country from which it was introduced. These Goffering Irons were finger-like tubes, ranging from one-quarter inch to one and one-half inches in diameter, mounted on heavy bases and heated by inserting a red hot bar or iron similar to a poker. The starched linen was grasped in both hands and pressed over the hot barrel which made a semi-circular crimp. Often two or more tubes of different sizes were mounted on the base. One type of goffering iron had closed tubes heated in the center by charcoal, Fig. 3. For general

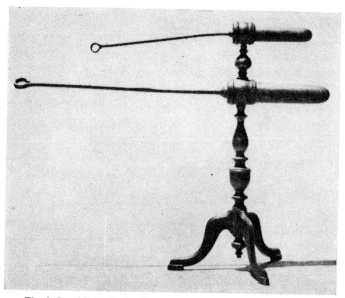

Fig. 2. Double Goffering Iron with snake feet — brass standard.

smoothing, there was the Box Iron—the fore-runner of the familiar Flat Iron. It was heated in the same manner as the Tally Iron—by inserting an iron heater through a door at the wide end, Fig. 4. Some of these were designed to be heated by charcoal.

Between the 16th and 19th centuries, most of the requisites for use in the home, as well as in the field, were made in the villages, so it was natural that the earliest flat irons were made on the blacksmith's forge. In common with most artisans of the past, the smithy took pride in his work and was a bit of an artist and a designer in his own field. Therefore, we find a great variety of handles on old irons—some quite decorative, with elaborate curls or twists where the handle is attached to the base. Some handles are high to keep them from getting too hot—some have shields to protect the hand of the user. They also vary in size and thickness. Sad irons are larger and thicker than flat irons and were used for smoothing heavy fabrics. The tailor's smoothing iron is called a "goose"— probably because its handle resembles the neck of a goose. (Incidentally, the plural of this word "goose" is "gooses.") Although gradually standardized in size and weight, the flat iron, together with the sad iron and the tailor's goose, continued in

*Fig. 3. Charcoal heated Goffering Iron. Height 12", maximum width, 11".
Found in Hamilton, New York.*

use until the advent of the electric iron.

When there seemed to be no possible way to improve the flat iron itself, new methods of heating the iron were devised. The Shakers built stoves with pyramidal sides against which many irons could be placed and heated simultaneously. When flat irons were heated on top of the kitchen stove, it was important to conserve the space directly over the fire. A cone of iron, made to fit into the opening of one stove lid, could heat your irons and still leave room on the top of the stove for other utensils, Fig. 5. Another unique heating gadget is a tray, the exact size to fit into the stove top opening when two lids and the divider were removed. The tray is divided into three sections with separate hinged covers to enclose the base of the irons only. Thus the handles were shielded somewhat from the intense heat of the stove, Fig. 6.

Fig. 4. Box iron in which was inserted a heated slug. Length 8", height 6".

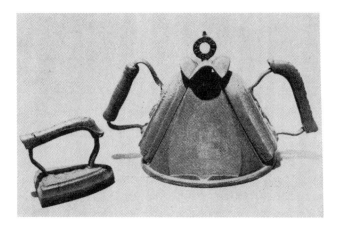

Fig. 5. Cast iron cone for heating four irons. Marked: "Reliable 93 S&T Co., Cleveland, Ohio."

Fig.6. Cast iron tray to heat three irons. Marked: "Wheeler and Bailey, Utica, N.Y. Pat. January 26, 1862 Number 9."

Fig. 7. Charcoal iron, patented 1852 by C. Bless and R. Drake.

Fig. 8. Geneva, Illinois, hand fluter. Patented 1860.

Fig. 9. Crimping iron with swinging handle. American. Length 7", width at heel 2½".

Fig.10. "The Original Knox" Patented 1870.
With photograph of SUSAN A. KNOX framed and set in the base.

.Fig. 11. "Diamond" oil iron. The Akron Lamp Mfg. Co

Fig. 12. Travel iron to heat over a gas jet. Patented 1888 by James A. Sharp of New York and James D. Suther of Brooklyn. Length 4½", width at heel 3".

By the middle of the 19th century, there were patents galore. To mention a few — In 1852, a charcoal iron was patented by C. Bless and R. Drake. It is quite high with a curved funnel at the front. Air was admitted through a vent at the back of the iron. On this vent is an oval impression of what is probably the likeness of Hephaestus—the Greek god of metal working, Fig. 7. Hephaestus is represented as heavily built and middle-aged with a beard and long hair, usually holding a hammer or tongs.

The "Geneva, Illinois Hand-fluter," Fig. 8, patented in 1860 was a great improvement over the old goffering iron. This fluter consisted of a heavy block of iron with a corrugated surface which could be heated in the fire. The material to be fluted was placed on the block and a rocker-shaped iron, also corrugated, with a wooden handle, was pressed over it. Another interesting fluter is shaped like a flat iron. The handle is attached to a corrugater lid hinged at the front of the iron, Fig.9.

A rather elaborate fluting iron, "The Original Knox," patented in 1870, Fig.10, has corrugated rollers, heated like the old goffering irons by inserting hot iron rods inside the rollers. The rollers are made on a heavy stand which could be fastened to a table and turned by a handle. The stand is prettily decorated and has a photograph of "Susan A. Knox" framed and set in the base.

A flat iron, with a handle that could be swung back on a hinge when the iron was heating, was patented by P.W. Weides of Philadelphia, Pennsylvania, in 1870. For ironing stiffly starched shirts and collars, a small oval "polishing" iron was made and patented by M. Maloney of Troy, New York, in 1876. The faces of these irons are etched in various patterns to keep them from sticking to the starched surfaces.

The Akron Lamp Mfg. Co. put "The Diamond" on the market. It burned oil and had a font at the back of the iron with a control to regulate the flame, Fig.11.

By the latter part of the 19th century, gas irons were in general use in cities and were made in various sizes and weights. This era opened up the country for travel, and there developed the need for a small iron which could be carried in one's luggage. One very unique example was patented in 1888 by James A. Sharp of New York and James D. Suther of Brooklyn, Fig.12. It measured 4½ inches in length, with a wooden handle. On top of the heating surface of the iron is a funnel-like opening and directly above this opening a small hole through the handle. The iron could be placed, handle side down, over any regular gas jet and heated to the desired temperature, lifted off and put to use.

Since time began, man has aspired to build that proverbial "better mouse-trap," and so it is with the household iron. Improvements are still being made and new patents applied for.

All irons pictured are from the author's collection.

Horse Whips

By MARY MOORE

January 1949

It is true that men had always cut themselves whips of hickory or white oak to speed up the ox- and horse-drawn vehicles. (Just as boys have cut themselves fishpoles.)

But in 1810 a man, said to have been actually named JO JOKES, first produced whips for commercial use. He, too, selected hickory sticks. But before attempting to sell them, he put on the end of each a lash made of horse hide. (It seems a bit strange that horses had to be speeded up by a forcible application of a piece of their own skin.)

A Mr. Socket is said to have been the first to have made a business of tanning white horse hair for lashes. And—as often— the Shakers of Lebanon, masters of craftsmanship, had a better way. The Shakers produced the lashes by stripping. (Stripping means taking off in layers—like peeling layers of bark from some trees.) Sometimes the lashes were smoothed and rolled in white soap.

As whips improved, the braiding of lashes was an important operation. Often this was done by the women of Westfield, Massachusetts, in their own homes. At one time Westfield made 90% of all the manufactured whips of the entire world.

Each day when mothers finished braiding the long pigtails of their little daughters and had sent them off to school, they could earn quite a bit toward cloth for a dress by simply braiding pieces of leathers. The bundle of completed lashes was taken to the store and traded out for credit. Today some of Westfield's leading men boast that the foundation of their family's financial standing was built on this piecework done by their grandmothers.

Not only in New England did women manufacture in the home. An elderly lady told me that in Pennsylvania on her father's farm she and her sisters always made the long lashes for the ox whips. They made them of old worn out leather boots which they cut into strips. In the end of each strip a slit was cut, through which the next strip was hitched. She complained that the end lengths were very short, but nothing that could possibly be used was discarded.

Years afterwards these little girls, with pigtails now pinned decorously about their heads, worked in the whip factories, either in the offices or along the production lines. Indeed, twenty years after the whip was first marketed, four hundred and fifty women were employed in the Westfield factories.

It is seldom that the construction of a whip is correctly described. It is difficult to convince anybody in New England that *his* grandfathers used *anything* but GENUINE WHALEBONE whips. It is a point of family honor, like passage on the Mayflower.

Perhaps masculine pride in quality made a man go to the harness shop where he believed only whalebone whips were carried. He asked for whalebone and the Yankee salesman only spoke of whalebone like ALL WOOL and PURE SILK. The *right* whale wore an upper plate. It was of soft springy substance, a springiness not only suitable for whips, but also an aid in confining feminine flesh to still narrower restricted areas. Soon the demand for these *right* whales exceeded the supply.

Whale oil for illumination went from 12¢ a gallon to 75¢, so some Yankees began making lard oil lamps in quantities. Down in Bridgeport, Connecticut, the Drs. Warner, who manufactured corsets, were experimenting with a Mexican cactus, as plentiful as the whale was scarce.

In Westfield, the industry that had originally paid 25¢ a pound for whalebone at the peak of its production in 1910 finally paid $10 a pound. $100 for a bundle of whalebone! So Westfield men began to substitute rawhide for the center of their whips at one-half the price of whalebone. This was called the "Twisted Whip." The green rawhide came from South America and India. Some of it from the water buffalo we see in our newsreels and zoos.

It was tanned, cut heavier at one end, twisted and allowed to dry. It was then unevenly round and about as big as a lead pencil. Through a machine it was rounded. (The scraps and dust from this process were salvaged to make glue. One firm was buying 100 tons of glue annually.)

Next came coats of filler, cornstarch or such—two or three coats.

Then an experienced workman took out the jumps. The whip at this stage might look straight to the average man, but to the experienced tester who took it in his hand and arched it, there might be jumps in it if it did not arch evenly. And he knew at once where the jumps were and with his hand planes removed them.

At last the whip was ready for the covering to be braided on. The braiding machine was a primitive affair. The whip was held above a barrel by means of a pulley and weight. Around the outside of the barrel were the bobbins. It took the time of several workmen to keep the bobbins moving. As the braiding progressed the whip could be pulled higher.

Beyond the tip of the whip an extra length of covering was braided and turned back into the body of the braid leaving a loop, called the keeper, to fashion the lash to.

One to three coats of varnish were applied depending on the quality of the whip. It took two days to dry each coat. You could not hurry old fashioned varnish.

A man hurried back to his factory with an idea and soon a whip company bought the patent for the shoelace braiding machine for $1,000, and the old barrel and pulley and weight were discarded.

Henry Hull thought of rattan sliced from tip to root and the sections glued about the core of whalebone or twisted rawhide. Soon just one plant was using 600 tons of it annually. Westfield's great factories and warehouses needed huge piles of lumber for packing cases. A single company used 3,000,000 yards of cloth and 75 tons of thread. Other factories were added: One making whip lashes, one or two making whip buttons, one or two making mountings, three making braidings or plaitings, three making snaps, five making wooden butts. One making whip-making machinery!

Production was soon so speedy in the plaiting or braiding room that at the peak of whip manufacturing it was impossible to watch the process closely. A former manufacturer likened that part of his factory to hundreds of toy electric trams running at full speed all over the place, so intricately and exactly was the work done.

A harness maker, until recently doing business on Westfield's Main Street, made the great leather cases in which samples were shipped. They looked like those in which symphony orchestras carry their harps. Smaller cases resembled our flat silver holders.

One firm had twelve salesmen. Another had fifty. One company offered 250 varieties. One carried eight different lengths.

As the motor age arrived, Westfield kept on making and selling whips, but when adjustment was finally forced upon them, many of the 200 or more factories went out of business completely. One has turned to the manufacturing of fishlines.

Horsehair Sieves

By MARY EARLE GOULD

September 1940

"When my father brought home a wire sieve there was great rejoicing. It was slow work sifting with a sieve of horsehair." Thus spoke an elderly lady during her reminiscences.

The hair for sieves was taken from horses' manes and cows' tails. Besides the black there was gray and an auburn shade. These hairs were sorted as to color and length and tied in bundles about two inches in diameter.

The hair mat was woven on a loom which was made for that special purpose. There is such a loom in the Farm Museum at Old Hadley, Massachusetts, which once belonged to two sisters living nearby, whose trade was making sieves for the market. It stands five feet high and is about two feet wide. The worker must have stood at her work or else sat on a high stool, for the threads were set up more than three feet from the floor.

In making sieves, the first step was to make the hoops. This was sometimes done at home but usually a cooper in the village was called upon to make them. Each sieve had two hoops, one fitting over the other like a box and cover. They were sometimes made in nests, like pantry boxes, and they ranged in diameter from three to seven inches. Larger sieves were made separately, measuring about fifteen inches across. The hoops were lapped and fastened with small flat-headed nails, sometimes of copper.

After the hoops were made and the mats woven, the process of putting the mat into the hoops was a separate task. Edward Andrews, in his pamphlet *The Community Industries of the Shakers,* tells of a sieve-binder on which the mats were finished. This binder was a stool with a seat or top of a particular size. Fitting over this was a cover of the milk-can variety but without a rim. The woven mat of horsehair was placed on the top of the stool and the cover pressed down on it and the edges of the mat were bound over and over with a stout thread. There must have been many stools of varying diameter to use for the several sized mats. Perhaps the smallest mats could be bound by contrivances held in the hand. History does not often enlighten us as to such details and we must usually speculate as to what methods our ancestors used.

After the mat was bound, it was clamped into the two hoops. Wooden pegs or copper nails held the two hoops together so that they did not come apart and the mesh was always very taut. One dear soul, presumably a woman, bound her mat with a bias strip of homespun and, because the work was done on her knees, the finished mat did not match the diameter of the hoops. Part of the bias strip shows inside the hoops and part shows outside. The horsehair gave away in the course of time but the binding remains, showing the finest of fine stitches.

In the nests of sieves, fine silk called bolting cloth was used quite often as hair, and sometimes very fine wire. These sieves were used in sifting herbs for medicinal purposes and such a set is in my collection, having been used in a doctor's office, and still showing the powder caught in the edges of the mats. The large sieves were used in household labor and one in my collection was woven with the design of a cross running through the center, of red and gray hair, with black as the background. There was scarcely any creation of the old days but showed an artistic leaning on the part of the worker.

Besides nests of sieves and the large pantry sieves there were double covered sieves. These were either round or oval and planned to keep the powder from flying about as it was sifted. Once such oval one was made with a mat of horsehair and one had bolting cloth. The crude, handmade, round sieve has a mat made from a thin gauze, still a different material from that which was generally used.

Remembering that everything came into the home in coarse form, we can well imagine the work of sifting. The mortar and pestle was used to crush the materials and then the sieves freed them from particles and undesirable lumps. We can sympathize with the "great rejoicing" when a wire sieve made its appearance.

The Old-Time Tanner

By Jared Van Wagenen

April 1949

In October or November the farmer killed a fat heifer or steer or cow to furnish the winter meat supply and preserved it by "corning" or drying and smoking, or otherwise utilized it according to the resourceful domestic science of that time. Then the hide, along with the skins of calves and the pelts of sundry sheep which had contributed to the family larder, were carried to the neighborhood tanner, who turned them into leather according to his primitive and yet skilled art. From them the household was shod and the teams were provided with harness. The master himself bestrode a saddle constructed of homemade leather, and not infrequently he went snug and comfortable in bitter winter days clad in garments of sheepskin with the wool unpulled but turned next his body.

In those days when all men wore boots that reached to their knees and all women were shod in stout and ample shoes, and the practice of splitting leather had not yet been devised; when there was not rubber or canvas or synthetic leather substitutes, it is evident that the tanner's art was relatively very much more important than now and that leather had a very much wider use than in our time.

Tanning was one of the pre-eminent rural handicrafts carried on in a multitude of small establishments so that every farmer had a tannery in his convenient vicinity. In 1845 there were 1,414 tanneries in New York State and there was hardly a township anywhere but had from one to three representatives of this most essential industry. The community tanning industry held its own better and longer than either the growing of flax or the local manufacture of wool. In 1855 the state still numbered 863 tanneries and strange to say these diminished in number only slightly during the next ten years, for there remained 820 in 1865 and not a rural county of the state but had several. It is evident that the Civil War brought about a temporary revival of flax growing, and I think it very probable that it had a like influence on the rural tanneries in that it kept alive a dying industry. Even as late as 1873 my own small and strictly rural county of Schoharie still listed 13 tanners in the local county directory. This, however, was the final stand of the currier and I think the last one of them has been extinct for more than a quarter of a century.

Most of these old-time tanneries were exceedingly primitive affairs, the whole labor force frequently being represented by one man who was both owner and manager. If two or three helpers were employed it was regarded as quite an establishment. Very often the proprietor was a farmer by vocation and a tanner on the side; in two local instances, within my knowledge, a third calling—that of harnessmaker—was added. Such multiplication of vocations was a perfectly natural and proper development of the handicraft age which decreed that a man must never be idle but must always have a job to which he could turn his hand. I cherish an old wallet or billfold made by one of these long-dead farmer-tanner-harnessmakers, and it is even yet a beautiful example of craftsmanship in leather. That was not the age of the machine, but it was emphatically the day of skilled and patient handiwork.

The tanner's art was one absolutely indispensable to our early agriculture. Rawhide might be tough, but it would stretch out of all reason when wet and would become hard and horny when again dry. Then, too, it would decay very quickly when moist and was in general wholly unsuited for boots or harness. Only when the gelatine of the skin had been chemically united with the tannin of oak or hemlock bark did it become adapted for the multitudinous uses to which leather is put.

The mechanical equipment of these pioneer tanneries was exceedingly primitive. Doubtless at a very early date the hemlock bark was crushed fine by pounding with a wooden mallet, but within the memory of any one now living the tanner had a bark-mill, which in its general construction was wonderfully like an overgrown, old-fashioned coffee-mill. It was turned by a sweep to which was attached a horse that marched around in a circle and thus crushed the bark. It is said that bark ought to be ground to the size of wheat kernels, but I believe this was an ideal seldom attained. One of the most shadowy recollections of my earliest boyhood is the rather patriarchal figure of Thomas Kilfoyl as he stood by his mill casting fragments of bark into it and urging on his patient and rather decrepit steed. He was a very fine old Irish gentleman and a maker of honest leather of good repute, but his skill and his business died with him and he left no successor. But I wish—I wish I might again hold converse with him of a summer afternoon.

Aside from his one-horse mill, the old-time tanner had no mechanical help except a few simple hand tools. His work was to a degree seasonable. In late autumn after the family beef cow had been slaughtered, the farmer brought her hide along with the calf skins and the sheep pelts which had accumulated during the year. Much, probably most, of the tanning was done "on shares"—that is, the finished leather was divided equally between the tanner and the farmer. Thus the tanner from time to time accumulated a load of leather which could be sent to some center of trade and sold for cash; by this means he was rewarded for his toil. When the farmer brought the skins, the tanner took his razor-edged tanner's knife and on one corner of the hide cut the initials of the owner so that a year later he might be able to hand him back his own. Then some of the poor, worthless parts of the hide—the legs, skirt and ragged head—were trimmed off, and these scraps, when bagged, were a saleable commodity to the tin-peddler, that once universal scout of trade now absolutely extinct in the rural community. Calf skins and sheep pelts were tanned whole but cow hides were split down the back line making two "sides."

The first step was getting rid of the hair or wool. Sheep and lamb pelts were moistened and stacked up in piles so that they would "sweat." This loosened the wool so that it could be "pulled." Almost a century ago, within a mile of my home, two Connecticut Yankee brothers, Sam and Lew Betts, not only tanned the pelts but also had a hat shop where, from the pulled wool, they made felt hats for all that section of the country—another example of how self-sufficient was the rural community in the days of which I write.

Cow hides could also be sweated, but the more usual plan was to put them in a vat with milk-of-lime which loosened the hair so that it would slip. But, having removed the hair with lime, it was of prime importance in securing good, soft, durable leather that the last vestige of lime be removed before proceeding any further; and to do this seems not to have been an easy job. The hides, after being scraped clean of hair, were transferred to another vat where they were soaked in a solution known as "bate," which was nothing more than a mixture of hen-dung and water. I do not pretend to understand the efficacy of this strange broth, but it was the recognized formula among old tanners and the universality of its use attests to its efficiency. After a period in this vat, the hides were very persistently scrubbed and rinsed in pure water and it is said that the character of this was very important, soft water giving much better results than hard.

Then the soft, plump hides were ready for their long immersion in the tanning liquor. Tannin is widely distributed in nature and there is a long list of vegetable tanning materials, to say nothing of the chemicals, which will do something of the same work. The old tanner could tan sheep pelts with the wool on by rubbing them on the flesh side with alum and salt and ashes, but on the whole he depended on the bark of hemlock and oak. Oak was regarded as better for harness-leather; but still, in our state, hemlock was far more widely used. The vats were commonly about six feet square and four feet or more deep and their tops were only a little above the ground-level, with narrow passageways between. I have been told that life was one long nightmare for the mothers of tanners' children because of a haunting fear that they would fall in a vat and drown—a dreadful, and in no way unlikely, accident.

In tanning, a layer of ground bark was put in the bottom of the vat, then a hide was spread on this, then a couple of shovelfuls of bark, then another hide until the vat was filled, when water was run in to cover the whole. From time to time the hides were looked over, re-piled and, if necessary, more bark added. In any case, they lay all winter in the liquor—a full six months or more. Usually it was about a year from the time the hides were brought to the tanner until the finished leather was ready for delivery.

Practically all the tanner's work was direct manual labor almost unrelieved by any mechanical helps; and part of it, notably "beaming"—that is, removing the fat and surplus connective tissue from the hides with the aid of the fleshing knife—was recognized as particularly hard and exhausting toil. There was a rather lengthy list of operations before the leather was finished. After removal from the tanning liquor, the hides were thoroughly washed and then dried by hanging over poles in a loft. It was important that the drying process should not be too rapid and yet with enough access of air to prevent molding. The need of loft space for this drying made it necessary that even a small tannery should be quite a sizeable building. Then the leather was scraped, beamed, rubbed with a mixture of tallow and neatsfoot oil, finally blackened on the grain (hair) side with lamp-black and then polished. I am told that modern tanning has almost wholly substituted various chemicals for hemlock bark, and that hot tanning liquor and other short-cut methods have reduced the time required to a tithe of the old period. But when we consider real leather quality—softness and strength and durability—there is at least a very general belief (in which I share) that the product of the old-time art was far superior to that produced under modern methods.

The pioneer tanner was engaged in a work which was essentially the product of complex chemical changes. He proceeded entirely by tradition and rule of thumb, and made no pretext of explaining what he accomplished; but, nevertheless, his was an art calling for good judgment and no small measure of technical skill. His calling was one where lack of conscientious care or an error of treatment would result in a greatly inferior product.

Making Barrels By Hand

By GEORGE L. MINER

January 1951

A mile from my farm in Hampton, Connecticut, is an old cooper's shop which turned out barrels, buckets, axe helves and other handmade implements from the early 1800's up to about 1890. Relics left in the old shop include a few tools, a huge caldron used in steaming barrel staves, and a cooper's draw-horse or shave-horse.

The late owner of this shop was Edward M. Fuller of Providence, Rhode Island. He had inherited it from his father and grandfather, and his boyhood days were spent at the Fuller Homestead of which the shop was a part. He was born in 1876. He had vivid recollections of the practice of barrel making in the 1880's and I will transcribe his narrative as closely as I can.

Ned Fuller's Story of Barrel Making

The cooperage business of my Grandfather Fuller was going strong when his children were little, but along in the 1850's the country coopers' shops gave way to the barrel factories and the handmade rum barrels fell behind in the competition. Grandfather continued to make barrels, but his chief product was buckets and farm tools such as bucksaw frames, rake stales, whip-stocks, axe helves and other wooden implements used by a farming community. There was a forge in the old shop and he turned out rough iron-work used in his farm wagons and tools.

He made good barrels. One of his customers was the big trading firm of Brown and Ives of Providence to whom he sold rum casks. He would work all winter making barrels, then when Spring came he would load them onto a big rack wagon with a yoke of oxen and cart them over the thirty-seven miles of country highway into Providence where he would unload them wherever his purchaser wanted them.

Usually the destination was the wharves on South Main Street toward Fox Point. Always however he stopped at the Hoyle Tavern to bait his cattle and to sleep overnight. When he sold his barrels he took the cash and spent the next day in the town of Providence buying groceries, dry goods and supplies the women needed and brought his purchases home for the use of his family for months to come.

As a boy I used to watch Grandfather working in his cooper's shop. His barrel making needed a certain amount of hand tools that are now almost forgotten. In the old shop are still left a few of the planes, a drawknife, or spoke-shave, and the old wooden horse on which grandfather sat to bevel his staves.

The staves were made of white oak which he bought from the Hampton saw mills in the rough. He planed the unfinished planks with a big jack-plane and worked them down to the thickness needed for staves. Then he took a thin wood pattern for a stave and laid it on the smooth plank and marked the outline in lead pencil. Then with a rip-saw he sawed out the staves; they were a little narrower at each end than in the middle.

Grandfather would then sit on a draw-horse or shave-horse as they were sometimes called, put a stave in the wooden vise in front of him and hold it tight by the foot pedal. With a small plane he could now shave down the square edges of the stave and thin the wood down towards the ends. Next step was to bevel the stave edges with a small block plane; the bevel was cut by eye and was rather a delicate piece of work in order to have the staves fit when placed together in a barrel—smaller on the inside than the outside of the stave edge.

There was a special tool [a croze] for finishing the ends of the staves so they would fit into the barrel head. The scant inch at the top and bottom of the staves where they projected above the head and below the bottom was termed the chimes. The croze reamed a V-shaped groove on the stave a little less than an inch from each end; into this groove the barrel head and bottom were edged and shaped to fit.

When a good supply of staves were all cut and planed and beveled they were ready to put together to make a barrel, but before bending them to barrel shape the tough oak had to be steamed to make it pliant. Grandfather had a steam box that fitted into his shop fireplace just under the lintel beam. That box measured about five feet long by a foot square and had a wooden lid—all of inch oak.

For steam he had a big iron kettle about 24 inches across with four legs on it. Two of these kettles are still in the workshop as good as ever. The kettle was hung on the iron crane in the fireplace by means of a strong wrought iron pot-hook, six or eight inches long, or longer if a shorter kettle was used. On the kettle was a lid of heavy block-tin into which was soldered a tin pipe about one inch in diameter. It took a pretty good fire in the fireplace to boil the water and make a head of steam sufficient to fill the steam box. The staves took some time to become pliant enough to bend. I don't remember just how long—perhaps two or three hours.

While the staves were steaming, the barrel heads were made ready. The heads were made in three pieces, the center piece had two straight edges and the other two pieces had one straight edge, like a pie cut squarely in half. The three pieces were laid on a bench, butted together and scribed in a circle of the diameter needed, then the circular edge was chamfered off to fit the V-groove that had been cut in the staves. This chamfering was done by a drawknife.

All was then ready for the assembly of the barrel. The coopers used stock rings of heavy oak; these were hoops made for temporary use; they were heavy and large and lasted for years. To assemble the barrel they took the bottom hoop and in it stood the staves up and slipped another ring inside the staves a little lower down than the outside hoop. Thus a slot was formed into which the staves fitted and roughly assumed the shape of a flaring barrel larger at the top than at the bottom.

The outside hoop was then driven down until it pinched the staves against the inside ring and held them firm. Then the barrel was turned upside down and a third temporary hoop was driven on to the bottom of the staves, making that end quite secure. Now the barrel could be turned back again with the flaring ends up, and moved over to the work bench.

The bench had a semi-circular curve cut in it big enough to hold the flaring staves. At the right of the semi-circle was a wooden roller with a ratchet and dog and with a crank and handle so as to act as a winch. On the left of the bench was an eye-bolt into which a three-quarter inch rope was hooked. This rope was led one turn around the flaring staves and over the roller of the winch. Turning the winch then drew the staves together, whereupon a fourth hoop was slipped over them and tapped down to hold them fixed.

The barrel was then ready to turn loose and could be handled as a barrel, though with no heads in it. It was up-ended and the bottom temporary hoop knocked off, and the three pieces of the bottom head were sprung into place. Permanent hoops were then driven on. These hoops were made of native ash or hickory, both of which woods were tougher and more withy than oak; they were cut with a lock or notch beyond which the two projecting ends were twisted under the main part of the hoop so it would hold fast. The temporary ring was then tapped off and the inside holding ring lifted out; the top was fastened permanently by two or oftener three hoops driven firmly on while meanwhile the three pieces of the barrel head were sprung into their groove in the staves.

Thus was made the rum or cider cask, ready to be filled from the bunghole. Barrels for apples and potatoes were left with the top head out and the top hoops loose, ready to head up when filled. It took some skill to make a watertight barrel. They were tight when first built and so long as they had rum or cider in them. When dry the seams opened up some, but quickly swelled when put into use again.

The Account Book of Terence Tracy
A 19th Century Cooper

By CHARLES REICHMAN

March 1989

A rare glimpse of coopery as practiced in the latter half of the 19th century is provided in the account book of Terence T. Tracy, a cooper residing in what was then the village of College Point in the town of Flushing, Queens County, New York. The account book is one of a number of local history gems to be found in the library of the Poppenhusen Institute at College Point. The account book had literally been wrested from a garbage heap moments before it was to be hauled off to the dump, by Robert C. Friedrich who contributed it to the Institute in 1962.[1]

Little is known at this writing about Tracy. He was born in Ireland in 1838 and died on Tuesday, February 25, 1879. He and his family resided in a house on 11th Street between Sixth and Seventh Avenues in College Point; it was one of four buildings on the street.[2] In the 1870 Census of Population he is recorded as being 32 years old, married and the father of three children, Kate 5 years, Alice 3 years and John 8 months. His wife was born, a year after her husband, in this country. The Census put the value of his real estate at $1,500; no figure is given for his personal estate.[3] Although hardly munificent, the value of his real property put him several financial notches above the majority of laborers and some craftsmen in College Point. At that time there was a marked dichotomy in income and real property ownership between the bulk of the village's inhabitants and its elite citizens residing in style in an enclave overlooking Flushing Bay, the East River and Powells' Cove.

Although Tracy's account book, which runs from August 1868 to July 27, 1875, shows its years — wear, water damage, sunlight, dust and its brief sojourn in a trash bin have taken their toll — most of its penned pages are still quite legible. Those badly faded are decipherable with a magnifying glass and perseverance and patience. The few pages that show the marks of mischievous little hands contain a separate history of their own, particularly of the older daughter, Kate, who used blank portions of the pages, initially, to practice her penmanship; then to write drafts of letters and ditties, perhaps of her own composition, and finally, to copy the texts of poems.

The account book runs to page 238 but pages 1 and 2 are gone — probably the result of the unhinging of the front cover when the spine deteriorated — and an undetermined number of pages appear to have been scissored out in the back. In between, an estimated ten pages are missing. Even so the vast bulk of the account book remains, making possible a coherent reconstruction of the kind of cooperage Terence

Tracy ran, who his customers were, what he charged, and what his annual income may have been over the eight year span.

Almost from the outset of examination of the account book, it became apparent that Tracy was primarily a "tight" or "wet" as opposed to a "slack" or "dry" cooper. The former made staved wooden containers chiefly for liquids; the latter provided less stoutly made containers for solids. It was not until a quarter of the way through the ledger that Tracy is also revealed as a "slack" cooper. However, the greater proportion of his work over the eight-year period was with vintners, spirits bottlers and dealers in medicinal, lubricating and whale oils.

Also evident from the first few pages was that most of Tracy's work —later determined by rough calculation to be 75 percent — was repairs and services rather than constructing wooden receptacles from the ground up. Coopery is defined in the dictionary as the work of the cooper, implying both restoration and original construction. In Tracy's account book it has less to do with the making of brand-new vessels than with the overhauling of old ones to usable condition for specified end-uses.

Among the "tight" receptacles Tracy was called upon to make, alter, or repair were barrels, casks, kegs and demi johns for whiskey, brandy, Scotch, sherry, sack, claret, port and other wines, syrup, castor oil, linseed oil, lubricating oil, whale oil and tallow. His use of the term "demi john" is a strange one, for it usually means a large, wicker-encased, glass or stoneware bottle. Perhaps he was merely decanting barrels into the demi-john containers. For gin he largely made pipes, huge casks closer akin to tuns than barrels in liquid capacity. The "slack" or "dry" containers he worked on were: barrels for sugar, flour, beans, dry apples, peaches, rice, coffee, bone dust, pecan nuts and nails. In addition he made or rebuilt casks for glucose sugar, prunes, plums, ingot copper, glue and madder. The root of the madder plant was used in the making of dyes and some medicines. Still other wooden containers included chests and packs for tea, cases for figs, nutmeg, licorice, bulbs, ingot copper and hardware and boxes for raisins and grape sugar. He also built firkins for butter. [A firkin is about ¼ barrel or 9 gallons; a tun is 252 gallons; and a pipe is about ½ tun.]

It is not clear whether the receptacles sent to Tracy's shop came broken down or whole. The fact that there are entries relating to coopering of "standing" or "empty" barrels may indicate that those not so noted were in "shooks" or broken down into component parts. Coopers called a fully assembled used barrel a standing or empty barrel; Tracy employed both terms interchangeably. When a barrel came into the cooperage knocked down or disassembled it was "shook." The following entry typifies the operation of coopering standing or empty barrels. Note that it also includes charges for materials required to make the barrels usable:

Coop 6 Standing Brls @ 25¢
3 iron hoops @ 25¢

8 wooden hoops @ 6¢

12 plugs @ 3¢

The entry below is believed to designate the reassembly of "shook" containers as well as the replacement, in the process, of damaged or lost parts:

Coop 4 brls Beans @ 6¢

4 hoops @ 4¢

1 stave @ 6¢

Whether Tracy actually started in business in 1868 has not been ascertained. There could have been an earlier account book. Assuming there was not, then it could be stated that he started off the business with a bang. His billings that year amounted to $934.36 and came from seven customers, four of whom accounted for nearly 90 per cent of the total. Actually, the billings were probably higher since, as previously noted, the first two pages of the account book are missing.

In the first six years of operation Tracy's billings increased annually, reaching a record high of $3,231 in 1873, a year of great economic upheaval. Though he apparently survived the panic of 1873, his billings the following year declined 22 per cent and his customer list shrank from seventeen to six. Since the accounting for 1875 ends with the July 27 entry (the rest of the binder's pages scissored out), annual totals are clearly not accurate and should be disregarded. What cannot be disregarded is that the meager billings for the seven months came from only two customers, a far from optimistic omen for the balance of the year. There is a good possibility too that his billings may have continued to decline because of technological innovations. By the mid-1870s the manufacture of barrels and other containers by machinery was already well advanced, signalling a brief and difficult future for village coopers.

Tracy's customer list over the eight years covered by the account book included more than 30 firms and upwards of 20 ships. Of the latter, six were identified as barks and four as brigs. The latter is basically a two-masted, square rigged sailing ship and the former a three-masted one. The term bark was also used to denote a small size sailing vessel. All the other vessels Tracy did business with were designated as ships. Whether these were also sailing ships or were steamers is not known, though some could surely have been of the latter type.

There is also no indication in the account book of the home ports of any of the vessels, though two were identified as bearing the Austrian flag, four the Italian flag and one the Spanish flag. Work was apparently done aboard the ships and not at the cooperage. Shipboard jobs almost always carried an additional charge for the number of men involved in doing the work and sometimes there was also a cartage charge tacked on, perhaps for the materials that had to be delivered to the dock or ship at anchor. Here is at typical shipboard entry for work performed on the brig *L'Amico*, Capt. Esposito:

```
  3 men 1 day @4.50   13.50
154 hd hoops @10¢    15.40
 50 brl hoops @10¢    5.00
 60 staves @6¢        3.60
  4 prune casks @1.50 6.00
    Cartage           2.00
```

Total billings to merchant ships were generally small, ranging from a low of $4 for the bark *Ercole* to over $200 for the Italian bark *Candida*. The entry for the bark *Ercole* was:

```
Coop    8 boxes of bacon @25¢  2.00
Strapping 6 boxes @25¢         1.50
Repacking 1 box                 .50
```

Note that there was no labor charge on this job. This was also true of the bark *Candida* entry, except that there were marginal notations of the number of men required — from one man working one-half a day on an $8.42 job to four men, presumably working one day, on a $70.64 job. Here are the entries for the bark *Candida* commission:

```
Coop 50 hds prunes @ 30¢   17.70[4]
      13 brls@ 18¢           2.34
     300 hd hoops@ 10¢      30.00
     100 brl hoops @ 6¢      6.00
      60 staves@ 6¢          3.60
       6 empty casks@1.50    9.00
Cartage                      2.00
```

Of the 30 or so wholesalers, trading companies and other land-based businesses comprising the greater part of Tracy's account book, only two spanned its eight years: Wilson & Co. and Thomas J. Daly & Co., successors to MacGregor & Daly. The former was a wine and spirits merchant and the latter a dealer in tea, sugar, dried fruits and spices and herbs. Billings to Wilson & Co. over the eight years amounted to just under $1,000; billings to the Daly company came to a whopping $7,600.

Thomas J. Daly & Co. was far and away Tracy's best customer. The aggregate billings to no other customer were anywhere near this figure. The closest was H.L.

Routhe & Son ($1,815) followed by Morton Bliss & Co. ($1,459), C.M. Fry ($1,343) and Benj. Rionda & Co. ($1,284). For the remaining customers, total billings ranged from under $25 for single jobs to three figure billings encompassing three or four years' activity.

One of the biggest jobs performed by Tracy for any customer, surprisingly, was not basic coopery. It entailed inspecting and tinning[5] 3,000 barrels of refined petroleum oil for Nolde & Degener, probably a fuel dealer. The work took from September 3 to September 8, 1868, and the rate charged was six cents a barrel. On September 8 there was also an entry, "Star and Daylight 1,500 casks" for which $180 was charged, for a total billing to Nolde & Degener of $360. Research in a number of books on coopering and speaking to a few authorities has brought no enlightenment on the meaning of "Star and Daylight." Could it mean transportation of 1500 of the oil barrels to each of two ships, the *Star* and the *Daylight*?

Tracy did not do business again with Nolde & Degener until April 1, 1869. This time he inspected 3,515 barrels of oil at three cents a piece, tinned the same number at the same unit fee and marked them at a penny apiece for a grand total of $246.05.

Nolde & Degener was not the only oil dealer Tracy served. Another was C. Heydecker, which handled mainly whale oil. For this customer Tracy in late May 1870 coopered a total of 47 casks of whale oil at 40 cents a cask. This work also required six splices at 25 cents apiece and one lead patch at 50 cents. In one of the few references in the account book to destination, it is noted that the whale oil was destined for I.A. Stember, a French concern. In connection with this shipment Tracy had to provide four affidavits (spelled "affidavids"). The cost for this, apparently laid out by Tracy, came to $14.75, broken down as follows: $5, lawyer's fee; 90 cents, New York County Clerk's fee; and $8.85, French consul's (spelled "council's") fees.

C.M. Fry, recorded above as Tracy's fourth largest customer in terms of billings, was a flour merchant and had done business with him earlier as a partner in Fry & Ramsey. Many of the Fry transactions might also have been performed off the premises of the cooperage, for the entries are usually followed by a reference to work locations, viz:

Coop 25 brls flour at 69 Front St.
Coop 25 brls flour at 117 Broad St.

The total price charged for the above two transactions was 50 cents. Obviously, at this price no cooperage could have been involved. Exactly what was done is not known and cannot even be surmised. Note should be taken that neither entry — including the rest of the like entries on the same page — included the cartage and labor charges characteristic of shipboard cooperage jobs.

"Taring" containers —determining the weight of a receptacle sans its content — was a service frequently performed by Tracy for some of his major customers, especially Thomas I. Daly & Co. Occasionally taring was combined with other services such as sampling, strapping, marking, identifying, and inspecting but mainly it was performed solo as the following entries show:

Tareing (sic) 13 packs of tea @30¢ $3.90
"   18 cases nutmeg @18¢   3.24
"   20 boxes figs @10¢   2.00
"   47 chest tea @30¢   14.10

How taring was accomplished and what tools or devices were used is, of course, not disclosed in the account book and is thus a subject for later study along with some of the many unanswered questions the ledger raises. It is hoped that when further investigation is undertaken, more information about Tracy and his cooperage will surface, particularly, the number and kinds of tools he used, the raw materials he stocked, and the extent to which new container-making machinery and technology was employed. On another level, it would be instructive to know whether a branch cooperage was maintained as part of his residence to service businesses and individuals in the College Point and nearby Flushing and Whitestone areas. The answers to these and other questions will shed light on how the business of coopering was transformed by the industrial boom that erupted in the United States in the wake of the Civil War.

NOTES

1. The author wishes to express his appreciation to Susan Brustmann, executive director of the Poppenhusen Institute, for permission to study the Terence T. Tracy Account Book over many weeks.

2. This address is at variance with a Pearl (street or avenue) address given for Tracy in Curtin's *Directory of Long Island, 1870-71*. Since there was no such street or avenue in College Point, a logical assumption is that it was the location of Tracy's cooperage in lower Manhattan. This view is fortified by the following: (1) It would help explain the relatively large number of ships with which Tracy did business; the ships would more likely have anchored off the tip of Manhattan than on the waters surrounding the tip of College Point. (2) It would better explain and certainly clarify, his customer roster; the firm names were not traceable to College Point plus the fact that, as will be seen, some coopering jobs were performed at lower New York City addresses.

3. Marshal's Schedules for Bayside, Flushing, 1870 Census of U.S. Population. Bayside was the then Post Office address for College Point.

4. There is no explanation for the difference of $2.70. Possibly an error or an unspecified labor charge or . . . .

5. This is thought to refer to the insertion of "bung tins" which are offered for sale in the 1884 Lang & Jacobs catalog of coopers supplies at 40¢ per hundred for the 2" to 2½" barrel size. Keg-sized tins were 35¢ per hundred.

The Wooden Leg of Gouverneur Morris

by SIGMUND EPSTEIN, M.D.

September 1940

At the time of writing, Dr. Epstein had been a practicing surgeon for over 50 years, and had contributed many articles to medical journals as well as to The Chronicle.

Today we are horrified by the thousands of motor accidents which each year take their terrible toll. But aren't we rather forgetting our great-great-grandfather who traveled on horse — the "safe" way? Considering relative sizes of populations, today's motor disasters seem as nothing compared to the once everyday hazards of the equestrian. Meaning no pun, in those days such incidents were taken in their stride.

Take the case of Gouverneur Morris, for instance. In spite of being a historical character and a considerable land-owner, — things we take somewhat seriously nowadays, — Gouverneur Morris, of Morrisania on the Harlem, was a good sport. He, as well, unfortunately, was the victim of what America was manufacturing by way of artificial legs back in the late eighteenth century.

In Philadelphia, in May of 1780, Morris, while still a young blade and man-about-town, was flung with much violence from his phaeton, as a result of his horses suddenly taking fright and deciding to run away. The consequence was, in a way, no more than that of the average Inquisition. The injury was to the left leg, and a dislocated ankle and several compound fractures,

Gouverneur Morris

which probably became infected, were sustained. He was confined to his bed for three months, and nursed in the home of one Plater, a Quaker member of Congress. The two physicians who were called to him recommended an immediate amputation as the only means of saving his life (among the things which they took somewhat seriously in those days were compound fractures and dislocations) and the dapper Mr. Morris cheerfully acquiesced to parting with his member. This, mind you, was a serious set-back for a glamour-conscious young man in a day when calves were more exposed than faces, and, consequently, more important. Patients are no longer so pliable. We doctors have to be careful nowadays, and the removal of a bunion is accomplished with less fortitude and more dextrous care. The leg was taken off below the knee, and the operation has often been lauded as being one of the most ill-advised and unskillful of all times.

The day following the awful event, Morris, after his indisputable courage of the yesterday, displayed again an admirable resilience, and niceness of disposition. To a friend and male Pollyanna, who came to console and assure him of this unforeseen cloud's silver lining, Morris grinned and said: "My dear sir, you reason so convincingly, and you show me so clearly the advantage of being without legs, that I feel almost tempted to get rid of the other one."

Actually, though, the fact of being without a leg, and, therefore, attached for his life's duration to a primitive wooden stump or pylon, hardly more than a rough oak stick, equipped with strapping for the leg and thigh and a wooden knob at the base, could not quite be said to have entirely impaired the trend of his existence, either politically or non-politically. In spite of definite evidence that he derived much comfort from simple efficiency, still it must be admitted that Morris entertained a fastidious hankering for the aesthetic, possibly stimulated by the look in a certain pair of brown eyes of French vintage. About the year 1790, he took considerable pains to replace the undeniably ugly and indisputably cumbersome wooden peg. A more pleasing affair of copper was ordered from a French artisan, who, at that time, was considered the last word in his profession. This was to be made to imitate and look like a leg with a foot on it, a cast of the stump, and of the sound leg having been furnished for guide. The procedure was gone through with much ado. Morris' own diary fashions us a sorry picture of what a skeletal deficiency could mean to a man of his day:

"April, Friday, 16, 1790.—. . .Genl. Morris and James call on me to inform of a Mechanic who can make wooden Legs very well. I desire that he may call on me ToMorrow. . ." (Showing his anxious haste to try anything that might improve his situation quite regardless of expense or effort.)

"Saturday, 17.— This Morning James Morris comes to breakfast with me and after Breakfast a Mechanic arrives who is to make me a Leg. Upon Examination of the Stump he says that I shall be able to take the Benefit of the Knee Joint. If this be so it will certainly be an Improvement but he acknowledges that the Machinery will be less solid than the simple Stick which I now Use.

"Tuesday, 20.—This Morning I go immediately after Breakfast to a Leg Maker and have my right Leg taken in Plaister of Paris as a Model by which to make a left Leg of Copper. By the Aukwardness of the Workman I am long detained and obliged to have a second Copy made: in fact he has not one needful Thing, which is a box for taking the Model by. Got a Model made of the Stump also, so as to prevent the Necessity of frequent Sittings to have the Cushions &ca. fitted. I am detained under these Operations until after four oClock.

"Saturday, 22.—...The Maker of artificial Legs also calls to try on mine. I doubt much the being able to make use of it."

And a month later:

"Saturday,17....Call on the Maker of Legs and try on that which he is preparing for me. It is very well made but will I think be useless to me."

This goes on and on. In August, 1790, he repeats disconsolately:

"Monday, 2.—.....The Machinist calls in the Afternoon with sundry-Appendages for my wooden, or rather Copper Leg, which will I fear prove useless to me after all..."

Poor Mr. Morris! Whether he liked it or not, his forked stump stick was still with him, if to the disappointment of his French lady friends cannot be said. It is known, however, that the inevitable happened: the good-looking prosthesis surely did not take the place, or give the comfort, satisfaction and security that his reliable old American carpenter-made support did. It is obvious that he never wore it. Nor did similar attempts made in London prove more fruitful. His picture shows him with his knee stretched out, firmly strapped to that homely good friend, his wooden leg.

The efficiency of this wooden leg had been tested by many years of trial. The new-fangled contraptions may have had poorly fitting sockets, the proper setting of which is a fundamental necessity in the artificial leg of today. A man of his size, moreover (for he was over six feet in height), requires heavy shoulder-straps in most of these prostheses, something never even conceived of by the old-time artisans. It sometimes takes over a year to learn how to wear an artificial leg, and there was no one then to give Mr. Morris an inkling of this. Many a World War [I] veteran in the great hospitals of this country, after an engagement overseas, left the hospital with his artificial leg strapped to his crutch. At the time of the episode of "a left Leg of Copper," Gouverneur Morris was no longer a young man, and his tissues (and skin) over the stump may have been intolerant to the bearing of weight. His mechanic too, in spite of his reputation and, probably, artistic training, seems not to have been so expert at his trade. Then the material itself—copper—is much heavier even in the form of a hollow cylinder, than the cotton wood or willow that we use today.

Mr. Morris' wooden leg may not have been the product of a highly skilled specialist but was probably the work of a good all-round mechanic, who acquired his knowledge in his father's shop. This solid, plain carpenter was typical of our early American workman. He did all the shop-work unaided and the results were not fancy but they were serviceable. Mr. Morris' leg shows evidence of good judgment in selection of woods, excellent skill in fashioning, planing and finishing an article of utility, competent to withstand much strain. It bespeaks comfortable padding, efficient and heavy straps, and good design. It was comparatively easy to carry around, was easier to remove, and served all sorts of purposes besides. Once, while being presented at the French Court in Paris, he appeared without a sword, a serious breach of court etiquette, but after all, no one could expect a man with a wooden leg to carry a sword, now could one? And then again during the Terror, when carriages had been abolished as being aristocratic, Morris, in no way willing to dispense with so comfortable a method of travel, drove through the streets, followed by stones and cat-calls of "An aristocrat!" He calmly opened his carriage door finally, and thrust out his wooden leg. "An aristocrat," he said. "Yes, truly, who lost his leg in the cause of American liberty." Whereat, to be sure, cat-calls turned to cheers, and once more the wooden leg of Gouverneur Morris had saved the day.

. . . and a final note . . .

July 1954

The New York Historical Society has been presented with the wooden leg of Gouverneur Morris by Mrs. Frederick Menzies of London, England. The leg will be on display at the Society's Museum, 77th and Central Park West, New York City.

Notes on Early Ohio Coffins and Their Makers

by RHEA MANSFIELD KNITTLE

August 1946

The earliest burial cases or coffins in use on the Ohio frontier were very primitive. In many sections of this territory a coffin could not be obtained until the arrival of a cabinetmaker to a settlement or small town. Frequently, this craftsman was the last worker to make his appearance in a village.

For some of the earliest burials, the body of the deceased was encased in bark, peeled from a large tree. Long strips were used for this purpose, provided the body was that of an adult. Many trees in the primaeval forests were of great height and girth. These strips were lined with grasses or moss and folded about the corpse, then rushes or withes, gathered from near-by swayles, bound the whole nearly together.

The second type of burial case consisted of slabs of native hard wood, such as black walnut. These slabs were dressed by use of a broad-axe and a drawing-knife, then they were held together in a firm manner by the insertion of wood dowels or pins. Few coffins of this type were in use prior to 1810.

As late as 1835, the coffin was made to order, cherry being the wood mostly in demand. Cherry boards were kept steamed and when required, were bent to the desired size and shape. The head of these cherry coffins were one-third narrower than the width across the breast of the corpse; and the foot was one-third narrower than the head. In some respects, these cases must have resembled the Italian burial-case, although I have found only one record in which glass was used, much of our earliest glass being green in color.

In many instances, when a poor woman could not afford to buy material for a shroud, "the neighbors substituted a clean shirt and white drawers for the shroud," and if a widow lacked the money for a coffin, the men and boys "gathered together, and with their broad axes, split in the middle the required length, then each half-section of the log was dug out in trough fashion, and the body was placed on a lining of moss laid atop a bed of leaves."

Many of the early burial services were very long, provided a circuit-rider was available, or perhaps a local minister "took charge." At the conclusion of the sermon and prayers the young girls in attendance "passed the mourning-cakes" and following the girls, came "six lithe youths with wine and whiskey." When these refreshments had been taken "The coffin then was bourne aloft on the shoulders of six or eight strong men who carried it to the grave."

The customary charge for a coffin between the years 1830 and 1835 is said to have been from $5.50 to $6.00. Frequently, farm produce was accepted in exchange. A coffin-maker in the Western Reserve Section of Ohio advertised that he needed beeswax as coffin barter.

There are comparatively few records giving the names of early Ohio coffin-makers; I am listing five of these accounts from my file:

In the year 1799: Alexander Hill, Marietta, Washington County, "coffin-maker" also was a carver of mantels and interior-trim.

1802: Richard McBride, Zanesville "made the first coffins in the Muskingum Valley."

1811: Daniel Baker (also a dish-timber man.) "Coffin Maker." Centerville, Licking County.

1817: Robert Moore, Stark County. "Coffin-maker also a tall-case- clock-maker."

1820: William Powell, "Bellefontaine, Logan Co. Coffin-maker, tall-clock-case-maker, and cabinetmaker." He made about 600 coffins and 10,000 lights for windows.

A Southern Wheelwright
of the Early 19th Century

by STEPHEN C. WOLCOTT

February 1934

The small numerals refer to the sketches of wheelwrights' tools on the next page.

To be truthful, it is necessary to date our shop as in the title, but I do not like to use the expression 19th century; it sounds too recent. To say about 1800, or even 1820, seems much further away. It is back to that period I wish to take you, and into a shop, some twenty by forty feet, still standing in good condition for all its age, this due to its heavy beams and rafters hewed from the great oak, gum or walnut trees, with the so-called turkey wing broadaxe; its heavy weather boards cut from the heart of virgin growth pine by the tiller type of pit saw,[1] and fastened with hand wrought nails or, in places, with wooden pegs; its cypress shingles, called here slabs or shakes, riven with the shingle frow[2,3] from blocks that had seasoned to just that proper condition for splitting to the best advantage; its brick chimney and large fireplace made of home-made bricks, the mortar made of that same oyster shell lime that has defied the weather for many years in all of our outside chimneys in the South. When first built, our shop had shutters only, as the sash were to come later when the door and sash clamps[4] and other tools had been made; these will also be used in making doors and windows for new neighbors.

Now that we have the shop closed in, let us consider what has been referred to—the making of clamps and tools. After the Revolutionary War, and especially between the years 1790 and 1810, the young people, having learned a trade, were pushing into new country. As with their fathers before, who also started with the wilderness, they had only such tools and implements as they had made at home, or could be spared from the home farm or shop. This necessitated making most of the tools needed. The tools usually deemed necessary by a wheelwright and taken on the adventure were a felling axe[5], a broadaxe[6], a tiller type of pitsaw[1], a frow[2], a few frame saw blades, both narrow, with small teeth[7] for circular work such as felloes would require, and wide, with large teeth[8] for ripping out boards, shafts, tongues, etc., a hammer,[9] a chisel[10] or two, an anvil[11], a bellows[12] and tuyere[13] or tire iron and a few planes[14] or at least some plane irons, also some bars and rounds of iron.

The first work will be to make up his benches, build his forge and set his anvil to be in shape to make up other tools and take on such work as he can do with his limited equipment. Before describing the tools he will use in his wheelwright work, and those

Sketches of Wheelwright's Tools

we describe will be only the large and noticeable ones that are essentially shop tools, let us consider his position in the community of which he is to be so useful a member. As has been said, he must be a blacksmith to properly complete his wheelwright work. Besides these two trades, he was always the local undertaker and made coffins as occasion required and to fit. (Throughout the South one finds this is still the case, or, if the old shop has been given up, the undertaker, and now the embalmer as well, has persisted.) One of the first pieces of work ordered will, no doubt, be some kind of a farm implement, a plow, harrow, wheelbarrow, cradles for grain, forks, rakes and handles for all farm implements and tools. Repairs to all of these, as to all kinds of household furniture and utensils, as well as making most of the simpler kinds, not excepting the high and low post beds and later the spool beds. He also helped out the local carpenter by making window sash and paneled doors. Owing to the few regular gunsmiths at that time, he took on this job, doing some excellent work, as I have seen, and from local report often being the best shot in the district. With his lathe he turned out bed posts, fence posts and post finials, wooden pump tops and spouts, butter moulds, potato mashers, wooden pestles, table legs, chair legs, spindles and braces, and for the children, tops and other toys.

Now that we have him sympathetic to both young and old, to housewife, farmer, carpenter and hunter, a man ready to smoke a pipe or tell a tale with any, maybe we can get him down to making the balance of his tools.

Such diversified work as his required a variety of tools, although at present we are interested only in the large ones used in the making of wheels. The first requirement would be three benches. The first will be a regular carpenter's bench, some two feet wide by ten feet long, with its bench screw or vise at the left end, with its top level with the top of the bench; a sliding panel under the front edge with peg holes and a peg for holding up the free end of a board clamped in the vise, an inch hole in top of the bench some six inches back of the vise for the bench stop,[16] and four inch holes scattered about the top for the holdfast.[17]

The next bench, known as a wheelwright's bench,[18] will be exactly like his carpenter's bench, except that the vise will extend about six inches above the bench top.

His second wheelwright's bench will be unique. So far as I can find out, this type of bench was used locally only. This bench, some two feet wide by six feet long, has a vise the same as the wheelwright's bench. Its difference lies in its having a V-shaped piece cut out of its top, starting some six inches to the right of the vise and extending to within a foot of the other end. This enabled the worker to stand inside the bench and in line with his work. This is known locally as a V-bench.[19]

Next to this bench, the most needed tools was a lathe,[20] with its Great Wheel[21] — as Dr. Mercer in his *Ancient Tools* calls it. A description of these is hardly necessary.

The lathe beds were usually ten to twelve feet long, for on them were turned many legs of our high and low post beds, and later various spindles and legs on the spool beds.

The turning tools[22] were very simple, usually home-made, and consisted of one or two gouges and one or two chisels; later as finer or more varied work was required, a few more of different sizes might be added.

At some shops the lathe and its wheel were placed inside the shop, but as a rule, and in this case, they were in a lean-to at the side.

These Great Wheels being so unique and confined mostly to the South, a word as to their use may be interesting.

The first criticism of them would be the need of another hand, and this is the reason they could be used in the South as against the one-man treadle lathe of the North, as there was always that extra negro boy handy, glad to make a penny or two. These wheels, sometimes operated by one man, sometimes with a handle on each side for two men, could spin a lathe with great speed, giving enough momentum to enable the turner to work the hardest and largest pieces of wood. Especially was this necessary in turning the large hubs used in old time log carriers.

Next in importance of daily use was the humble chopping block and felloe horse. [The felloe was a segment of the rim of the wheel.] This was usually saved from a large tree that had been cut for weatherboards; carefully selected from that part of the tree where a branch some six inches in diameter had grown from the side; it was considered a great prize if he could find a piece with such branches coming off nearly opposite each other. These branches were cut off some three or four inches from the block and a hollow made into the block just above the stump. With such a block some twenty inches or more in diameter and two feet high, our wheelwright had not only his chopping block but his felloe horse,[23] for in these holes could rest the felloe while its inside curve was worked out with an adz.[24]

Under one of the benches stand two logs on short legs, with most of the top center cut out. Like many of the practical and efficient tools of our early artisans, these two belie their looks, for not only is ease of operation permitted, but comfort to the operator combined in these hub mortising blocks[25] or horses. After a hub is turned down, it is marked for the positions of the spokes, and it is then wedged in one of these horses.

With the operator seated at one end, he is at the right height and place to work to the best advantage. This work required a steady hand, so most of it was done by the best workman. After the hub was set, the proper dish of the spokes was determined; with that angle in mind and gauge adjusted, holes were bored with a nose auger[26] straight through toward the center until the center hole was struck. This center hole was about an inch in diameter from end to end of the hub, for the purpose of seasoning. After all spoke locations were bored, mortising chisels[10] were brought into action until it was practically finished. There remained smoothing up of the sides with a firmer chisel and cleaning out the corners with the bruzz or corner chisel.[29] After the first hole is finished, the hub is turned over that the opposite mortise may be cut, the first cut allowing the borings and shavings to drop through. This operation was continued until all the required mortises were finished.

The mortising horse having been returned to the seclusion of the bench, the hub was clamped as tightly as possible on the iron spindle sticking straight up from the center of the wheel block.[31] This was true in our shop, although for the sake of honesty and clearness it is only right to say that some wheelwrights used a wheel pit. This was a pit about six inches wide by four to six feet long, and four feet deep. When this was used, the hub was securely clamped at its center.

The spokes were begun some two years ago, when they were riven out with the spoke frow, or in our early shop the shingle frow, having it on hand. Now, properly seasoned, they are roughly chopped into shape with a hatchet on our chopping block; then with a draw knife, while being held in the vise, they are further shaped into being. With a tenon or hand saw, or lacking these the frame saw will do, the shoulder of the tenon at the foot is formed.

To put the final touch or finish to our spoke, it is clamped in a spoke or tool clamp[40]—a piece of wood some four feet long and three inches square, with a four-inch upright piece shaped like a triangle fixed at one end, and one of the same shape, but movable, at the other. This latter one has an adjustable threaded iron piece through it, facing the other triangular piece. Between these two pieces is held the spoke, while with a spoke shave[41] it receives its finishing touches.

Mr. Pierce used a spoke gauge or template for laying out his spokes. I am not sure this was true with other wheelwrights.

With the spokes ready and numbered for each hole, a sledge hammer[32] or maul[33] at hand, comes what to me would have been the crowning operation of wheel making. First starting the spoke into its hole with gentle taps until you find it steady, then with a little moisture on the palms of your hands you swing onto the maul, and with a great drive, force the spoke maybe a half inch, then checking with a gauge and another blow, a check, and you drive it home. True, your spoke is somewhat battered at the top, but that is the reason you left it a few inches long, and gives you the use of the next tool, the laying out frame.[34]

This frame is nothing more than two pieces of three-inch by four-inch oak, six feet long, held together about twelve inches apart by cross pieces at the ends. In the center toward each other from the sides are two pieces of iron, each threaded through the long side pieces, making them adjustable, to hold different size hubs. With the wheel held as above described, the wheelwright, using a heavy marking gauge,[35] marks the length of each spoke, cuts it off, checks the dish of each spoke, and forms the tenon on the end that will go into the felloe.

We have been so interested in the larger tools, we have neglected those required to make the spokes and felloes, although we have used the one and are about to use the other.

The felloes, having been roughly cut out with the frame saw[7] and seasoned for two years, are now brought into the shop. The first operation is to get a flat surface. This will be done by working one side of the rough piece down, first with the jack plane and then with the trying plane or large smoothing plane. The trained eye was usually capable of developing this surface; if not, two trying sticks were used. Having one side true, it was easy to shape the other, and it was then possible to lay out the exact size and shape from a pattern—many sizes of which hang over the bench. The piece was left some four to six inches long, to be cut off later when fitted over the spokes in our wheel.

With the pattern properly marked out, one end of the felloe is put in a hole in the side of the chopping block, hence the name felloe horse, and with an adze[24] cut down on the inside to the line, then into the bench vise, and with a hatchet and adze worked down on the outside, the latter having been marked out with a marking gauge[35] from the finished inside, unless, as stated, a template was used. In forming the tenons on the spokes, Mr. Pierce at first used a saw to form the shoulder and a draw knife to cut away the surplus wood, the size of the tenon being constantly watched and gauged with a piece of hard oak having a hole of the proper size, called a tenon gauge. As soon as he had time, and properly seasoned wood, he made hollow augers[49] for this purpose. These were blocks of hard wood three or four inches in diameter of the same thickness. A hole was bored through this block, and a V-shaped cut made in its side for the full length, and into the center hole. This cut, about three-eighths of an inch wide at the center hole, became wider toward the outside. Into this cut was fastened a narrow knife-shaped piece of iron, sharp on the inside. With handles on each side of this block, it was possible to form spoke tenons easily and quickly.

The wheel was taken from the laying out frame, and was again placed on the wheel block, or possibly held in the bench vise, where the felloes were placed in position, marked for length and position of spokes, then sawed off, dowel holes marked at each end, and with a dowel bit, holes made to take these bits of wood. These were small round pieces of hard wood, usually maple or oak, three-eighths to half inch in diameter, and one to one and a half inches long, made in the shop with a dowel former. After fitting the felloes, boring the tenon holes with a nose or shell bit, the felloe is forced into place. This is a piece of work for the man of experience. He has two spokes to go into each felloe, as well as fitting the dowels into the preceding felloe. In this job he is greatly assisted by the spoke dog,[52] a stout piece of wood some sixteen inches long and about an inch and a quarter thick, having a piece of iron eight or ten inches long, three-quarters of an inch wide, and a quarter of an inch thick fastened loosely near one end, so that one end of the iron, turned up as a hook, projects an inch or two beyond the end. By placing the stick against the spoke already engaging the felloe, and the hook over the obstinate one, he is able to force the spoke into its hole. The wheel now goes to the blacksmith, but our readers being already tired of details—as I am—we can leave the blacksmith work for another article.

There remains, however, one more operation and, therefore, one more tool—nothing more than a heavy oak plank, some three of four inches thick, with a ten-inch hole in it, and some eighteen inches square, mounted on sturdy, strong legs, and called the hub boring block.[25] After being properly ironed, the wheel is placed with the hub in this hole, and with an auger and gauge to begin with, a hole is worked in the center of the hub for the axle. This hole is finished with the long—some two or three feet—hub reamers,[56] or rimers as they were called, made of iron curved to a half circle and sharpened on one edge, and tapered from one inch at the bottom to two or three inches at the top. These, of course, came in different sizes for the various types of wagons and carts, also the rear wheel would need a larger, heavier axle than the front one.

With good luck and few interruptions, Mr. Pierce could make a wheel in a day, so for the time being, let's give him time for a pipeful.

How Tinsmiths Used Their Tools

by JOHN H. DEMER

December 1973

At the time of writing, Mr. Demer was an Assistant Curator at Cooperstown.

Tinning: The art of coating other metals with tin.
Tinplate: Ironplate coated with tin by dipping it into a molten bath of the latter metal.

Knight's *American Mechanical Dictionary*, Boston, 1878

In the nineteenth century, tinplate was among the most useful and versatile of construction materials. Using hand tools and simple machines, tinsmiths fashioned tinplate into a seemingly endless variety of articles including domestic utensils, containers, and architectural elements such as roofs and gutters.

Although products of nineteenth-century tinsmiths are many, evidence of the process of tinworking seems to be limited. Two studies, Margaret Coffin's *American Country Tinware* and Shirley DeVoe's *Tinsmiths of Connecticut,* discuss form, style, decoration, and sale of tinware, but fail to adequately discuss the step-by-step processes, the craft of making tinplated objects. Even heavily-relied-on sources such as encyclopedias of Denis Diderot, Johann Krunitz, Charles Tomlinson, and Abraham Rees, fail to discuss in detail how tinsmiths crafted objects. Rather, their attitudes are that the

craft of tinsmithing was common knowledge. If so, then what was common knowledge in the eighteenth and nineteenth centuries is largely forgotten today.

One might think that, in learning the "art and mystery" of his trade, the tinsmith was well acquainted with his tools, machines, and craft, and that the tinsmith passed his knowledge on to succeeding generations. Oddly, there is evidence that this assumption is false. As L. Broemel wrote in his preface to a sheet metal worker's manual in 1918:

The author, having spent a lifetime with the tinsmith and sheet metal worker, during that time has visited many shops the country over. It is surprising to observe how limited is the knowledge of a large percentage of apprentice and advanced sheet metal workers regarding the essentials of machine and tool construction and application. Therefore, the subject of sheet metal working machinery and tool construction, their uses and application, is treated in these pages in an extensive way.[1]

Because of a lack of material illustrating how tinsmiths used their tools, there is a need to document the processes used in making various objects of tin. Essentially there are three methods by which a tinsmith can make tin objects by hand tools or by machines, which include stamping, hammering, and spinning. Hand tools, such as stakes, are templates over which the tinsmith shapes tinplate by using manual pressure. Machines are also templates, but rolls and rollers shape tinplate when turned through gearing by the tinsmith. Stamping machines shaped through uniform pressure on a blank applied by a heavy die, unlike machines which required the metal to be turned.

Hand tool shapes can be traced to eighteenth-century encyclopedias such as Diderot and Krunitz. Machines were invented in the decade following 1800 by Eli Parsons and Calvin Whiting of Dedham, Massachusetts, but gained their greatest popularity after 1819 when Seth Peck, a toolmaker from Southington, Connecticut, began his extensive manufacture. As Edward Hazen wrote in his *Panorama of Professions* in 1839:

On examining almost any vessel of tin ware, it will be perceived that, where the parts are united, one of the edges, at least, and sometimes both, are turned, that the solder may be easily and advantageously applied. It will also be discovered that iron wire is applied to those parts requiring more strength than is possessed by the tin itself. The edges and handles are especially strengthened in this manner.

The edges of the tin were formerly turned on a steel edge, or a kind of anvil called a stock, with a mallet; and in some cases, this method is still pursued, but this part of the work is now more expeditiously performed, by means of several machines invented by Seth Peck, of Hartford Co., Connecticut. These machines greatly expedite the manufacture of tin wares, and have contributed much towards reducing their price.[2]

Parsons and Whiting also claimed credit for inventing stamping machines for tinware, but stamping was used in Europe in the eighteenth century to produce furniture brasses, so their claim is not so important for stamping as it is for inventing tinworking machines.

To understand the craft of the tinsmith, one must first become acquainted with his tools and their operation. Accordingly, the writer has compiled a series of illustrations and diagrams that hopefully show some, but by no means all, of the tools at the disposal of the tinsmith and how the tinsmith used many of these tools. First illustrated are the hand tools that were basic to all tinsmiths' shops. Next, tinsmith machines and their uses are illustrated, and, finally, the writer briefly mentions stamping machines and their application.

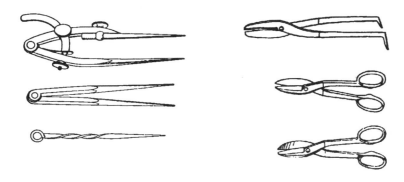

Figure 1. Single Leg Extension Divider (top),Compass (middle), Scratch Awl (bottom). Unless otherwise indicated, all illustrations are from L.Broemel.

Figure 2. Bench Shears (upper), 1819 Original Pexto Straight Snip (middle), 1819 Original Circular Snip (lower).

Using the divider, compass, and scratch awl (Figure 1), the tinsmith traced a thin pattern line on sheet tinplate. As Broemel noted, each point was sufficiently sharp to make a clear line. Of the tools, the awl was most used, because after cutting templates from tinplate, the tinsmith could quickly and easily scratch around them to make a design for an identical object. After outlining the object on tinplates, the tinsmith was ready to cut the shape, and had a variety of hand and mechanical shears to choose from.

Among the most basic tools of the tinsmith, shears enabled him to cut the desired pattern from flat tinplate. (Figure 2). Bench shears were fastened to the bench, as the name implies, through the tang and a hole in the bench plate. The tinsmith used bench shears to cut heavy sheet metal—20 gauge or less. After 1850, floor shears became more popular and largely replaced bench shears for cutting heavy metal.

On the other hand, straight and circular snips have continued to be of prime importance to tinsmiths, and enable smiths to cut and trim a variety of patterns from tinplate. The shears in the middle of Figure 2 have a straight blade and are particularly suited for cutting straight edges. The shears on the lower end of the same illustration have tapered cutting edges and sloping sides. They are best suited for cutting circles. Hawk-billed shears (Figure 3) have sharply pointed blades that enable the

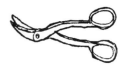

Figure 3.

tinsmith to cut around sharp corners without buckling the metal. Hawk-billed shears are also suited for detailed work as they are well-balanced and easy to control.

After cutting tinplate to the desired shape the tinsmith who did not own machines used a variety of bench stakes (Figure 4) to shape tinplate into tinware. The tinsmith put his bench stakes into a heavy metal bench plate (Figure 5) that held the stakes secure during the shaping.

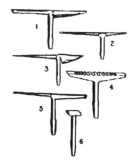

On the beakhorn stakes (Figure 4-1), the tinsmith could make right-angle bends on the square end, and tapered round bends on the other end. Using the beakhorn stake, the tinsmith could shape square pans on one end, and tankards and coffeepots on the other end. On the candlemold stake (Figure 4-2), the tinsmith could make not only candlemolds, but also small cups and other objects having diameters of less than two inches.

Figure 4. Bench Stakes.

The tinsmith used the funnel-shaped end of the blowhorn stake (Figure 4-3) to shape funnels, tapered covers, and flared pans. The other end he used for less tapered cylinders. The different-sized indentations in the creasing stake (Figure 4-4) allowed the tinsmith to crimp, make grooves to receive wire, and bend wire for pail handles.

Figure 5.

The needlecase stake (Figure 4-5), represented as being the same size as the candlemold stake but actually much smaller and thinner, was used not only for needlecases but also tapered cylinders of thin diameter, such as coffeepot spouts. On square stakes (Figure 4-6) the tinsmith fashioned objects requiring right-angle bends, such as pans.

The double-seaming stake (Figure 6-1) permitted the tinsmith to make a series of watertight seams in a vessel, as shown in Figure 7.

The creasing stake with horn (Figure 6-2) is a variation of the creasing stake mentioned earlier and its use is virtually the same. The writer is not familiar with coppersmithing tools and cannot comment on the purpose of the coppersmith's square stake (Figure 6-3). Using the hatchet stake (Figure 6-4) with its beveled edge, the tinsmith could bend an edge for strength or prepare a rectangular object such as a deed or candlebox to receive a wire. Finally, a bottom stake enabled the tinsmith to form a double seam on a vessel with a narrow diameter. Other stakes not shown, such as the mandrel stake, enabled the tinsmith to shape gutters, while the conductor stake enabled the tinsmith to shape stove and other piping.

Tin "knocking" was needed for sharp angles, but firm hand pressure only was needed to form cylinders, especially those having diameters greater than two inches. Generally speaking, when the tinsmith formed a cylinder using bench stakes, he used only his hands to shape the tinplate. When the tinsmith formed an angle using bench plates, he usually used a hickory mallet to shape the tinplate.

Figure 6. Bench Stakes

Figure 7.

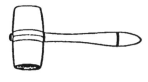

Figure 8. Hickory Mallet.

The most basic and useful hammer to the tinsmith was the hickory mallet (Figure 8). With a broad face and polished end grain, the mallet was well suited for the task of bending tinplate, whose surface coating of tin was easily scratched or marred. Using gentle taps, the tinsmith could gradually but positively shape tinplate with his mallet and stakes.

Raising, or, as they were sometimes called, bumping hammers (Figure 9), enabled the tinsmith to form curved moldings for cornices, and rounded covers for tubs and pails. To raise a piece of tinplate, the tinsmith placed the flat metal over a depression in lead or wood such as in Figure 10.

In the process of raising a cover (Figure 10-A), the tinsmith first lightly worked the outer edge of the circle into a shallow dish by tapping the edge and turning the circle, as in Figures 10-1 and 10-2. Next, the tinsmith repeated the procedure more toward the center

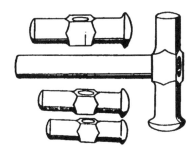

Figure 9. Raising Hammers.

of the circle, as in 10-3 and 10-4. Then the tinsmith lightly formed the center of the cover, as in 10-5 and 10-6 by tapping the center of the circle. If there were any pronounced hammer marks, the tinsmith could smooth them by placing the core on a round head stake and tapping where needed.

Other special purpose hammers included the setting hammer (Figure 11-left) and riveting hammer (right).

Of tinworking machines, the simplest and most basic was the folder (Figure 12). The folder allowed the tinsmith to turn the edge of the metal to form a lock or receive a wire.

To set up the folder, the tinsmith first turned the gauge hand wheel until the gauge indicated the width of bend the tin-

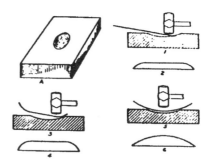

Figure 10. Process of Raising a Cover.

Figure 11.

ner wanted. Then he tightened the gauge locking screw with the key wrench. Next, the tinsmith adjusted the wedge screw to the left for a closed lock, or to the right for a wire lock. Then he tightened the wedge screw with the key wrench. Finally, the tinsmith located the adjustable stop at the desired degree of bend, which could be from 10 degrees to 120 degrees.

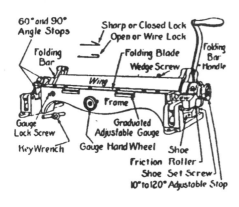

Figure 12. Bar Folding Machine.

After setting up, the tinsmith inserted the sheet metal between the folding bar and jaw, and pulled the forming bar handle, making the bend.

A brake, or open throat folder (Figure 13), like the bar folder, turned an angle in sheet metal, but the brake was more flexible. The jaw of the folder, though adjustable, was closed, and could accept metal for only limited fold widths, usually not more than one-half inch. The bar folder was useful for small work but awkward for large. On the other hand, the brake had an open throat and could bend a long piece of sheet metal.

Like the bar folder, the brake had adjustable gauges, called pin gauges, which the tinsmith could set at a predetermined width. Also like the bar

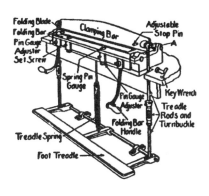

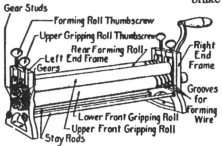

Figure 13. Open Throat Folder, or Floor Brake.

Figure 14. Shapes Formed on the Cornice Brake.

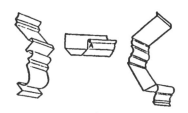

Forming Machine with Solid Housings.

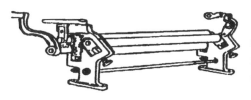

Forming Machine with Slip Roll.

Figure 15.

folder, the brake had stops, which the tinsmith could set at predetermined bend angles. But the important difference between the bar folder and brake was the brake's use of a foot-operated clamping bar which held the sheet metal in place while the tinsmith formed the bend. The term "brake" comes from the clamping bar that holds or brakes the metal during the bend.

There were different types of brakes, including the cornice brake, which, through the use of various shapes, allowed the tinsmith to bend cove and beaded moldings. (Figure 14).

Forming machines (Figure 15) are sometimes called rolls, and form flat metal into cylinders. Broemel claims that no other machine in a sheet metal shop was used more than the former. Using the former, the tinsmith could shape cylinders such as lanterns, coffeepots, gutters, and stovepipe.

Setting up either the forming machine with solid housings or with slip roll was essentially the same. First the tinsmith adjusted the upper and lower gripping roll thumbscrews to insure the sheet metal to be shaped would smoothly and uniformly pass between the gripping rollers. Next, the tinsmith would adjust the rear forming roll by turning the forming roll thumbscrew. Raising the rear forming roll decreased the diameter of the object to be formed while lowering the rear forming roll increased the diameter of the object to be formed. By

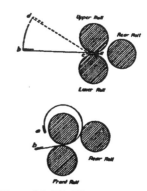

Figure 16. Side View of the Forming

turning one side of the forming roll thumbscrew more than the other, the tinsmith could shape a tapered object, such as a tapered coffeepot or a flared pan. Grooves in the gripping and forming rolls allowed the tinsmith to shape tinplate that had previously been wired.

After setting up, there were no set rules for shaping tinplate. The tinsmith experimented and adjusted until his cylinders were of the desired diameter. The forming machine with solid housings required less skill to use but removing a narrow cylinder was often awkward. The forming machine with a slip roll allowed the tinsmith to slide the finished cylinder off the forming roll, which pivoted and opened. The diagrams (Figure 16) help illustrate how the forming machine works. In each figure, the rear roll is the forming roll and the front rolls are the gripping rolls. The figure on the top shows the sheet metal on insertion, the figure on the bottom shows the turned cylinder ready for grooving.

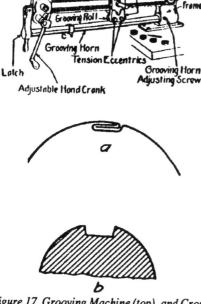

Figure 17. Grooving Machine (top), and Cross-Section of a Grooving Horn and Countersunk Pipe (bottom).

After bending locking seams on a folder and shaping a cylinder on a former, the tinsmith closed a seam on a grooving machine (Figure 17). Though there were many

Figure 18. Section of Pipe before Grooving (left) and Section of Pipe with Outside Seam (right).

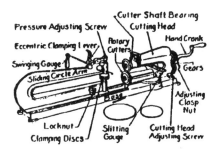

Figure 19. Short Horn Grooving Machine.

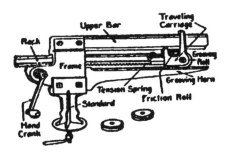

Figure 20. Circular Shears.

types of grooving machines and different types of seams, the process for grooving was similar among all machines.

To set up, the tinsmith would attach a flat-faced grooving roll for a countersunk seam, or a grooved roll for a seam located on the outside (Figure 18). To operate the grooving machine, the tinsmith cranked the traveling carriage to the frame, inserted the seam to be grooved on top of the grooving horn, then cranked the traveling carriage across the seam, from the frame to the end of the grooving horn. The length of the cylinder that could be grooved was limited by the length of the grooving horn. But by reversing the pipe, the tinsmith could groove a cylinder twice the length of the grooving horn.

A short horn grooving machine (Figure 19), similar to the grooving machine in the Winterthur collection, apparently could be used only to groove seams located on the outside of a pipe, according to Broemel, because the traveling carriage was too light to flatten a seam in countersinking.

Circular shears (Figure 20) enabled the tinsmith to quickly cut accurate circles of sheet metal to be used as bottoms for vessels and cans. Though there are different types of circular shears, their methods of operation are similar. Basically, a piece of sheet tinplate held between two clamping discs was cut by two vertical rotary cutters turned by a hand crank.

To set up the rotary shears, the tinsmith loosened the locknut and slid the circle arm to the desired diameter, measured by a gauge on the circle arm. Circles could be cut from diameters of 2¾ inches to 22 inches. After setting up, the tinsmith cut the circle by turning the hand crank. By turning the cutting head adjusting screw, the tinsmith could compensate for wear in the cutting blades.

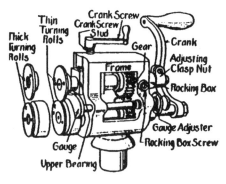

Thin Turning Rolls
Thick Turning Rolls
Crank Screw
Crank Screw
Crank Screw Stud
Gear
Crank
Adjusting Clasp Nut
Frame
Rocking Box
Gauge Adjuster
Rocking Box Screw
Gauge
Upper Bearing

Figure 21. Turning Machine.

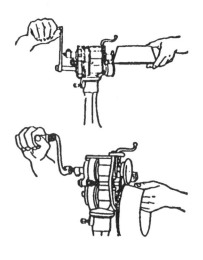

Figure 22.

Ring and circular shears were a variation in basic design, with the cutting blades meeting at an angle. The angle apparently facilitated the cutting of inside circles.

Turning machines (Figure 21) were used to bend the edge of sheet metal to receive a wire. The tinsmith used the turning machine to turn edges on cylinders or irregular shapes. He used the folding bar to turn edges on straight edges, but could not use the bar folder to bend a curved edge.

To set up the turning machine, the tinsmith first attached either a thick or thin turning roll, depending on whether he planned to insert a thick or thin wire. Next, the tinsmith adjusted the gauge which determined the width of the bend. Then he inserted the edge to be turned on the lower roll and lowered the upper roll by turning the crank screw. After setting up, the tinsmith turned the hand crank and the rolls made a slight bend on the edge. The tinsmith worked the metal through the machine several times, each time tightening the crank screw slightly, and tilting the metal into a bend. After bending, the tinsmith inserted the wire.

The diagrams in Figure 22 show how a tinsmith uses the turning machine to bend a wire edge on a flaring article. On the top, the tinsmith has set up and is making his initial turn. On the bottom, the tinsmith has made a succession of turns, each time tightening the crank screw and raising the pan until the groove is deep enough to receive the wire.

The wiring machine (Figure 23) was used in conjunction with the turning machine to seal the bent edge around the wire. Its setup and operation were similar to the turning machine.

To set up the wiring machine, the tinsmith first adjusted the forming gauge at the proper width by turning the forming gauge worm screw gear. Next, the tinsmith tightened the crank screw until the fit was snug, as indicated in Figure 24. To wire, the tinsmith turned the crank,

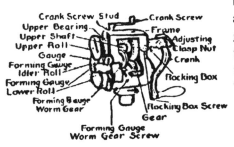

Figure 23. Wiring Machine.

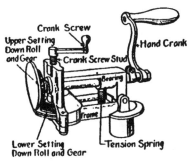

Figure 25. Setting Down Machine.

closing the seam. As was true with the turning machine, several tightenings of the crank screw were required to complete the wiring.

Tinsmiths used the setting down machine (Figure 25) in conjunction with the burring machine. The burring machine formed a double seam, the setting down machine sealed it, as shown in Figure 26. On the left (A) are edges turned by the burring

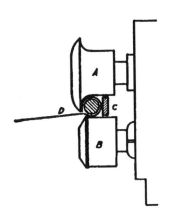

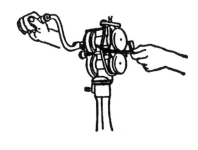

Figure 24

Figure 26.

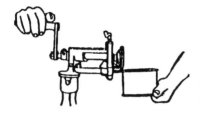

Figure 27

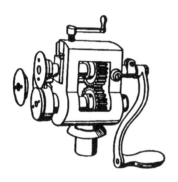

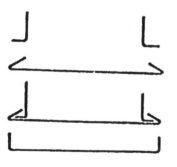

Figure 28. Burring Machine (top), and Three
Steps Needed to Attach a Bottom to a Cylinder
Using a Burring Machine (bottom).

machine; on the right (B) are edges that have been closed by the setting down machine.

To operate the setting down machine, the tinsmith placed the burred edges of the vessel, bottom up, between the setting down rollers, applied pressure with the crank screw and turned the hand crank to close the seam. A variation of the setting down machine, not shown, had inclined rollers, which enabled the tinsmith to close a seam with the bottom either up or down.

How the tinsmith used the setting down machine is illustrated in Figure 27. The two rollers are pinching the seam shut.

Tinsmiths called bending two pieces of metal over one another, as in the diagram on the bottom (Figure 28), "double seaming." The burring machine turned an edge on the cylinder and a similar edge on the bottom that were joined as on the bottom figure in the lower diagram. A double seam was clean and watertight.

To set up the burring machine, the tinsmith adjusted the gauge for the desired width and raised the upper roll. To operate the burring machine, the tinsmith placed the metal against the gauge on the lower roller and tightened snugly the upper roll. Then he turned the burr, increasing the pressure of the upper roller and increasing leverage on the metal, as shown in the diagram in Figure 28.

The double seaming machine (Figure 29) was used in conjunction with the burring machine and setting down machine. The diagrams in Figure 29 show what the double seaming machine did after the setting down machine sealed the seam (A).

The double seaming machine first bent the seam on a 45 degree angle, then bent the angle parallel with the side of the vessel, as shown below in Figure 29, B and C.

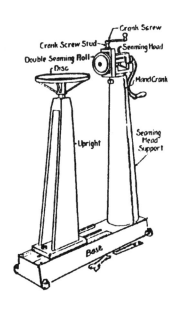

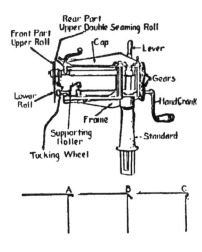

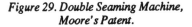

Figure 29. Double Seaming Machine,
Moore's Patent.

To operate the double seaming machine, the tinsmith slid the vessel over the lower roll so the bottom was flush

Figure 30. Variation,
Double Seaming Machine

against the face and the lip rested on the supporting roll. Then the tinsmith moved the lever to the left, which put the upper roll in a position to turn a 45 degree angle (B). The tinsmith turned the angle, then moved the lever to the right. Moving the lever to the right brought the lower tip of the upper roll (marked Front Part Upper Roll) in contact with the raised part of the seam. The tinsmith closed this seam by gradually applying pressure to the crank screw (not marked) as he turned the work with the hand crank. Moore's Double Seamer could double seam vessels with a diameter of 4½ inches and a throat of 15½ inches. A variation of a double seaming machine similar to the one in Winterthur's collection is shown in Figure 30. Like Moore's, the horizontal disc double seaming machine required two operations to form a double seam. On the horizontal disc double seaming machine, the bottom of the vessel was placed on the disc, perpendicular with the roll. On Moore's double seaming machine, though, the bottom of the machine was placed over the lower roll, parallel to the face.

A simple machine to use was the beading machine (Figure 31). By beading a vessel or a strip of sheet metal, the tinsmith strengthened it or made it more stiff.

To use the beading machine, the tinsmith first attachd a set of bead rolls, chosen for appearance rather than for specific use. Ogee beads, for example, were used mostly on stovepipes, while triple beads were used mostly on coffeepots and other vessels. A

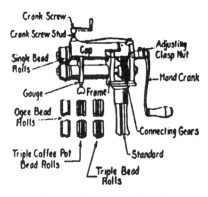

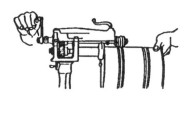

Figure 31. Beading Machine   Figure 32.

single bead would be used where necessary. Next, the tinsmith set the gauge at the desired width, inserted the vessel between the rolls, tightened the crank screw, and turned the bead, as shown in Figure 32.

Material illustrating stamping machines, especially for stamping tinware, seems limited. The illustration (Figure 34) from Rees' *Cyclopaedia*, published before 1834, shows a level view of a die stamp used in plated ware (Figure 34-1) and a top view of the same die stamp (Figure 34-7). The process of stamping appears obvious. A weighted die (d) drops down tracks (a) and forces a flat piece of tinplate into the shape of the female die (3). Another variation of a stamping machine (Figure 33) combines turning as in the machines, and male and female dies as in a die stamp, to make a wide variety of container covers. The covers are actually spun, much the same as britannia wares were spun. Thus the press was neither a true die stamp nor a true machine.

By using the tools and machines illustrated, tinsmiths in the nineteenth century were able to produce a wide variety of goods such as those shown in

Figure 33. Spaulding's Patent Tinners' Press. From Dover Stamping Company, 1869.

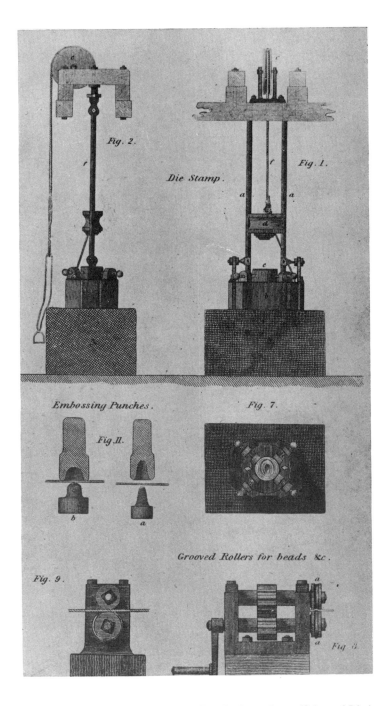

Figure 34. Die Stamp. From: Abraham Rees, The Cyclopaedia; or Universal Dictionary of Arts, Sciences, and Literature (Philadelphia: Samuel F. Bradford, n.d.) Plates IV, see "Plated Manufacture."

the catalogs of the Dover Stamping Company of 1876. Though Hazen and others claimed that the use of machines and stamping equipment lowered tinware prices, figuring the degree of savings is at best difficult. Too many related factors, such as the inability to compare exact objects of the eighteenth century with those of the nineteenth century, production, and inflation, make comparisons inconclusive.

For example, in the Winterthur library is a *Book of the Prices for Making Tin-Ware*[3] published by the tinworkers of Philadelphia in 1796. The price book is important because it lists prices of tinware before the invention of machines. But the only entries that can be compared with nineteenth-century catalogs are those for coffeepots. Many other items such as candlemolds, colanders, egg slicers, graters, and bread pans are listed according to how many objects the tinner could make from a sheet of tinplate, which measured 10 inches by 14 inches. They are not measured according to volume or inches, as are objects manufactured in the nineteenth century. Also, prices in the Philadelphia price book are those the tinsmiths charged, where prices in the nineteenth-century catalogs include costs of production and the wholesalers' markup. Thus, to compare production in the eighteenth and nineteenth centuries would be difficult using only price lists and catalogs. Tinware manufacturing company account books are needed for more valid comparisons, and they are difficult to locate. Finally, inflation is an important factor that cannot easily be accounted for in the span of eighty years between the publishing of the Philadelphia price book in 1796 and the St. Louis Stamping Company catalog in 1876.

Subject to the above, we have listed the prices of coffeepots made in 1796 and prices of similar pots made in 1876.

Philadelphia Prices, 1796:[4]
Coffeepot, gallon, per doz. $1.60
3 quart, per doz.   1.20
1 quart, per doz.   .80
1 quart, per doz.   .65

St. Louis Prices, 1876:[5]
Coffeepot, gallon, per doz. $4.50
3 quart, per doz.   3.75
2 quart, per doz.   3.25
1 quart, per doz.   2.25

In dollars, coffeepots made in 1876 were more expensive than those made in 1796, but as mentioned above, variations of Federal money, production rates, and inflation are not taken into account in the 1876 figures. Both price lists are for finished coffeepots, but in 1876 a tinsmith could order pre-cut pots from a stamping company and assemble them in his shop. An advertisement for unassembled coffeepots is shown in Figure 35.

Thus, by buying unassembled coffeepots, a tinsmith could save 40 percent of the cost of a finished pot, in addition to saving the additional costs of shipping a dozen bulky assembled pots.

Though it is difficult to document lower production costs due to use of machines and stamping equipment, contemporary sources claim costs were lower, and until there is additional documentation this writer can only agree with those who manufactured machines and sold stamped wares. Additional research will probably support their claims. Documenting the craft of the tinsmith, especially those who used machines and stamping equipment, is equally important, because the use of tinsmith's machines has become a dying art. Perhaps rather than looking for eighteenth-century sources illustrating the craft of tinsmithing, researchers might do well to concentrate

Figure 35. Unassembled Pots. From: Catalog of the St. Louis Stamping Company, 1876, p.63. Prices added by this writer.

on little-known and long-forgotten sources published in the nineteenth and early twentieth centuries, such as Broemel. There seems to be a revival of interest in traditional tinsmithing, especially among younger craftsmen. Again, perhaps this new generation will make the knowledge of their craft as well known as it apparently was in the late eighteenth century.

BIBLIOGRAPHY

Book of Prices, of Journeymen's Wages for Making Tinware. Philadelphia, 1796. Joseph Downs Manuscript and Microfilm Collection, manuscript number 55.506. Winterthur Museum, p.4

Broemel, L. *Sheet Metal Workers' Manual.* Chicago: Frederick J. Drake & Co.,1918.

Catalog of the St. Louis Stamping Company, 1876, supplemental price list for January 1, 1877, p.11. Copy in Rare Book Room, Winterthur Museum Library.

Coffin, Margaret. *History and Folklore of American Country Tinware, 1700-1900*. Camden, New Jersey: Nelson Press, 1968.

Daboll, Nathan. *Schoolmaster's Assistant: Being a Plain Practical System of Arithmetic: Adapted to the United States*. New London, Connecticut: Samuel Green, 1806.

DeVoe, Shirley S. *Tinsmiths of Connecticut*. Middletown, Connecticut: Wesleyan University Press, 1968.

Dover Stamping Co. 1869. 1869; rpt. Princeton: Pyne Press, 1971.

Flower, Philip W. *History of the Trade in Tin*. London: George Bell, 1880.

Hazen, Edward. *The Panorama of Professions and Trades; or Every Man's Book*. Philadelphia: Uriah Hunt, 1836.

Krunitz, Johann George. *Oeconomische Encyclopadie, Oder allgemaines System der Land-Haus-und-Staats-Wirthschaft, in alphabetischer Ordnung*. Berlin: J. Pauli, 1828.

Rees, Abraham. *The Cyclopaedia; or Universal Dictionary of Arts, Sciences, and Literature*. Philadelphia: Samuel F. Bradford, n.d.

St. Louis Stamping Company, 1876. Catalog and supplement published by the company.

Tomlinson, Charles. *Illustrations of Useful Arts & Manufactures*. London: Society for Promoting Christian Knowledge, 1859.

FOOTNOTES

1. L. Broemel, *Sheet Metal Workers' Manual* (Chicago: Frederick J. Drake & Co.,1918), p.7.

2. Edward Hazen, *Panorama of Professions and Trades; or Every Man's Book* (Philadelphia: Uriah Hunt,1836), p.274.

3. *Book of Prices of Journeymen's Wages for Making Tin-Ware*, Philadelphia, 1796. Joseph Downs Manuscript and Microfilm Collection, manuscript number 55,506, Winterthur Museum, p.4.

4. *Ibid*.

5. *Catalog of the St. Louis Stamping Company, 1876*, supplemental price list for January 1, 1877, p.11. Copy in Rare Book Room, Winterthur Museum Library.

The Bygone Cobbler

by JARED VAN WAGENEN

March 1947

The community tanner and cobbler have disappeared together. Both of them represented the same phase in our industrial development; each of them was to some extent dependent upon the other and both of them are only a memory. It is true that in our larger towns and cities we still have cobblers of a sort. Sometimes the shop is called the "Electric Shoe Repair Parlors" and commonly these places are manned by Greeks or Italians, who, if necessary, will mend a rip or put on a "live rubber heel" or a neolin halfsole while you wait. But the genuine craftsman of three generations ago, who, out of home-tanned leather and hand-twisted flax, wax-end and wooden pegs, and over a last which he himself whittled out of a block of maple, made boots and shoes—he survives only now and then as an ancient, time-worn man.

I am sure that of all our community handicrafts, there is none more primitive or more universal or representative of a wider range of technique in a one-man business than is the ancient and honorable art of the old-time shoemaker. He was once found not only in every township, but literally in almost every school district.

The census of 1845 counted the workers and establishments in several lines of rural industry, but for some reason failed to enumerate the shoeshops. The census of 1855 reported 1,467 such establishments, and ten years later there were reported only 515. As a matter of fact the true number was very much greater. The enumerators were either very careless or else did not bother to count individual cobblers. Much better evidence can be gathered from the various county directories and gazeteers, a large number of which were published both before and after the Civil War. The usual plan was to sell subscriptions to the directory in return for which the subscriber's name would be listed, together with his business. In *Child's Directory of Schoharie County* for 1872-73, it appears that there were no less than 96 men who were willing to pay their money in order that they might be listed as "custom" shoemakers. By the same token we had 131 blacksmiths. Without question my own county was merely typical of all the rural counties of the state. That was 75 years ago. Since then the very last of the cobblers has laid aside his tools forever and of all that goodly company of blacksmiths, only an insignificant fraction remain.

In the United States Census for 1860 is an article in which the author moralizes upon the rapid development of the factory system as applied to shoe manufacture, and he writes with great satisfaction that it is only a question of time until the cobbler with his "bench and kit" will be as extinct as the "hand card" or the "great and little spinning wheel" (meaning thereby the wheel for wool and the smaller wheel for flax). Time has vindicated his judgment for his prophecy has been almost literally fulfilled.

The shoemaker was such an absolutely indispensable member of the community that he appears very early in the history of all our settlements. Thus in the second year of the Plymouth Colony arrived one Thomas Beard with his tools and a stock of hides. The London Company not only accredited him to the governor of the colony, but also promised him a yearly salary of ten pounds in addition to his earnings.

The Yankee inventor seems always to have had a peculiar flair for shoemaking. Almost from the beginning Lynn was a center of the business, and until recent years the practical supremacy of Massachusetts in this particular industry has never been questioned. In 1860, three counties of that state produced more than one-third of all the factory shoes manufactured in America.

The Use of Wooden Pegs in Tapping

The use of wooden pegs to fasten the "taps" to the "uppers," while at one time almost universal for the footwear of both men and women, was by no means the earliest method. Indeed it is said that wooden pegs were first used in New England about the year 1812—or according to another, 1818. At any rate, their use spread very rapidly. Time was when each cobbler laboriously whittled his own, but shoe-peg factories were soon established. In 1841, a factory at Laconia, New Hampshire is said to have produced nearly all the pegs used in this country. They were made from black, yellow or white birch and from hard maple. The daily production of pegs exceeded fifty bushels. They did not long have the field to themselves, however, for in 1850 the Census returns for Massachusetts showed a total production of 17,800 bushels of pegs worth $12,900. This seems very little money for such a tremendous output of pegs, but it was the day of cheap labor and abundant raw material. In 1855 New York State reported 15 shoe-peg factories, but ten years later only three survived. Cobblers bought their pegs by the quart in several different sizes, and the price for that unit of measure was only five or six cents.

Advent of the Front Lacing Shoe

It is said that the idea of a shoe open in front so that it could be laced up with a shoe string originated with a Yankee cobbler about the year 1791. There are still living old men who can remember when boots were not made "left and right"; both were made over one last. Then if the owner would be careful and conscientiously transpose his boots from one foot to the other each time he put them on, they would never "run over" on the heel or wear out the taps unevenly. My father frequently told me of this careful economy of his boyhood.

I think it safe to say that up until at least the close of the Civil War, the average man and woman of the farm went shod in the product of the local cobbler's art, and even after factory shoes were generally introduced, there were many conservative men who steadfastly refused to wear "store boots."

There is no question but that in those years when the Homespun Age was in flower in our state, literally thousands of cobblers sat on their low benches and industriously tapped-tapped-tapped or drew out the long waxed ends from morning to night in order that our farm people might be shod. There was hardly a crossroads hamlet but had its representative of that trade, of which St. Crispin is the patron saint. The last man who ever practiced his craft in my community was one Fred Martin. Very dimly I remember him. The house where he worked still stands and the big window with sixty panes of glass, built of this unusual size so that as he sat by it his bench might be flooded with light.

I think it will be proper that I should here speak of his unfortunate venture into agriculture. Through many years, by patient industry, he had accumulated what seemed to him a little competency, perhaps $2,000 or thereabouts. Then, in his folly he must needs leave his craft which he understood and buy a farm about which he knew nothing. He purchased a poor hill farm at an inflated price just about the time that land values began their long decline. On the land he was helpless, and his equity was soon wiped out, and he was forced off the farm—a sadder and a wiser, and unfortunately an aged and broken man. This back-to-the-land foolishness is responsible for too many tragedies such as his.

The Itinerant Shoemaker

The shops of these old-time shoemakers were very frequently in the living-room of the house, but there were also itinerant workmen who with their bench and kit went from farm to farm and wrought beside the fires of their employers. If it was a large family, the cobbler might stay a couple of weeks before all the members of the household were properly outfitted. The arrangement seems strange to us, yet it was very widely adopted and probably had a sound basis. In the days when there was very little contact with the outside world, the traveling cobbler, especially if skilled in the art of pleasant gossip, was welcomed both as an employee and as a guest. Then, too, inasmuch as his clients were all on hand, it was easy to cut and fit and try as the work proceeded. My father lived long enough ago to remember how the country tanner was scolded and importuned to hurry and get the leather finished because the shoemaker was expected.

At the Pioneer Home on the State Fair Grounds in 1926, might be seen Mr. John Mulberry of Potter's Hollow, actually making a pair of boots after the exact methods of long ago. Mr. Mulberry is my authority for the statement that at an early date the shoemaker whittled out his own lasts. Sometimes when he established himself for a genuine siege of family cobbling, he measured the biggest foot of the family and then made a last to match it. This last having served its purpose, he proceeded to shave it

John Mulberry of Potter's Hollow, making a pair of boots after the exact methods of long ago.

down and shape it to compare with the next largest foot, and so on until every member of the family, down to the six-year-old boy, would have had his footwear built over the self-same last, finally reduced to a small fraction of its original size. It is at least an interesting tale.

The coarse working boot of the farmer was made of cowhide, and in the boyhoods of the oldest living men cost not less than $3.00 a pair. The cost was much less, however, when the leather was furnished.

For Sunday and town-meeting wear, the well-to-do farmer had a pair of fine boots made of calfskin. On these the maker sometimes lavished wonderfully patient and skillful care. I have seen a pair of very old calfskin boots which had an inside lining of leather sewed to the outside by stitches that went partly through the outside, but never pricked entirely through or showed any indication of the seam. To do this required craftsmanship of a very high order. A pair of these fine boots cost $5.00 in the days when that was a large sum of money, and they were expected to last the careful owner for many years.

John Mulberry, before referred to, tells me that in his prime he could make out of the raw three coarse boots—a pair and a half a day; but it is well not to forget that old men remember the maximum rather than the average day's work.

The cobbler at his task sat on a very low bench hardly eighteen inches above the floor, and true to his trade, his seat was made of a sheet of leather stretched over a round hole. He sat hunched over his work, and some of the time with the half completed boot held against his chest. For this reason his calling was commonly alleged to be an unhealthy one, and I find one writer using this as an argument for the introduction of factory methods. As a matter of fact, I think most cobblers did grow stooped and round shouldered with the years. Still, I remember that some time ago there died at Lowville, N.Y., one David Cronk who had not only attained the great age of 105 years, but also had the honor of being the last surviving soldier of the war in 1812. He was given a public funeral, and his body lay in state in the City Hall of New York while 20,000 people filed past his casket. He was by occupation a cobbler, but it can hardly be alleged that as the result of his unhealthful calling he was cut off prematurely in the flower of his youth.

The Cobbler's Bench

The antique hounds have long been spoiling the rural neighborhoods of spinning-wheels and nearly all the other reminders of the handicraft age, but I have not yet heard that they have taken to collecting cobblers' work-benches. I judge that there still remain literally thousands of them stored away in sheds and attics about the state. This bench was four or five feet long. At one end was the round, sunken cobblers' seat, while the other end was divided up into a large number of different compartments to hold and keep separate the different sized pegs and nails. His tools were simple and inexpensive, but a full "kit" had a surprisingly large assortment. There were many sizes and shapes of awls, and there were three or four thin knives, one or two of them curved and all kept as keen as a good razor. For this purpose there was always at hand a fine whetstone and a strop. His wax-ends and his ball of shoemaker's wax were within easy reach. At his knee stood his "sewing-horse" whose jaws clamped and held his sewing while both hands were employed in punching the holes with his awl and drawing through the threads— a stitch entirely different from that employed by a woman sewing cloth.

In order that he might be prepared for all kinds of customers— for men with number 10 feet and for women, and for little boys and girls—he had a wooden rack where his lasts were arranged in orderly rows from which he picked out the one he judged would best fit his client. Up until say seventy-five years ago, he sewed with a hog's bristle waxed fast to his thread, but later he had steel needles with various curves adapted to his particular job. The "tap" (never "sole" in his speech) was fastened to the uppers with pegs. With his short, stout "pegging-awl" he punched a hole for each peg, set the peg in place and then drove it home with a single smart blow of his broad-faced hammer. If it was necessary to strike the same peg twice, by that token he was emphatically a botch and no cobbler. Against his wall hung tin or wooden patterns which he laid on the leather, and then with his keen knife cut out the peculiar shaped pieces, which when crimped and drawn down into position became the "upper" of a boot or shoe. In the corner were flung pieces of leather and rolled up hides which the farmers had brought him and which later would be shaped into footwear for their needs, while over all was the pleasant aroma of tanner's oil and hemlock bark.

The invention of a machine that could at one operation make its own pegs and drive them, and the introduction of a sewing machine which could stitch leather, marked the beginning of the end of the individual craftsman. Of them we must write in the phrase of long ago, "They rest from their labors, and their works do follow them."

The 'Mysterie' of a Cordwainer

A brief essay on the technical particulars of shoemaking in the 18th century.

By D.A. SAGUTO

March 1981

Mr. Saguto, a 'Cordwainer' since age 16, is the Historical Footwear Advisor for the First Virginia Regiment, Inc., a non-profit educational corporation. His collection includes not only the antique tools of his trade, but an extensive collection of technical works on this subject, dating from the 18th century to the present.

Preserving and fostering both interest and knowledge of the shoemaker's trade, Mr. Saguto has displayed his trade and tool collection in several special exhibits at the Smithsonian Museum of History & Technology, as well as before groups such as the Potomac Antique Tool & Industries Association and at many historical sites.

All the lawful secrets of the said Fellowship and all such things as at any time of assembly shall be lawfully in communication amongst the Fellowship at the Common Hall you shall keep secret and not disclose the same to any not of that Fellowship and especially to any such person whom the same in any wise doeth not concern or touch, or to any manner of person whatever.

So help you God.

[A portion of the "Freeman's Oath" of the London Cordwainers Guild, circa 16th century.]

How truly gratifying it is indeed to find that despite such ominous oaths as the one above and the subsequent passage of years, glimpses into the particulars of the Cordwainer's trade have managed to survive. How often have we sat, pondering the exact application of the "long-handled-whatzits" and "double-edged-gizmos" that pop up in almost every collection of tools? The men who fashioned them, and most who used them are no longer with us, and with their passing has gone a great wealth of information, as well as the epoch of the hand trades.

In this vein, it is my purpose here to animate the inanimate, and thus provide the vital link between the tools and the purposes to which the tradesman's hand applied them by briefly describing the processes employed by the shoemaker, or Cordwainer of 200 years ago.

In the following, I will outline the making of an ordinary man's welted shoe from start to finish. However, this outline is not intended as a detailed instruction, rather, a simple narrative, which along with the accompanying figures should give the reader a basic insight into the labors of a Cordwainer, not to mention the basic tools and their applications.

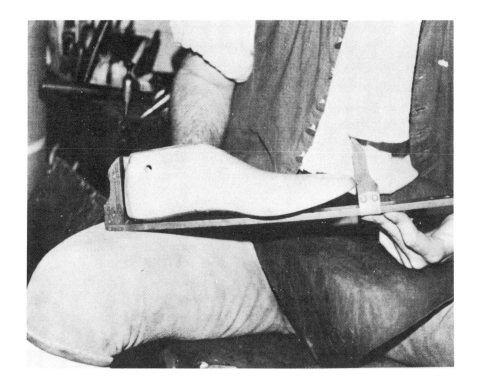

Fig. 1. The size-stick is shown here being used to measure a last to illustrate how it is operated when measuring a foot.

Measuring the Foot

Once the particulars of construction, style, etc., have been ascertained the feet must be measured. The length measurement is taken with the size-stick. This instrument is a sort of sliding wooden ruler, on which is marked out increments of measure, or sizes. [The oft touted one-third-inch per size, or "barleycorn" system of foot measure, was not commonly used until the last quarter of the 19th century. The size increments of original 17th and 18th century size-sticks vary from one-quarter-inch per size to three quarters-inch per size.] At one end of this device is mounted a fixed upright, which is placed behind the heel of the foot to be measured. Over the ruled portion of the stick, a mobile upright is affixed, which is slid along until it contacts the tip of the longest toe (see Fig.1).There were several varieties of these, some telescoping, some folding, and others one piece, as shown here. After the foot is removed the final position of the mobile upright indicates the size. A certain amount is added to this to allow for the toe shape of the shoe. [Shoes with pointed toes require an allowance, to prevent cramping the wearer's toes. Square, or rounded toe shoes require less of an allowance.]

The circumference measurements of the feet are taken at several points with an unmarked paper, or parchment tape, and small snips, or marks are made on this to indicate the various measurements.

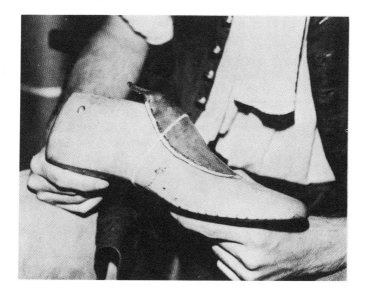

Fig. 2. The selected last is shown here after having been fitted-up with an instep-leather, which has been tied on with a thread ligature.

Fitting-Up the Last

Since the order of the day was the straight-lasted shoe, a suitable shoe-last must be either made, or selected from stock. Since this one last will have to serve either foot, it must be fitted-up to answer the larger of the two feet. The instep portion is left quite under in measure and a leather shim, called an instep-leather is selected to make up the difference. This piece is placed on the last and secured by a thread ligature, so that it

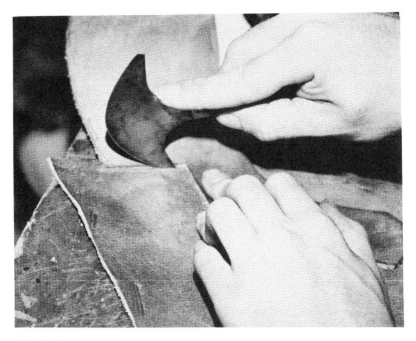

Fig. 3. *The angular cut that forms the side-seam is shown here being cut into the vamp with a round-knife, using the quarter for a guide.*

can be easily removed later on (See Fig.2). Since the shoe is built tightly around the last, the removal of the instep-leather creates the necessary looseness so one can pull the last out easily. If the last still does not quite answer the foot's measurement exactly, some wood can be rasped away, or the last built-up with bits of leather to suit.

Patterns and Clicking

The patterns used to click, or cut out the vamps and quarters, are only rough-cutting templates and have neither the side-seams, nor tongues cut to them. [The vamp is the portion of the uppers which covers the front of the foot; and the quarters, the two pieces covering the sides and heel of the foot.] These are cut directly into the pieces of leather later on by eye, to suit the last. The clicking is done with a round-knife, while the leather is lying on the cutting-board held upon the knees.

After the vamps and quarters have been roughly cut, the pieces are stretched forcefully between two pairs of lasting-pincers. This removes any excess stretch that could lead to the uppers becoming deformed, and pulling out of shape during the later operations. The stretched pieces are now cut to answer the patterns again. The position of the tongue and side-seam is now marked, and cut into one of the quarters. This quarter is now used as a guide for making the required cuts in the vamp (See Fig.3).

Closing

The completely clicked uppers are now ready to be sewn together, or closed. A thread is made-up from several strands of linen to the required thickness, then waxed, and a boar's bristle is spun onto either tip. This is the well known shoemaker's end, or waxed end used for sewing. The quarters are first to be closed up the back. For this, they are placed together edge to edge over the closing-block, and held in position upon the left thigh with the stirrup (See Fig.4). The sewing is begun at the top of the heel-seam, and is finished off with a special knot-stitch. For this a small round curved awl is used called a closing-awl. The vamps are now sewn to the quarters along the side seam, commencing at the extreme front of the quarters where they will overlap the vamp. This overlapping portion is closed with a stabbing-stitch made with a straight awl, passing completely through from one side of the leather to the other. For this sort of stitch the work is held in the clam (See Fig.5). Once past this overlap, the stabbing is discontinued, and one picks up round-closing to the end of the seam. This is the name of the same stitch that was used for the heel-seam. It enters the surface of one piece and curves down through its thickness, exiting right through the cut edge. From there it enters the cut edge of the other piece and curves up, exiting through the latter's surface. In this way all of the stitches are formed on the outside of the uppers, with none on the

 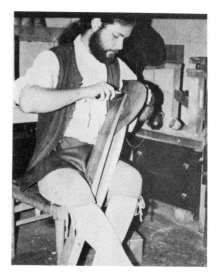

Fig.4. The workman is shown here round-closing a seam of the uppers. Note how the pieces are draped over the closing-block and held in place with the stirrup, which passes over the work and under the left foot.

Fig. 5. The over-lap of the quarter and vamp is being stabbed in this shot. For this sort of stitch the work is held in the clam, which in turn is tightened between the workman's legs, leaving his hands free to stitch.

inside to chafe the foot. This seam forms a double row of stitches, one to either side of the two edges, butting the pieces together edge to edge.

In addition to these seams, various stiffening pieces called linings are sewn to the inside of the uppers to help them keep their shape in wear. These are sewn in with a whip-stitch that passes only partially into the surface of the leather. An awl and thread is the preferred method, but for this operation a steel needle was occasionally used. [Garsault, Diderot and the English works of the 19th century, all mention a steel needle, very much like the modern "glover's needle." It is described as being exclusively used for whipstitching and very fine closing. It has a diamond-shaped sharp point, unlike the blunt harnessmaker's needles, used by many today in place of bristles.]

Making

Under this heading, formerly, was included the remainder of the manufacture, i.e., lasting, sewing, stitching, heel-making, etc. Now that the uppers have been closed and

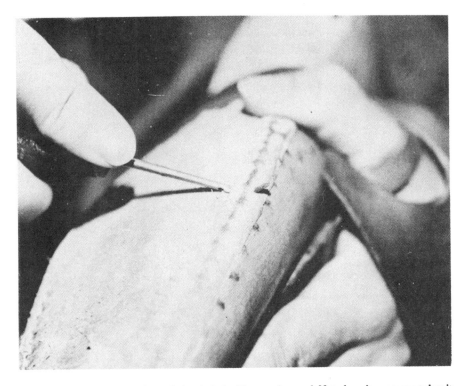

Fig.6. Here the insole is shown being holed with a sewing-awl. Note how its passage takes it only partially through the thickness of the leather as it pierces from the channel outward.

lined, one prepares the first, or insole. It is upon this sole that the foot will rest, and to it are sewn the welt and rand.

The lower edges of the uppers are likewise sewn in with this seam all around, being sandwiched between the welt and insole. The insole is cut and moistened, and then stretched and tacked to the bottom of the last at several points. This operation, naturally enough, is called blocking. Once in place, the insole is rounded, or trimmed close to meet the edge of the last. The cut edge of the insole is now shaped into a sloping bevel called the feather, which runs around the entire edge of the insole. Within the border created by the feather, a shallow channel, or V-shaped groove is cut all around to recess the sewing-stitches. After the channel has been cut, the next procedure is the holing of the insole. In this operation the holes are made that are to receive the sewing-stitches. An awl, similar to the curved closing-awl, though larger, is used, which is called a sewing-awl. With this awl the holes are pre-punched all around the insole, running from the bottom of the channel and curving outwards, exiting through the middle of the feather (See Fig.6). In this fashion the seam only nips partially through the thickness of the leather. When the insole has been completely prepared, the uppers are now lasted, or mounted over the last with the lasting-pincers.

With the uppers in position over the last, they are pulled tight "to the wood" inch by inch with the pincers, and held secure at each pull with a lasting tack, driven in with the hammer portion of the same tool (See Fig.7). In this fashion one works one's way around the entire last, pulling obliquely, or straight, depending on the particular area being lasted. After the uppers have been lasted and made smooth, one takes note of the position of the side-seams, and then makes a mark across the face of the insole at a

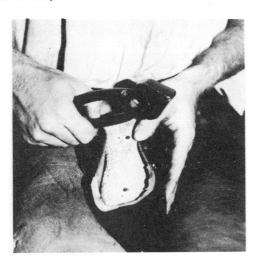

Fig.7. The lasting-pincer is shown here being used to drive in a lasting-tack at the spot where the uppers have just been strained with the same tool.

certain distance in front of them. This mark will be used as a guide for the front edge, or breast, of the heel. At this point in the making, the rand and welt are cut and made ready to be sewed. The rand is made first, and is cut from uppers leather in the form of a strip, approximately 1¼ inches wide, and long enough to run around the seat, from the heel-mark on one side, to the heel-mark on the other. [The term seat is used to designate the portion of either sole that will lie beneath the heel of the shoe]. The welt is now cut from stouter hide, and is roughly one inch wide and long enough to run around forepart from heel-mark to heel-mark. In other words, the rand and welt will butt together, end to end, right at the heel mark on both sides forming a lip to which the sole will be stitched. One now prepares a sewing-thread, which is done in the same manner as the closing thread, though made much heavier, and usually of hemp, to the tips of which are attached boar's bristles. The lasted shoe is now placed upon the left thigh and held down firmly with the stirrup.

The rand is sewn in first, commencing at one end, and continuing around the curve of the seat, removing the lasting-tacks as one sews. When making the sewing-seam, or in-seam as it is sometimes called, one runs the sewing-awl through the hole made during the holing and pierces on through the uppers, exiting through the rand or welt, as the case may be. The stitches of the sewing-seam thus formed lie in the channel in the insole on one side, and on the under face of the rand or welt on the other. When the rand has been sewed, the shoe is turned around and the welt is slipped into position underneath the stirrup to hold it in place. The sewing is continued in the same manner removing the lasting-tacks one by one; the seam ending at the first stitch made in the rand (See Fig.8). In this way a continuous seam is made, which will serve as a foundation holding the shoe together, by uniting the uppers, welt, and rand to the insole. Since this seam is made with only one thread, it usually runs around two fathoms long from bristle to bristle.

With the sewing finished, the next step is to trim the excess leather down to the seam, and turn the rand over, and brace it across. [In bracing, the two loose folded-under edges of the rand are drawn in and secured by lacing them back and forth across the insole with a thread, much in the same way one would lace up a boot.] The waist, or narrow portion between the seat and forepart, is now fitted with a shank-piece, which will stiffen the shoe through this area, keeping it from sagging down in front of the heel, as well as preventing the sole from bending anywhere other than across the wider portion of the forepart. The shank piece is usually fashioned from leather, and fits within the sewing-seam, where it is pasted and pegged into place.

The remainder of the forepart is now made level between the ridges of the sewing-seam, by filling it with either leather shavings mixed with paste, or a piece of thick felt cut to shape.

The second, or outer-sole, is now ready to be stitched to the welt. This sole, having been cut out earlier, was moistened and hammered upon a smooth stone to compact it

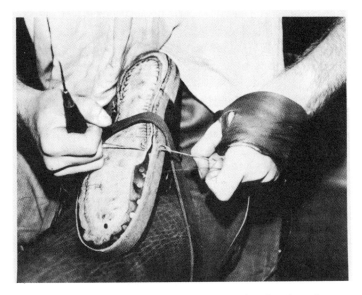

Fig.8. *Here the workman is in the process of sewing the welt*
Lasting-tacks still remain around the toe, which has yet to be sewed.
Note how the workman uses the knob-end of the awl-haft in one hand and
the manicle, or hand-leather, on the other, to protect his hands
from being cut by the thread as he pulls his stitches tight.

and make it harder wearing. [For this operation the French "Cordonnier" used a block of oak, into the sides of which concave ovals had been hollowed. By hammering the sole over the hollow of appropriate size with the end of the hammer handle, the sole was not only compacted, but at the same time given a cupped shape, allowing it to be placed on the shoe without further shaping.]

To begin this operation, the under side of the insole, rand and welt are smeared with starch paste, and likewise the under side of the sole. The sole is now positioned and pressed fast to the bottom and secured temporarily with a few tacks. The welt is now beaten down flat to meet the outsole, and the outer folded edge of the rand is likewise hammered into shape. All around the face of the sole a channel is marked off, at certain distances in from the edge, continuing around just below the welt, and ending where it does at the heel-mark. This channel is created by making an angular slice into the face of the sole with a knife, which is then pried open with the channel-opener. A stitching-thread is now made, a bit lighter than the sewing-thread and a stitching-awl is selected. The stitching-awl, also known as the French Awl, is curved like the rest, but its point is square and shaped like a vertical chisel. This awl makes a narrow slit, rather than a round hole, which allows a greater number of stitches to the inch to be made. For the stitching, the shoe is strapped to the left thigh again with the stirrup. The stitching-awl is placed first on the welt, and then run through it into the sole, piercing on through the bottom of the channel, where the stitches will be made (See Fig.9). In this way the sole is stitched to the welt from end to end, and when completed, the flap created by the channel-cut is smoothed back down in place over the stitches to hide them from both wear and view. The edge of the sole is now pared quite close around the welt, very near to the stitches, with a knife. At this point in the production, the heel is made. Heels of both leather and wood are used, and each one is made in an entirely different fashion. The wooden heel always has a leather heel-cover, sewn in like a rand, though wide enough to cover the sides of the wooden heel-block. The heel-cover, rather than being folded under onto the insole and braced, is folded down over the sides of the heel. In this case though, a leather heel is to be made.

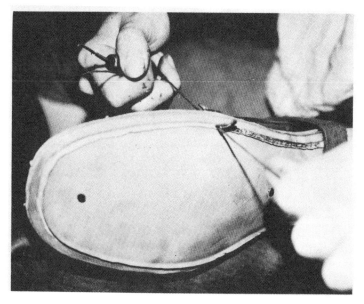

Fig.9. The manner in which the sole is stitched to the welt is shown here. Note how the stitching-stitches form at the bottom of the channel, the protective-lip of which is barely visible at the lower right of the figure.

A split-lift is now cut; this is a strip of stout hide that is bent into a horse-shoe shape and fitted around the seat. The inner edges of the curve so created are skived thin. The split-lift is wedge-shaped in cross-section and is put in place around the seat to make it level and flat, since the sole is slightly convex in this area. The split-lift is pasted onto the sole and secured by a few wooden pegs. A very thick thread is now made up, and with the heel-awl the split-lift and sole are stitched to the rand. The heel-awl is a very large version of the sewing-awl, but less curved. It is run down from the crevice between the quarters and the folded edge of the rand, piercing quite through the lot and exiting through the face of the split-lift. The stitches so formed, sink down into the crevice when they are pulled tight, where they are hidden. This sort of rand treatment is called a blind-rand. [Stitching the seat via the blind-rand is only one of several ways shoe heels were built. The seat stitch usually maintains the same path, but can enter the folded edge of the rand instead, as well as being extended to stitch the entire multi-layered heel to the shoe, rather than simply pegging it, as described here.] Once the seat is stitched, the other heel layers, or lifts are pasted and pegged one upon the other on the firm foundation of the split-lift and stitched seat (See Fig.10).

Fig.10. The workman is shown here pegging on the top, or tap-lift of the heel. Note how the stirrup is used to hold the shoe across both thighs, by passing it beneath a stick held under the feet.

With the heel built, all that remains is to remove the rest of the tacks and trim the edges of the sole and heel, after which they are burnished, along with the under-side of the outsole. This is done with various wooden and bone sticks. The edges, as well as the seams of the uppers, are now coloured with black ink and finished off with a gum solution. The last is now ready to be removed, so the thread ligature that holds the instep-leather is now cut and pulled out. The instep-leather can now be drawn out with the lasting pincers. Once removed, the last is slipped or withdrawn, by hooking the last-hook through the hole provided for this purpose at the rear of the last. The insides of the shoe are checked over and any peg points are bruised down with the hammer and cut out with a peg-knife. All that remains to be done, is to trim the width of the latchets or buckle straps, and fit the required buckle, and the shoe is fit to be put on.* (See Fig.11).

*The reader will notice that in my discourse I have only been referring to one shoe, rather than the pair. During the hey-day of the straight lasted shoe this is how they were made, that is to say, one at a time. Since only one last was used for both shoes, the operations that required the shoe to be mounted over the last had to be completed for one shoe so that the last could be withdrawn and used for the second shoe. Usually the uppers are clicked and closed together, but when the making takes place they part company. When the first shoe is finished, the insole is blocked onto the last and all of the subsequent operations repeated. The trimming and finishing of the soles and heels usually does not take place until the second shoe comes off of the last, so they can be mated perfectly, but has been included in the description purely for the sake of completeness.

Fig.11. The finished product — an ordinary man's welted shoe of a style typical of the 1770's. All that remains is the fitting of a shoe-buckle.

BIBLIOGRAPHY

James D. Devlin, *Guide to the Trade. The Shoemaker, Part I,* Lon. 1839.

Denis Diderot, *L'Encyclopedie,* Vol. III, Chapter: "Cordonnier et Cordonnier-Bottier", Paris, 1763

Francois Alexandre de Garsault, *L'Art du Cordonnier,* Paris, 1767. (Descriptions des Arts et Metiers).

C.H.W. Mander, *A Descriptive & Historical Account of the Guild of Cordwainers of the City of London,* Lon. 1931.

Martin of Taunton, *The Art of Cordwaining,* 1745, annotated by John Thornton in the *British Boot & Shoe Institution Journal,* Vol. 15 No. 11, 1968.

John O'Sullivan, *The Art & Mystery of the Gentle Craft,* Lon. 1830.

John F. Reese, *The Art & Mystery of a Cordwainer,* Lon. 1813.

John H. Thornton, "The English Shoe Size Scale", *B.B.S.I. Journal,* Vol. 4. 10. 1951.

ACKNOWLEDGEMENTS

The writer wishes to express his thanks to:

Mr. Ernest Coleman for the photographic work for this article.

Miss June Swann, Keeper of the Shoe Collection, Northampton Central Museum, Northampton, England, for photocopies of several of the above listed works.

Mr. John H. Thornton for the copy of the *B.B.S.I. Journal* containing his article: "The Art of Cordwaining."

A Letter to the Editor

September 1984

I thoroughly enjoyed D.A. Saguto's beautifully written article *The 'Mysterie' of a Cordwainer*. In the article he mentioned that in an early day the same last was used for both shoes.

When I was a very young man in the U.S. Cavalry, they decided, as an economy measure, to use up old stores and we were issued boots, dating back before the turn of the century. They were black with a fairly square toe and were made to fit either the left or right foot. The instructions with the boot told us to stand in pails of water or in a stream until the boots were thoroughly wet. Then we were to walk in them and they would take the shape of the foot.

The leather of the boots was fairly heavy, heavier than present day riding boots, and the sole was quite thick. The heel was broad and slightly higher than ordinary shoes.

I wore these boots for about two years and enjoyed them for they were comfortable and excellent riding boots.

The regular issue foot gear for the Cavalry by that time had become a hobnailed shoe worn with a leather faced legging on the side toward the horse. This was to prevent your clothing from wearing out.

No one will believe me about these boots being made for one foot. I am glad to see that the Master Cordwainer brings the subject to light.

Incidentally, the U.S. Army had a wonderful surgeon by the name of Dr. Munson. I believe he was a major. He had a tremendous collection of feet in plaster at Walter Reed Army Hospital in Washington, D.C., but, unfortunately, these were all thrown out at the beginning of World War II because of the need for space.

Munson designed a shoe which we called "the Munson last" that the Army still uses. I contend that the Army's present day black shoe, the dress shoe, is the most comfortable shoe ever made.

The Clogmaker

By DAVID G. PERCH

December 1975

One afternoon late last spring, I had a chance to experience a once-in-a-lifetime sight. For two hours I watched a Dutchman go through the steps of making wooden shoes by the same method and using the same tools as those illustrated in Diderot's *Encyclopedia* of 1763. On arriving home I set about putting down on paper the steps

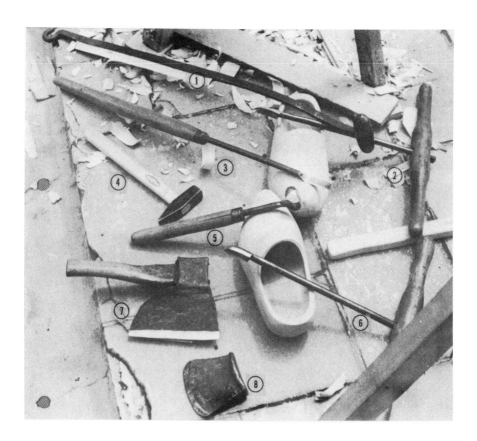

Figure 1. (1) bench knife; (2) first bore; (3) long crooked knife; (4) hammer; (5) crooked or bottom knife; (6) second bore; (7) small broad axe; (8) wedge.

this craftsman went through. The following day, armed with a camera, plenty of film, a notebook and pen, I returned to the place where he would be demonstrating his craft. There I set about taking photographs of all the various steps he went through from the start to the finished product — a pair of clogs. I watched him work for the better part of three hours and had the opportunity to talk with him. The name of the sabot maker is Wim Holmer. He is Holland's best known clogmaker, and comes from a long line of clogmakers. He travels around the world demonstrating his craft and was brought to Ottawa as part of the annual tulip festival activities. He was good enough to answer my questions and read over my notes to make sure I had the steps in their proper sequence and had the correct explanation of techniques employed and the names of the various tools used.

Figure 2

The wood used in the manufacture of wooden shoes is either poplar or willow. Logs of poplar, which Wim was using, are stripped of their bark and cut into proper lengths, depending on the size of shoe to be made. They are worked green; no drying time is allowed whatsoever. Nine tools and two horses are the only equipment necessary for the job at hand (Figure 1).

Once the logs are cut to proper length, they are split lengthwise in either four or six pieces, depending on the diameter of the log. The wood is split using a heavy four inch steel wedge and a hammer. These rough split pieces are trued up using a small broad axe, similar to the axe used by coopers, with a ten-inch-long offset handle. (The poll of the axe is also offset.) I was told that this allows the craftsman to see exactly where he is chopping. This axe is also used to rough-shape the block. It cuts away the excess wood from the toe portion and rounds off by the heel, the piece that will have the opening where the foot will enter, and the heel notch (Fiqure 2). (A pair of clogs are usually worked together.) The blocks are then given a more decisive shaping using the shaving knife (or bench knife as we know it). (Figure 3).The bench knife is mounted

Figure 3

Figure 4

on the piece of burl having three legs. The side facing the workman is straight-sided and has the knife mounted so it will cut about three inches in from the straight side and parallel to it. Once the shaping has taken place, the two shoes are mounted in a horse at an an angle of 150° from perpendicular and wedged tight, with the toes pointing downward (Figure 4). Using the spoon auger with the pointed end, or first bore as it is called, three holes are made in the block. The first hole is bored to a depth of approximately one inch straight down, as is the second. The third hole (Figure 5) starts in number two hole but in the direction of the toe of the shoe, using the round-nosed and wider spoon auger (Figure 6). A channel is opened up between holes one and two. This channel is gradually widened and worked in towards the third hole, widening it out (Figure 7). The long crooked knife (Figure 8) is used to open out the third hole even

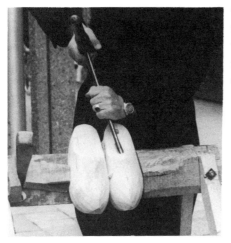

Figure 5

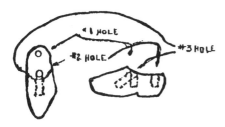

Figure 7

wider and to remove the excess wood for the instep and the fore part of the sole. The tool is worked three ways: first by holding it high on the handle and rotating it in a conical motion; second by grasping it with the right hand just below the handle and the left hand just above the cutting portion of the blade and moving it from side to side. A third cut is effected by reaching into the opening towards the toe and pulling the tool towards the heel.

The crooked or bottom knife (Figure 9), resembling a question mark, is used to smooth off the bottom. The second bore finishes the inside opening at the heel, bringing it to proper shape and width. Again, the bottom is touched up with the bottom knife and the inside for the forward half of the foot with the long crooked knife. Chips are removed periodically with whatever tool is being used at the time. The finishing touches are given the shoe with a small sharp knife. This is used to remove the sharp corner around the inside opening of the shoe (Figure 10). The ruler is used only to determine the inside length of each shoe; otherwise all measurements are done with the experienced eye of the craftsman. Shoes are made in half-centimeter increments (23, 23.5, 24, 24.5, etc.).

Tools were sharpened with powdered emery and a long tapered wooden stick. The only time a sharpening stone was ever used was when the tool was knocked over by a careless spectator. All of Wim's tools were old except for the hammer and wedge, and they were definitely hand-forged by some blacksmith of a bygone day. He told me that replacements are next

Figure 8

to impossible to obtain, so he, therefore, takes extra special care of all his tools. The two pegs on the left hand side of the horse are used to hold the tool while it is being sharpened. The tool is placed between these pegs and held tight by pressure on the handle from the thigh. Before the days of machine-made clogs, a skilled workman could turn out between eight and twelve pairs a day, depending on the sizes made. The average time taken to complete a pair of clogs is about an hour and a half.

It is thought that the wooden shoe or clog originated in the south of France, but it has been worn by all kinds of Dutchmen since the middle of the fifteenth century. Farmers, construction workers, butchers, dike workers, brewery workers, masons and even suburban weekend gardeners buy five million pairs a year. The clog was carved by hand until after the First World War, when machinery took over the clogmaking industry. Now Holland has close to 400 fully mechanized clogmaking establishments that can turn out a pair every few minutes. Thanks to men like Wim Holmer with their primitive tools and the knowledge handed down in each generation from father to son, their craft is not forgotten. It is still with us, even in an age so eager for mass production of quantity, not quality.

Figure 9

Figure 10

James Wilson
A Pennsylvania Chairmaker

By DONALD L. TUTTLE and JOYCE BAUDUIN
Illustrations by Linda Scharf-Jones
December 1979

The Wilson Chair Shop

Chairmaking in small shops in eastern Pennsylvania reached its peak shortly after the middle of the nineteenth century and before the close of the century had given away completely to the competition of mass production and large scale operations. In western Pennsylvania, however, the fruits of volume production and the latest styles took longer to disseminate, and small shops like the one James Wilson operated at Taylorstown, Buffalo Township, Washington County, Pennsylvania, for nearly six decades, continued to thrive, serving the needs of the local community.

The chair of choice manufactured in James Wilson's shop appears to have been the plank-bottom, "common" arrow back chair, about three feet high, four arrows in the back, rounds or stretchers on each side and at the back, below the seat, and with a shaped stretcher in front. Wilson's process of finishing chairs seems to conform to a general pattern, excepting some variations he found helpful or time saving. Usually the raw wood was primed with venetian red mixed with oil. Over this was applied a finish coat of various colors, or grained to simulate more expensive rosewood or mahogany. Decorative designs were usually applied by means of paper stencils and free-hand striping on the seat and posts. Stencils of Wilson's which have survived were cut from account book paper. Because of the relatively fragile nature of the paper, many of his designs are prick-punched to facilitate copying and making new stencils. Extant examples of Wilson's chairs and his patterns preserved in his shop show he also made Boston rockers, high chairs, child's rockers, settees and various other plank-bottom chairs with lyre and fiddle-shaped splats. It appears Wilson never made rush-seat chairs more often found in New England and the Western Reserve of Ohio. One author contends plank-bottom chairs were fairly limited to makers who had learned chairmaking in Philadelphia and there became skilled in making this type of chair. Family records show Wilson was trained in Philadelphia; this may account for his penchant for the plank-bottom style.

James Wilson was born on May 9, 1826, in Buffalo Township, Washington County, Pennsylvania, of second-generation Irish parents. In 1846, he began an apprenticeship in Philadelphia, studying chairmaking and ornamental work for a number of years. A page from Wilson's account book for this period now in the Meadowcroft Village collections records the following notation for his ornamenting chairs:

August 3rd 1849
Mr. Thomas B. Boyson
To ornamenting Chairs

2 French	$ 2.62½
2 Settees	.75
6 Doz. Common	3.75
1½" Scroll Top	2.25
	$ 9.37½

January 31st 1850

6½ Doz. Common Chairs	$ 4.03½
5½ Scroll Tops	8.75
One Rocking Chair	.25
	13.03
	8.87½

October 4th/49

5 Dozen Common Chairs	$ 3.12½
6 Arm Chairs	1.50
6 Scroll Back	.75
18 French	2.25
2 Common Rocking Chairs	.25
2 Settees	.75
2 Polka Chairs	.25
3 Settees	.62½
	9.50
	9.37½
	18.87½
	10.00
	$ 8.87½

Oct. 4th Received of
    Thomas B. Boyson 10.00 Dollars

It is unfortunate that the account book does not offer us more on his period as a young apprentice in Philadelphia — it was heavily cut later to provide paper for stencils and little else remains but some pages of copybook exercises in ornamental penmanship. A number of rococo ornament designs with a strong Philadelphia Chippendale influence among his papers hints that he trained with a master furniture maker. And the fact that Wilson's tools include several firmer and mortise chisels stamped "James Cam," a Sheffield toolmaker reported as out of business in 1833 (while Wilson was still in his childhood in western Pennsylvania), suggests that he may have acquired them from relatives years later when he came to Philadelphia.

By 1852, Wilson had returned to Buffalo Township, married a Rebecca Wilson of Bridgeport, Ohio in 1855, and a year later purchased a 100' x 124' lot in Taylorstown already containing a single story brick house and a weatherboarded log house one and a half stories high. The log "house" became James' chair shop which stood on Lot #40, Main Street, until it was moved to Meadowcroft Village and restored in 1969.

It is probable that the small log building had been built originally as a shop and never intended to be a residence. Constructed in the 1830's it contains no fireplace and probably was always heated with a stove. Chairs were constructed on the first floor which contained a large foot treadle lathe and two long work benches. A ladder was used on the end of one bench to reach the loft where chairs were painted and decorated. The ceiling of the loft is covered with stenciled patterns, sign letters and freehand designs which James Wilson, and later his son DeForrest, who worked with him on house and sign painting, tried out before finishing a piece of work. A small storage shed attached to the rear of the shop provided a place to work out wood in the rough.

The fancier styles of Philadelphia chairmakers seem to have influenced Wilson very little. His way of life probably left him with virtually no leisure in which to study. Perhaps by choice, but probably from the necessity to augment the family income, he worked at other trades during the slack seasons: housepainting, signpainting, graining of woodwork, decoration of the interiors of local churches, and occasionally coffin-making occupied his time, in addition to his chairmaking business.

And there were other factors that had a bearing on his production. One was the prospective local market, including his own and other families in Taylorstown (many plank-bottom chairs and settees still in the possession of older Taylorstown families can be traced by patterns and styles to Wilson's handiwork) and also a larger market in Washington and Claysville on the National Road, within accessible hauling distance by horse and wagon. Maple or hickory could be cut from nearby woodlots; poplar trees for splats and planks purchased from a sawmill. Powdered paint pigments such as venetian red, yellow ochre and lamp black, were probably purchased in Claysville; Wilson's letters and receipts indicate he purchased his gilding powder and graining and signmaking supplies from several supply houses in Chicago and New York. Several copies of a monthly trade magazine, *The Chicago Paint and Varnish Journal* for the

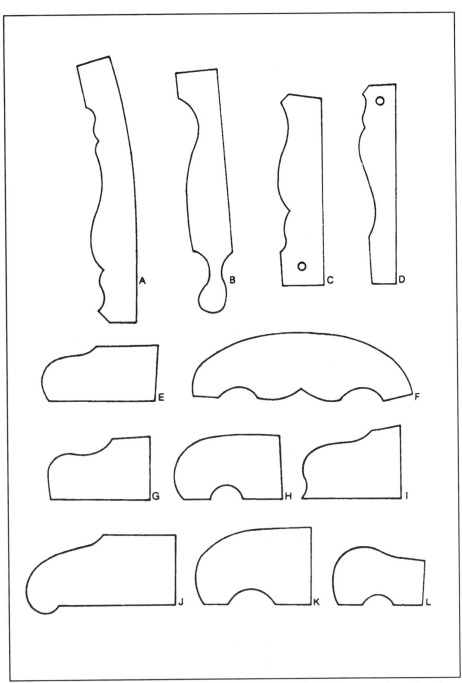

Fig.1. James Wilson Chair Patterns, made of pasteboard. A-D, rocker arm and outside splat patterns. E-L, chair and settee back crest patterns. All approx. ½ actual size.

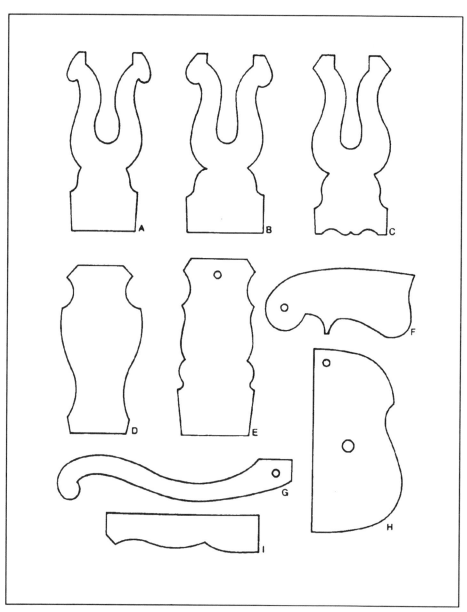

Fig. 2. *James Wilson Chair Patterns, made of pasteboard or wood. A-C, lyre-shape splat patterns. D-E, urn-shape splat patterns. F, rolled curve template, 11½". G, rocker arm pattern, 19½". H, plank-bottom seat pattern. I, outside splat pattern. All approx. ⅛ actual size.*

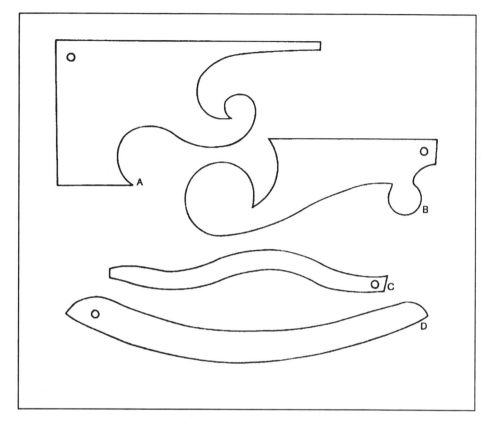

Fig. 3. James Wilson Chair Patterns, made of wood. A-B, compound and rolled curve templates. C, chair arm pattern, 21½". D, rocker pattern, 28¼". Approx. ⅐ actual size.

years 1888 to 1891, preserved with Wilson's papers are full of articles on the "Harmony of Color," "The Painters Library," "How to Grain Burl Walnut," "Mixing Striping Colors," and discourses on the advantages of joining a painters' union or the merits of fireproof whitewash and mechanical graining machines. It appears Wilson succumbed in the case of the latter; his papers include an instruction pamphlet for the use of the "Adams' Air Cylinder Graining Machine," a wondrous patented pneumatic device for graining woodwork that cost in 1868 — net cash — $100.

Although the contents of the chair shop were sold out before the shop was moved, Meadowcroft Village was fortunate in being able to buy back a majority of Wilson's tools in 1977. These tools, amounting to over 100 items, include five bow saws, a felloe saw, chisels and gouges, a pair of calipers, gimlets and awls, lathe chucks and other lathe parts, a mortising brace jig, draw knives, gilding equipment and a number of tool

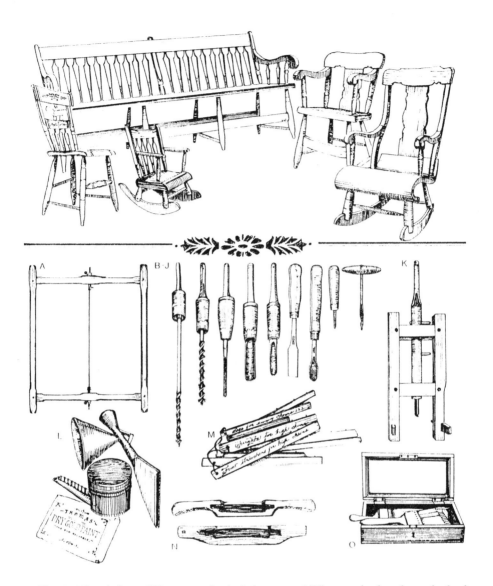

Fig. 4. (Above) James Wilson arrowback chair, settee, child's arrowback rocker, splat-back chair, Boston rocker. Meadowcroft Village collection.

(Below) Selections from the tools of James Wilson, Taylorstown, Pa., chairmaker. A, felloe or chairmaker's saw. B-J, (left to right) spiral bits, spoon bits, firmer chisel marked "James Cam," lathe gouge marked "James Cam Sheffield," brad awl, gimlet. K, mortising brace jig with stopped spoon bit. L, gilder's funnel, shellac can, and powdered gilt packet. M, pattern sticks. N, spokeshaves. O, graining kit.

handles and chair patterns. The complete collection, along with several examples of chairs made by Wilson, is on display in the restored shop.

James Wilson's tools and equipment are much like those of any rural carpenter, with a few exceptions. The use of the axe, bowsaw, the treadle lathe, draw knife and plane is obvious. Included is a set of chisels and gouges used in making grooves for slats, as mentioned earlier, some of them bearing the toolmaker's mark of James Cam, Sheffield, England, 1787-c.1833. For drilling holes in the legs of stretchers, Wilson used a brace and a bit or a handmade mortising brace jig fitted with scalloped-edge pewter ferrules. A set of "pattern" sticks was also used. These are sticks of hardwood, squared and marked for the type of chair being made. The patterns have pins inserted, protruding so that when the proper pattern was placed against a new stretcher or upright and tapped, the points left marks which served to guide Wilson where to drill holes.

Unusual among the saws in the collection is a felloe saw or a chairmaker's saw. This type of saw is a lighter and shorter form of the frame or veneer saws usually found among cabinetmaking tools. Mercer illustrates this saw and states it was used in England by chairmakers for cutting straight or slightly curved pieces.

James Wilson, chairmaker, like everyone else, worked at his craft, accumulated wealth or a parcel of debts, raised a family, was subject to taxation and when the time came, died. When his will was probated, he left an estate consisting of one house and lot in Taylorstown valued at $800 and personal property valued at $3000. To his son, DeForrest, he bequeathed "my Watch and all the tools in the shop."

BIBLIOGRAPHY

Beers, J.H. & Co., *Commemorative Biographical Record of Washington County*, Pennsylvania; 1893.

Kebabian, John S., ed., *Joseph Smith Explanation or Key to the Various Manufactories of Sheffield, with Engravings of each Article*, EAIA, 1975.

MacDonald, Wm.H., "Central New Jersey Chairmaking or the Nineteenth Century," *The Chronicle*, Vol. III, Nos. 6 and 7, January-April, 1946.

Mercer, Henry C., *Ancient Carpenters' Tools*, Bucks County Historical Society, Doylestown, PA, 1960.

Washington County Court House Deed Book 3R, p.127.

Washington County Court House Will Book, No. 34, pp.567-568.

Wilson, James, MS Collection, Meadowcroft Village, Avella, Pennsylvania.

Meadowcroft Village was developed by Albert and Delvin Miller as a memorial to their ancestors, who settled there in 1795, and to perpetuate their rural heritage and that of the Scotch Irish in Pennsylvania. The Village is a collection of old buildings, restored and filled with the appropriate furniture, antiques or tools. It is three miles west of Avella in Washington County, Pennsylvania.

Burning Brands or Marking Irons

By JOHN R. GRABB

September 1978

At the time of writing John R. Grabb was a retired letter carrier, engaged in cabinet making, and an archivist at the Ross County Historical Society, Chillicothe, Ohio, where he did historical research.

Through past articles in *The Chronicle* we have noted the importance of identifying marks or stamps on tools, devices, etc.[1] Such articles have dealt mainly with makers' marks on the fore end of planes and the touch marks of metal objects forged by blacksmiths.

Silversmiths, potters, gunsmiths, and other artisans have marked their finished pieces. Myriads of identifying marks have featured combinations of initials, names, geometric designs, or figures of men, animals and plants.

Serious collectors are challenged as they seek to date and identify their newly acquired pieces by marks uncovered in cleaning them.

This essay will mostly concern the marking of wooden objects by the use of the burning brand—sometimes called "marking" or "branding" iron. The word "brand" is from Old English, akin to Old High German *brant*, and means a mark of a simple, easily recognized pattern made by burning with a hot iron to attest manufacture or quality or to designate ownership.[2]

The word "brand" also became known as a grade of goods or the products of a certain maker. In past articles we have noted that wood planes marked "Scioto Works" were the second grade or brand of goods made by Ohio Tool Company.

Remember the thrill we received as youngsters at Christmas when we received a "brand new" sled, doll, wagon or pair of shoes? In this instance, the indication was that the gift was fresh from the storekeeper or manufacturer and had never been used.

The use of brands or branding goes back to antiquity. We notice it first in the Old Testament[3] when we read of the "Brand of Cain." "The Lord set a mark upon Cain, that whosoever found him should not kill him." This was the badge of the manslayer.

The Chinese penal code is based upon enactments for which a remote antiquity is claimed, and the earliest system of punishment is ascribed to the "Emperor" Shun (2255 B.C.), who is said to have established the "Five Punishments" that were in vogue to the end of the Chow dynasty (255 B.C.), viz.,(1) branding on the forehead.[4]

Branding was one of many varieties of corporal punishment. It was used not only to chastise, but to identify and stigmatize criminals. The letter used generally identified the crime committed. The practice was common in medieval and modern Europe. It

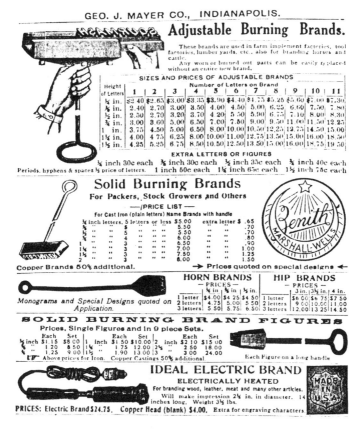

Figure 1. Geo.J.Mayer Co., Indianapolis, ca.1924.
Manufacturer of marking devices of all kinds.

persisted in England until abolished in 1829. In colonial America, petty criminals and runaway slaves were frequently branded, but the practice was discontinued before the American Revolution.[5]

While this authority says "the practice was discontinued," there is record of an instance here in Chillicothe, Ohio, in the year 1804, in which a criminal was branded. A man was tried for murder, but the jury found him guilty of manslaughter only. He was sentenced "to be burned in the palm of the left hand, with a hot iron, so that the letters M S shall be plainly marked thereon, and pay the costs of prosecution." This man's name, of all things, was John Brandy.[6] At that date Chillicothe was a frontier town and the capital of the new state of Ohio.

It appears that the use of hot branding irons directly on the hides of live cattle came into general use with the beginning of the great western cattle drives, where stock of several owners was commingled and driven to railheads for shipment east.

Because of the many tales written about the cowboy and the spring round-up of stock to be branded, when we see the term "branding iron" we generally associate it with the hot searing of live animals with the iron to mark ownership. In the eighteenth and early nineteenth centuries, such brands were little used on the hides because the hides were too valuable to be defaced in such manner. Burning brands were used, though, on horns and hoofs.[7]

Generally the smaller stock, such as hogs and sheep, were identified with ear cuts, described as a crop, a slit, an underbit, a hole, a swallow fork, etc.[8]

From laws long established in England and followed in America, we find that owners of brands were required to register them with county or township officials, and we note this:

All those who wish their Brands or marks recorded will call on
William Niblick, at Mr. John M'Dougal's Store.
Chillicothe, May 19, 1802.[9]

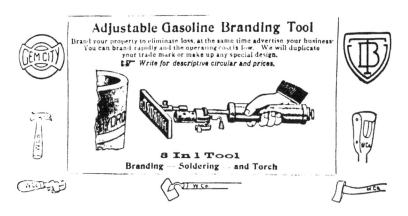

Figure 2. Geo. J. Mayer Co., Indianapolis, ca. 1924. Remember the brands on telegraph and electric poles? Naugle Pentrex was the brand on many such poles. The gasoline branding tool was probably used to apply such brands.

County officials themselves sometimes required a brand to identify public property. In 1802, it was "Ordered that the inspector for Ross County provide at the expense of the County a branding iron with Ross County on it, at full length & the letters M.W.T. underneath."[10]

Shortly after the settlement at Chillicothe began in 1796, John McDougal opened the first store. He advertised in a Wheeling, Virginia (now West Virginia), paper that he was "prepared to furnish the very best whiskey, and other things required. . ." Monongahela whiskey from around the Pittsburgh area was much prized by the early settlers, who imbibed freely of it. You can be sure that the distiller, anxious to advertise his brand, prominently branded his whiskey kegs.

John McDougal was also a heavy buyer of pork for which he executed "Pork Notes." These were agreements by which hog raisers promised to deliver so many pounds of fat hogs between the 10th of December and the 15th of January. McDougal would give "Cash and Merchandise in advance for good Pork."[11] He also advertised that "A contract will be entered into the building of Orleans Boats."[12]

Orleans Boats were flatboats, as much as 90 feet or longer, which were built of white oak in the winter season and loaded with the rich produce of the Scioto Valley—mainly pork, beef, salt, corn, flour, whiskey, skins, pelts, and ginseng. Upon the arrival of the spring freshets the loaded boats floated off on the long voyage down the rivers to New Orleans. Such ventures were fraught with danger on the flood-swollen rivers and many wound up wrecked. The remains of one such wreck found on a gravel bar in the Scioto River 52 years afterwards contained a couple of barrels of mess pork that was distinctly branded McCoy & James.[13]

Figure 3. F. HANGS—12TH DIST. O. Frank Hangs was a Chillicothe cigar maker in the 1870's. He used this brand for his boxes of 100 cigars. The lettering is 3¼" across and is cast iron. It was applied on the bottom of the boxes. Handle is 18" long.
The S E E D S brand is bronze. Letters are 2 ½" wide. Whether it was used to mark boxes of seeds or whether it was the name of a man is unknown. Author's collection.

In the early 1800s, Chillicothe was an important frontier town where artisans made all sorts of wooden articles besides pork barrels and lard kegs. Craftsmen stood behind their work and were proud to brand such products as cabinetmaker's case pieces, chairs, smith's bellows, ladders, looms, churns, well buckets, tubs, washboards, measures, pumps, apple and salt barrels, tobacco hogsheads, beer and powder kegs, tool handles (Figure 2), wooden planes, cigar boxes (Figure 3), and many other wood products.

Chillicothe had four known spinning wheel makers: Joseph Hopkins, Joseph Lemun, Henry May, and James Howard, and they may have branded their work, but the writer has never found a marked wheel with the name of any of them. Joseph Lemun's spinning wheels were "justly esteemed the best in the state."[14]

Weavers were busy at their looms supplying the settlers with coverlets, rugs and coarse cloth. The shops of Evarard Harr and M. & W. Simpson turned out weavers' reeds. The latter were proud to announce "all reeds sent from the shop will bear the brand M. & W. SIMPSON."[15]

You might wonder where a branding iron might be obtained in territory so far removed from a metalworking center such as Pittsburgh. In 1818, James Brown and

Figure 4. The burning brand of John Crouse was a huge affair weighing 18 lbs. Overall length is 31". John Crouse made a fortune on the flour and whiskey that he shipped in barrels and marked with this brand. From 1798 until 1820, in Ross County, Ohio, he operated a mill and distillery works on Kinnickkinnick Creek near where it emptied into the Scioto River.
Collection Ross County Historical Society, Chillicothe, Ohio.

Figure 5. I. [John] CROUSE. The head is nine inches wide with one-inch letters.
A number of brands could be made with it on one heat.
Crouse believed in doing things big, with no mistaking of his mark.

son set up a shop in Chillicothe where they ". . .established a manufacture of sundry articles of Hardware suitable to the country. . ." Branding irons were included in a long list of edge tools, scale beams, etc., which the Browns advertised.[16]

Chillicothe was a supply station for General Wm. Henry Harrison's troops during the War of 1812. Local miller and distiller John Crouse shipped a large quantity of both wet and dry goods during that era (see Figures 4 and 5).

In 1823 the Navy Commissioner's office advertised in New York, Baltimore, Philadelphia, Alexandria, Norfolk, Kentucky, St. Louis and Chillicothe newspapers to ". . .receive proposals for furnishing the use of the Navy of the United States, 1500 barrels of Beef (Figures 6 and 7) and 1000 barrels of pork. . .These provisions must be of the best and most approved quality, well salted, and salt petred. The barrels must be of seasoned heart of white oak, and fully hooped. . ."[17] All of these barrels would have been branded to identify the shipment when opened up and "duly inspected."

By 1819 steamboats were in general use on the Ohio River, laden with all sorts of boxes, cases, and barrels. Such shipments received rough handling and were often stacked on deck where they were subject to the elements. Paper labels were not practical under such conditions so branding and stenciling were mostly used to identify the shipper and consignee.

Often these wooden barrels and cases were reused and the original brand or stencil was removed with the box shave.[18] Oldtime storekeepers always had a box shave among the tools used in their stores.

A teamster who worked at the mines in Arizona in 1869 wrote about his experiences in hauling explosives.[19] "The heavy boxes were branded 'Nobel's Blasting Oil' and when handled and jolted, there sifted out of the cracks of the boxes a fine white powder, like flour."[20] It's clear that no chances were taken with paper labels coming off or

Figure 6. Left to right. RYE FLOUR. This all iron brand is 7 ½" wide and 32" long. S. DISBROW. The copper head is 5½" wide. Handle, 22". PRIME BEEF. *The bronze head is 7½" wide. Handle, 24". The letters are burned rather thin from long use; perhaps it marked shipments to the army or navy.* Collection of Dard Hunter, Jr.

Figure 7. All these brands have long iron handles to dissipate the heat when not in use. The smaller brands contain the names of T. G. CHASE, F. NOURSE, S.O. GEBHARD, A. S. MARTINI, and DEAN BURT. All are iron. They are photographed atop an Ohio & Erie Canal lock stone at the Ross County Historical Society. Collection of Dard Hunter, Jr.

stencils being smeared. Recognition of such dangerous contents had to be positive.

In 1884, a New Mexico correspondent of the St. Louis *Globe-Democrat* wrote about *The Branding Iron as the Basis of Fortunes.*[21] He told of the gangs of rustlers who stole Mexican stock and drove them to the San Carlos Indian Agency. "Sometimes when they stole American stock, they had to get to work and remove the brands. The plan was simple enough. A piece of blanket was taken and wrung out after being dipped in water. A common frying pan was heated nearly red hot. The wet blanket was applied over the brand and the red hot pan pressed hard against it. The steam generated scalded the hair clean off, and the job was done. In a few months the hair grew again and a new brand was put on. There again the fellows had a kind of branding iron with which they could change a number of brands."

I once heard a Texan boast of a man in Presido County, Texas, "who started in only five years ago with two old cows and a branding iron and is today worth $100,000." I told him, "I knew half a dozen men in New Mexico who started in with nothing but a branding iron and are today worth $200,000." The branding iron has laid the foundation of many respectable fortunes, both here and in Texas.

In the nineteenth century, marking irons were peddled on the streets of London and we find this in Tuer[22]: "The old cry of 'Marking Irons' has died out. The letters were cast in iron, and sets of initials were made up and securely fixed in long-handled iron boxes. The marking irons were heated and impressed on a proof of ownership.

Hence ladders, bellows, tubs and pails,
Brooms, benches, and what not,

Just as the owner's taste prevails,
Have his initials got."

Germans made up a large part of Chillicothe's population in the late nineteenth century, and how the men loved their beer and cigars. There were quite a few German cigar makers who plied their trade in shops attached to their homes. They put up their hand-rolled "segars" in mahogany boxes and branded them with their names and revenue districts as the law required (Figure 3).

Devotees of early trades are familiar with the stamps used to mark wooden planes and also the marking hammer used to whack an identifying mark on the ends of railroad ties, large timbers, and logs. These devices worked well in compressing the end-grain and left marks that have endured for centuries. Such stamps are not satisfactory on lateral surfaces as they cut and distort the wood fibers. This is where the branding iron does such a good job in imparting an indelible mark. Early house joiners and cabinet-makers often used the wood chisel to cut Roman numerals in their various framing pieces.

NOTES

1. See Mary Earle Gould, *The Chronicle*, Vol. XIV, 1, pp. 4, 5. Also Rockwell Gardner, *The Chronicle*, Vol.XV, p.16 and others.

2. Webster's *Dictionary*.

3. Genesis, IV, 15.

4. *Encyclopaedia of Religion and Ethics*, James Hastings, ed. Scribners, New York, Vol. 4, p. 269.

5. *International Encyclopedia*, Vol. 3, p. 219.

6. Williams Bros., *History of Ross and Highland Counties, Ohio*. Cleveland, 1880, p.70.

7. See R.A. Salaman, *Dictionary of Tools*, p.106. Also see "Horn Brands" in Figure 1.

8. *Book of Records*, viz., Marks, Brands, Private Roads, &c., Union Township, Ross County, Ohio, John Crozier, Township Clerk, 1816.

9. *Scioto Gazette*, May 22, 1801.

10. *Minute Book of the Court of Quarter Sessions of the Peace*, session of the 4th Tuesday, Sept. 1802. Chillicothe was then the capital of the Northwest Territory.

11. *Scioto Gazette*, Nov. 20, 1809.

12. *Ibid.*, Oct. 9, 1809.

13. John McCoy & Thomas James were the first to pack pork west of the Alleghenies and they opened the second store in Chillicothe in 1798. Much of the pork was destined for Havana and other West Indian Islands.

14. *Scioto Gazette*, Nov. 13, 1806.

15. *The Supporter & Scioto Gazette*, Dec. 20, 1823

16. *Scioto Gazette & Fredonian Chronicle*, April 3, 1818.

17. *The Supporter & Scioto Gazette*, August 9, 1823.

18. See R.A.Salaman, *Dictionary of Tools*, p.180.

19. *Scioto Gazette*, March 30, 1881.

20. Nobel's Blasting Oil was no more nor less than nitroglycerine. That was the name it was used under for a time after its discovery, or rather, its practical application to blasting, by Alfred Nobel, in 1863.

21. *Scioto Gazette*, Oct. 22, 1884.

22. Andrew W. Tuer, *Old London Cries and the Cries of Today* (London, 1885). For more on these peddlers, see Grabb, "The Knife Grinder," *The Chronicle*, Vol.30, 3, pp.37-40.

The Paper Industry, 1810-1860

By CHARLES I. FOSTER

July and September 1947

Mr. Foster was the Director of the Worcester, Massachusetts, Historical Society.

As the great presses thundered away printing books by the millions, and newspapers multiplied, the country's appetite for paper grew in proportion. It is interesting to note the efforts of the paper mills to meet this increasing demand for their product, for the art of papermaking was a slow and laborious process.

The American paper mills met this challenge by the gradual mechanization of their handicraft; they adapted machines to perform each operation in much the same way as it had been done previously by hand and throughout the whole long process maintained certain handicraft characteristics of their art. They used machines and yet the manufacture of paper refused to become mechanical.

In the early days of the art when only manpower was used, the papermaker pounded cotton and linen rags in a mortar with a pestle until they disintegrated into fibers. He mixed these rag fibers with water in a vat and into that vat he dipped his mold, a wooden frame about twenty inches wide by thirty inches long, covered with a wire mesh or screen and surrounded by a removable wooden edge or "deckle" which served to retain on the mold the desired quantity of the "stuff." As the papermaker lifted his mold from the vat, he agitated it gently to distribute the stuff evenly over the wire and held it long enough to permit the water to run through the screen and leave on the wire that meshed layer of fibers which was to be a sheet of paper. Then he removed the deckle, handed the mold to his helper, the "coucher," placed the deckle on a fresh mold and started over.

The coucher had before him a piece of felt of the same size as the mold which he received from the vatman. Against this felt he pressed the fiber-coated side of the mold and then gently withdrew the mold. As the fibers clung to the felt and left the wire of the mold, a sheet of paper was born. Over the newly formed paper the coucher placed a fresh layer of felt and repeated the operation as the vatman handed him another mold with its burden of fibers.

Drying this sodden paper was indeed a tedious process. When the pile of alternating felt and paper reached a convenient height, the workers lifted it to a press for a gentle squeeze which removed some excess water and matted the fibers more firmly together. Then it was necessary to peel each tender sheet from its mother felt, place it on another felt, give the pile a harder pinch in the press and repeat the operation. When the still damp paper was judged to be sufficiently strong, the papermakers piled it sheet to naked sheet for another, but gentler, nip in the great press, the start of another repetitious process: between each increasingly severe squeeze the craftsmen carefully peeled the tender layers apart and turned them upside down, changing their order, so that the two sides might be finished alike. The final step in drying was to take the pile of paper into a ventilated loft and hang swatches of four or five sheets over poles or ropes, there to stay for weeks, if necessary, to dry and season.

The first step in mechanization of papermaking was taken long before the period with which we are concerned. Stamping mills, which consisted of ironfaced rams pounding in armored tubs, had replaced the mortar and pestle in shredding the rags to fibers. In turn, a marked improvement over this stamping mill resulted from German-Dutch competition early in the eighteenth century, when the Dutch found their wind-driven stampers unable to compete effectively with German waterpower. The Dutch devised an oval tub partially divided longitudinally by a central fin. Between this central support and the edge of the tub they mounted a cylinder about twenty-six inches in diameter, armed with iron bars which bore with a regulated pressure against a similarly armed bedplate in the bottom of the tub. By this means a mixture of rags and water could be circulated around the tub with a continuous shredding and hydrating of the fibers. This "Hollander" or beating engine is still in use today with only minor improvements.

An effective device indeed, the Hollander could also be used for washing the rags by easing the pressure of the cylinder against the bedplate and increasing its speed. Invented in 1750, the Hollander reached America in 1775 and generally replaced the stamping mills here during the following five years.

The only other step in mechanization which had been introduced by 1810, the time when our study opens, was a power-driven rotary duster for removing loose dirt and lint from rags. Meanwhile, the color and quality of the product had been improved by the practice of putting bleach into the washwater. This not only whitened the stock but softened the fibers to make easier beating.

The process of papermaking as it functioned in the United States in 1810 may be better understood if we study a typical mill in operation at that time. For this purpose, the Burbank mill in Millbury, Massachusetts, built in 1800, will serve admirably. This was a two-vat mill of the size usual for those days; that is, it had two vats at which master papermakers dipped their molds. A twelve-foot water wheel supplied power to drive two Hollanders for washing and bleaching, a rotary duster and a grindstone on which to sharpen the bedplates of the Hollander.

We may readily visualize the daily work of the mill. Ten or twelve women sat at large boxes each of which was armed at one corner with a sharp scythe on which they cut dirty cotton and linen rags to bits, freed them from bone and metal and dropped them into the box, a foul and dangerously unsanitary task. The filled box went to the duster which whirled the rags about to free them from loose dirt and lint. From the duster, the rags went to the Hollander engines for washing and bleaching. The beating roll of the Hollander was set for light pressure against the bedplate and high speed to circulate the water and rags fairly rapidly about the tub. Here the bleach started its work to remove as much as possible of both dye and dirt stains. Incidental to washing, the Hollander did a certain amount of beating with the result that by the time the washing was finished the rags no longer resembled woven cloth but were mangled to a mass of stringy fiber called "half-stuff."

There was, of course, some method of draining the wash-water from the half-stuff. Our records are not clear, but it seems probable that draining rooms were in use by 1810. These were brick-lined chambers with a bottom outlet in which the half-stuff was left for a period of weeks while the water leached away and the bleach continued its work of whitening and softening the fibers.

From the drainers, the half-stuff went back to the Hollander engines on its way to becoming "stuff." This time the beating roll was set snugly against the sharpened bedplate for grinding, mashing and hydrating the fibers. The half-stuff, mixed again with water, moved sluggishly around the tub and under the Hollander roll for fifteen to twenty-four hours. Toward the end of this period, the papermaker often squeezed a handful of stuff seeking to feel with his fingers that greasy texture which would tell him that the beating was finished.

With beating, mechanization in the process of papermaking ceased in 1810. From the Hollander, the stuff passed to the vat where the vatman added water, stirred to keep the fibers suspended and went to work with his molds. Operating two vats, the production of the Burbank mill was 230 to 250 pounds per day and that limited output was supposed to provide wages for about seventeen workers as well as some return on the capital invested.

Faced with the rapidly mounting demand for paper as the people of the country placed more and more faith in the power of the printed word, the paper mills sought to overcome two severe obstacles to increased production: cotton and linen rags formed the only supply of raw material and they were scarce; the latter half of their manufacturing process was entirely unmechanized and laboriously slow. Often three to four months elapsed in the conversion of rags to paper.

The search for a new and plentiful raw material was fairly frantic. There were experiments with various grasses, cabbage stumps, corn husks and stems, nettles, ferns, bagasse, cotton stalks, wild indigo and banana plants. Many of these fibers made paper of sorts, but not economically nor of a quality to compete with rags. The single success in this field was registered by Matthias Koops in England who made good paper of straw in 1800. By 1827 we had started to use this new raw material, not for printing and writing papers, but for paper boards with which to bind books and make boxes.

In spite of experiment and search, rags continued to be the only supply of fibers for printing and writing papers in America until after the Civil War. People brought rags to the printer where the wagon from the paper mill picked them up, and bell carts roamed the streets of our cities, but the domestic supply failed to meet the situation. Importation provided the necessary supplement to keep our mills going. In 1829, the mills of Massachusetts consumed 1700 tons of rags, a large proportion of which came from Italy and Germany. Difficult as the raw material problem proved to be, the historian cannot but give thanks, as he examines the tough and durable documents of the period, that the papermakers were obliged to use rags.

While the mills scratched feverishly for more plentiful supplies of fiber, the other main obstacle to greater production, the mechanization of the latter half of the manufacturing process received quite as much attention with more successful results. The first move to replace the vatman by a machine came in the early years of the of the nineteenth century with the invention of the cylinder machine by Dickinson in England. This engine employed a drum, faced with mesh wire, which rotated in the vat. Suction applied to the interior of the drum drew fibers to the wire surface and as the drum rotated, the layer of meshed fibers were peeled from the drum, a continuous web of paper. In 1816, Joshua Gilpin successfully adapted the Dickinson principle to a machine which went into operation in the Thomas Gilpin mill near Philadelphia where it did the work of ten vatmen. In the years 1818 to 1839, many American mills installed these machines, a number of which were manufactured by John Ames and Isaac Burbank of Worcester, Massachusetts, but their adoption was not very rapid, for as late as 1829 only six of sixty Massachusetts mills had papermaking machines of any type.

Probably one reason for the relatively slow adoption of the cylinder machine was the fact that it was better suited to the manufacture of paper board than to the lighter weights which the country required for its literature. Today that process is used exclusively for making container and card stock.

But the same period which saw the development of the cylinder machine brought an invention which completely solved the problem of making and drying light weight paper in a continuous web at high speed. As much as any other single factor, the Fourdrinier machine made possible the mass production of literature for the common man.

The most important feature of this device was a long, endless belt of wire mesh on which the water stock from the vat flowed in a continuous, regulated stream. Power applied to the rollers which sustained this wire belt rotated it steadily as it carried away its burden of watery fiber. Traveling on the edges of the wire mesh to keep the stock from spilling over the sides were two endless belts of heavy leather which took their name, "deckle straps," from the frame of the vatman's mold which served the same purpose. The Fourdrinier wire traveled far enough and slowly enough to permit most of the water to drip through the wire and the sheet of paper to form. At the end of the wire belt, the wet layer of paper was strong enough to carry its own weight a few inches to what was appropriately called the "couch roll," which in turn supported it to pressing rollers for further extraction of water. The Fourdrinier was a step in mechanization comparable in importance to the Hollander. Even in its early, crude form, it could accomplish the day's work of two vatmen and the two couchers of the typical mill in about an hour.

The development of this useful machine, according to findings of J. Wallace Tower, was the product of cooperation between Frenchmen and Englishmen during the Napoleonic wars. In 1799, Nicholas Louis Roberts took out the original patent in France. Two years later, Leger Didot purchased the Roberts' patent and, with John Gamble, took it to England where Byron Donkin worked out its difficulties to make the device practical. By 1803, the machine was in operation under a patent granted to Gamble by the English government. The next year, Henry and Sealy Fourdrinier, stationers and manufacturers, purchased the interests of Didot and Gamble and managed to put a second machine into operation. In 1806, Henry Fourdrinier became the sole proprietor of the English patents for the machine which bears his name.

Efficient as it proved to be, the Fourdrinier machine did not reach the United States for over twenty years after it went to work in England. The reason for the delay may probably be found in the fact that it was a complicated piece of engineering, difficult for our crude shops to manufacture and costly to buy in a period when our mills did not command much capital. Of course, the idea came before the machine. For example, the Eagle Mill in Dansville, N.Y., had its own adaptation working in 1824, a wooden contraption thirty-six inches wide by twenty feet long which carried the stuff over a woolen blanket in place of the Fourdrinier wire. In one respect the Eagle Mill was quite advanced as the paper web traveled from the blanket over four heated driers. The steam-heated drying cylinders had been added to the English machine only a year before.

Finally, in 1827, Henry Barclay brought the first Fourdrinier machine from England to America and installed it at Saugerties, N.Y. In addition to the steam-heated drying cylinders, several other improvements had been added by that time. A woolen jacket on the couch roll helped to pick the paper from the end of the Fourdrinier wire; a second press roll meshed the fibers more firmly; a light roller, called the "Dandy" revolved on the wire near its dry end and impressed the soft paper with laid watermarks to make it resemble the hand-made product more closely and under the wire were suction boxes which pulled water through the mesh to permit higher machine speed.

After 1827, our adoption of the Fourdrinier was rapid. The next year saw a second machine set up in the mill of Joseph Pickering in North Windham, Connecticut, and with this working model at hand, our New England mechanics could solve the problem of manufacturing the Fourdrinier. Phelps and Spafford, later the Smith and Winchester Manufacturing Company, built the first American Fourdrinier in 1829.

It is interesting to note that this important date roughly coincides with the intro-duction of the steam-driven printing press. With mechanization completed, the American paper mills were prepared to feed the country's hungry printing plants which, in turn, supplied a nation's hunger for the printed word. Here is a clear example of the interdependent nature of our economic, social and intellectual life. These aspects of our history are as closely interwoven and meshed together as the fibers of the paper which gives expression to so much of it.

In spite of this mechanization of its various processes, papermaking remained distinctly an art rather than an exact science. The craftsmen simply used more efficient tools. Once installed, the heavy and expensive Fourdrinier machine remained a fixture in the community for half a century and more while the secrets of its peculiarities passed from generation to generation. The long reign of the Fourdrinier has been protected by the fact that since its birth in 1799, no new principle of paper manufacture has appeared to render it obsolete. There have been minor improvements such as the introduction, in 1830, of tubular rollers under the wire mesh to draw water from the forming sheet by capillary action. This enabled the machine to further increase its speed. In the same year, the marked direction of grain in machine-made paper, a characteristic which distinguished it from the hand-made product, was partially overcome. It was caused by the natural tendency of the fibers to turn lengthwise in the current of stuff as it flowed onto the wire. To cross these fibers more effectively, the shopmen added a crank which gave the wet end of the machine a gentle lateral shake similar to the motion of the vatman as he lifted his filled mold from the vat.

But no refinement in the process has ever reduced papermaking to mechanical exactitude. As the requirements of art and industry have increased the variety and quantity of paper production, the skill of the papermaker and his intimate knowledge of his machine have been so much the more necessary. In this instance, the craftsman has remained a craftsman.

Bookbinding in Colonial America

By C. CLEMENT SAMFORD

January 1954

John Saunders, bookbinder, took the freeman's oath in Boston in 1636 and purchased a shop, presumably for his business in 1637. The first product of record from an American press to need the services of a binder, *The Whole Book of Psalms*, more generally known as the Bay Psalm Book, was not published at Cambridge until 1640. Nothing more is known about Saunders or his activities. George Parker Winship suggests that he probably had been connected with some printing establishment in England, since binding and printing had always been closely allied. It is reasonable to assume that he may have been associated with the early printing venture in Massachusetts Bay where his experience would have been valuable.[1] Throughout the colonial period, printing and binding were usually carried out in the same premises. As trained hands in either craft were scarce, it was often necessary for one to aid the other. Records of early printing offices cite numerous instances of this. Franklin, in his *Autobiography*, says that Keimer (his employer) employed Hugh Meredith and Stephen Potts, among others — that "Meredith was to work at press, Potts at bookbinding, which he (Keimer) by agreement was to teach them, though he knew neither one nor t'other."[2]

The materials essential to binding were mostly readily obtainable in colonial America. Since the earliest days, boards of wood or pasteboard, leather or vellum, glue, paste, pack thread for bands and linen thread for sewing, have been the basic materials necessary for most bindings. "Scabbord" (scaleboard-thing, birch or oak) was widely used in the early days in place of pasteboard which had to be imported for covers.[3] There was no problem about leather, as tanning was one of the earliest colonial industries.[4] Sheep, calf, and the skins of deer and other animals were widely used for a variety of purposes including bookbinding. Among the earliest tanneries were those in Virginia in 1630, and Lynn, Massachusetts, a few years later.[5] In 1640 a Massachusetts law required that hides should be carefully removed and promptly taken to the tannery, and fixed penalties for the home tanner who produced an inferior leather.[6] By 1734, the report of the Lords of Trade stated that "a great part of the leather used in the country is . . . manufactured among themselves."[7]

Seventeenth century colonial books bear evidence that this leather was not always a finished product. It was often rough, and at best inferior in finish to the European leather. It was serviceable, however, and the many uses to which it could be put, combined with a very adequate supply of raw skins, promoted a rapid growth of this industry. By the first decade of the nineteenth century, the annual total amounted to twenty million dollars.[8]

Flax for thread and vellum for parchment were also produced. Flax was grown in Massachusetts, Connecticut and probably Virginia as early as 1640.[9] Linen was one of the early cloth products. Not much vellum was used on colonial bindings compared to leather coverings in spite of its relative cheapness. There is evidence that some copies of The Bay Psalm Book, and possibly some of Eliot's Indian Bible were bound in vellum, though it was most frequently used on account books and ledgers.[10] Local vellum was used by the Maryland government in 1704 for engrossing the laws at a price of 18 pence a skin. By comparison, Franklin paid, not much later, three and four shillings for calf skins for binding operations.[11]

Certain other supplies such as morocco, milled binder's board, tools for decoration, etc., were not locally produced and had to be imported at considerable expense.[12]

Paper, also, was largely imported until well along in the eighteenth century.[13] In 1664, John Ratcliffe, the Boston binder, complained that "I find by experience that in things belonging to my trade, I here pay 18s. for that which in England I could get for four shillings, they being things not formerly much used in this country."[14] Conditions were slow to improve in a country not essentially industrial in character, and the colonial binder often reverted to economic practices common in fifteenth century Europe. Temporarily out of boards, he would utilize waste paper, pasting the separate sheets together to make covering material. Left over printed matter for which he had no further use was ideal for this purpose, and consequently, many early American bindings contain fragments which bear much the same relationship to early colonial imprints as the early Gutenberg fragments bear to the beginnings of printing. Ten different William Bradford imprints were recovered from the binding of a later volume, two of which had previously been unknown. *A Collection of the Governor's Several Speeches*, printed by James Green, in the Maryland Historical Society Library, was entirely recovered from a binding of the next year's session-laws, its pages pasted together for binder's board.[15] The John Carter Brown copy of *The Compleat Laws of Maryland*, printed by William Parks in 1727, contains an extra title page announcing the inclusion of the Charter of Maryland in the volume. For some reason, the charter was never printed, and the rejected original title page was pasted down on the boards as a lining in this copy.[16] That future discoveries of great historical importance may be made in this fashion is highly probable.

Early colonial bindings were generally plain and utilitarian. Only such materials as were at hand were used freely. The craftsmanship was often clumsy and ornamentation was almost nonexistent. Consequently, the average book bound in early colonial America consisted of little more than a scrap of leather drawn over the boards, often without any paring of the turned in edges. Backs frequently were not rounded, headbands did not exist for the most part, and leather thongs or pack thread on which the sections were sewn were frequently sunk into grooves to produce a smooth back. Lettering, if it occurred at all, was generally only a scrap of paper pasted on the spine, hand written in ink. Much of the leather used was rough calf of local manufacture, though account books were often bound in vellum and sometimes sided with paper, frequently marbled.

There were exceptions, however, as certain binders even from the earliest times occasionally used morocco imported from England and worked out more or less elaborately tooled designs in blind or gold leaf. Notable among these bindings are those of John Ratcliffe and Edmund Ranger of Boston in the seventeenth century, and later, William Parks of Maryland and Virginia. The craftsmanship of the emigrant binder was always superior to that of the workman trained in the colonies, and wherever extra pains were taken, evidenced by sewing on raised bands, general care in forwarding, the use of morocco and good calf, sewn headbands and gilt decoration, the craftsmen probably had served their apprenticeships in England or Scotland. Even so, the best work done in colonial America was not on a par with the average of the leading European centers, and a critical comparison would be unfair. Considering the limitations under which the colonial craftsman was forced to work, the absence of the patronage of royalty or wealth, the isolation from superior sources of supply, the economic and political factors present in the new country, etc., the amount of comparatively fine work which was produced is very impressive.[17]

The emigrant binder naturally followed the prevailing practices of the mother country in both technique and style of decoration. The "Cambridge style" had been very popular in England and is consequently found on the majority of the more elaborately decorated colonial books. The average book was decorated more simply. Raised bands were conventionally outlined by fillets, and the board edges were almost always decorated by lines or a narrow flower roll. A single or double line might be run around the front and back boards. Often, the only gold used was on the board edges, the rest being "blind."

Some form of the basic panel design was found throughout the colonies. Boston and other New England bindings make constant use of it. The early Philadelphia bindings are typical, as are also, almost without exception, the Maryland and Williamsburg bindings of William Parks. Two striking examples of this style are the *Charter of the College of William and Mary*, Williamsburg, 1736, printed by William Parks (John Carter Brown Library, Providence, R. I.), and Franklin's *Cato Major* of 1744, (American Antiquarium Society, Worcester, Mass.), a presentation copy to Thomas Clap, President of Yale. Sometimes other decorative schemes were used. A fillet or decorative roll around the boards, with small fleurons at each corner and a central ornament — a larger fleuron or a design built up from small tools is fairly frequent. Mather's *A Call from Heaven*, (Alderman Library, University of Virginia), bound by Ratcliffe in green morocco is an especially attractive example of this style. Occasionally an allover design is encountered, built up of many small tools, but these are exceptional. These more or less elaborate examples of blind and gold tooling serve to show that the colonial binder could on occasion produce work of a high calibre, being limited mainly by the demands of his market. That this work was actually done in English America is supported by a consistently inferior technique in handling the tools, when compared with the more finished efforts of Continental binderies.[18]

Samuel Willard's *Compleat Body of Divinity,* (Boston Public Library and Columbia University Library), printed in 1726 by Keeland and Green, is probably the first colonial book with gilt lettering on the spine.[19] However, books continued to be bound without lettering or labels until well into the nineteenth century, except for the work of the emigrant binders. Their books, from the early part of the eighteenth century, carried labels, generally of red leather, and their spines were frequently decorated.[20] Gold was not widely used before the Revolution, and was generally confined to double fillet lines around the boards, or bordering the bands across the spine. Gold tooling, however, had been done in the seventeenth century by Ratcliffe and Ranger, among others.[21] Inventories and accounts of this period list gold leaf letter stamps, and various gilding tools, as well as charges for books bound "gilt." It is generally impossible to say how simple or elaborate these "gilt" bindings were. References to the use of gold are too numerous to list completely. The day books of William Hunter and Joseph Royle of Williamsburg, for example, list many purchases of gold leaf in quantity, and many charges for binding "gilt"; as do also the accounts of Franklin and others.

The ornamental tools used by colonial binders were mostly identical to those in use by English and French binders of the period. An extensive examination of colonial bindings has been made for this study, and the decorative tools are traceable as far back as the sixteenth century. As the goldsmiths of the various centuries and nationalities were responsible for the original designs, and as duplication was possible only through re-cutting of new tools by hand, slight variations and elaborations can be noted, which often enable us to establish within reasonable limits, not only the date of an unidentified binding, but its locality, and sometimes even its binder. The tools used in colonial America by the emigrant binders could be brought in by them duty free, but as it is not known that any binder's tools were manufactured in the colonies prior to 1768, it has been assumed that local craftsmen had to import them at considerable expense — a circumstance which would certainly limit their supply, and consequently a chance to develop their skill in using them. There were goldsmiths in the colonies in considerable number at the time of the revolution, and some as early as the late seventeenth century. While it is not probable that any of them specialized in cutting binder's tools at a time when the binder himself had to serve in several capacities, it is possible that an occasional smith turned out work of this type. The first mention that is known concerning the actual making of bookbinder's tools is the following advertisement in the *Pennsylvania Chronicle* for April 18, 1768:

"James Smither, Engraver, At the first House in Third Street, from the Cross Keys, Corner of Chestnut-Street, Philadelphia, performs all Manner of Engraving in Gold, Silver, Copper, Steel, and all other Metals — Coats of Arms, and Seals, done in the neatest Manner. Likewise cuts Stamps, Brands, and metal cuts for Printers, and ornamental Tools for Bookbinders. He also ornaments Guns and Pistols, both engraving and inlaying Silver, at the most Reasonable Rates."[22]

It was in this year that Abel Buell, of Killingworth, Connecticut, was making his first experiments in cutting and casting type.[23] Perhaps further research will be able to establish that native craftsmen were able to supply the colonial binder with some of his tools at an earlier date.

Much of the product of the early American press did not call for an elaborate binding. Session laws, assembly proceedings, pamphlets, sermons, and the almanacs, were frequently only sewn and covered with a piece of paper "drawn on" — pasted to the end papers, front and back. This cover was ordinarily either plain or blue, or sometimes marbled paper. The paper known as "Dutch Gilt," decorated with gold stamped animals, Biblical characters, or other designs also was used. The John Carter Brown copy of the *Charter of The City of New York,* printed by John Peter Zenger in 1735, is covered with an unusually well preserved example of this paper. These books and pamphlets, essentially very simple and hardly to be described as "bound," provided a cheap, bright contrast to the more common dull brown calfskin so common to the period, and were evidently very popular.[24] Colonial Williamsburg owns some excellent examples of small blank books, ledgers, and "pocketbooks," as well as alphabets and ledgers, covered with marbled paper in this manner. Cloth as a binding material was occasionally used during the eighteenth century. These cloth bindings are not to be confused with the elaborate velvet and embroidered ones from an earlier date in England and on the continent, but were of canvas, linen, or calico-like material. The Government of Maryland commissioned William Parks to bind some of the early laws in canvas in 1729.[25] Franklin's account for July 16, 1732/3 lists the following item:

"Dr. do (Thomas Hopkinson for) covering with linen 1/6."[26]

A letter from Thomas Longman of July 21, 1772, to Henry Knox, Boston bookseller, speaks of a shipment of books from England in canvas bindings. Isaiah Thomas also made use of cloth in his early days at Worcester.[27] While the use of cloth certainly was not widespread, it is interesting to note the beginnings of a practice which in the nineteenth century revolutionized the printing and binding industries.[28]

We do not know precisely what were the hours of labor in the colonial printing office. That the hours were long and the wages "something less than munificent,"[29] though the journeyman printer had always been one of the best paid craftsmen, is amply attested by Franklin in his *Autobiography*. Daylight was largely depended upon to set the printing office hours, as composition by candle light was neither popular nor practical. In England, it had been the custom to work up to eighteen hours a day as shown by the following statement of James Watson of Edinburgh. He attributes the poor quality of Scottish printing to "the little esteem we have for Press-Men, and the narrow prices given them." He continues:

> *"The Dutch, who, it must be acknowledged, are the neetest*
> *Printers in the World, have different Thoughts of them: They give*
> *larger wages to good PressMen than to Compositors: They will not*
> *allow a PressMan to work above Eight or Nine hours in a day, lest by*
> *working much they work not well. But here and in England, he that*
> *works Seventeen and Eighteen Hours is recon'd a choice workman:*
> *And indeed there is a Necessity for working much, their Wages are so*
> *small . . . For my Part, I'd rather give a Crown a Day to a good Press-*
> *Man, who brings Reputation to my work and preserves my Letter,*
> *than Eighteen Pence to one who must certainly destroy it by careless*
> *and base Working."*[30]

The bookbinders' wages during the colonial period are not so easy to determine. Records for journeymen printers' remuneration are fairly clear, however, and as the binder so frequently served in a dual capacity, or vice versa, it may be assumed that their wages were on much the same level. In general, printing office records for wages and charges for printing and binding show very little variation for the entire colonial period. A document in Benjamin Franklin's hand and endorsed by Isaiah Thomas gives the following price scale for printing and journeymen's wages:

"Prices of Printing Work in Phila. 1754
Books per Sheet

Compute Journeyman's Wages at Press and Case, treble the Sum, and that is price per Sheet for the Work. If you find Paper, allow yourself at least 10 per cent in the price of it. For Pamphlets of 3 Sheets, and under, 'tis best to agree at so much a Piece. Compute the Price by the above Rules, add the Paper, then add for folding and stitching 6d. per Quire; devide the whole Sum by the Number to be done, and if the Cost of each Book be above 3d, call it 3d½; if above 3d½, call it 4d. &c. and fix the retail Price at ½ or a 3d. more, as may be found most convenient.

Single Advertisements, of a moderate length, 5/- in the Gazette, small and middling Advertisements at 3/ the first week, and 1/ per Week after, or 5/ for three Weeks. Longer ones to be valued by Comparison with the foregoing; as if 20 lines be a middling Advertisement, Price 5/ for 3 Weeks, one of 30 will be 7/6d. &c. judging as near as you can, by the Sight of the Copy, how much it will make.

Blanks for Offices, ½ Sheets, No. 300 and upwards,

Printing 1d. a Piece.

Broadsides Ditto 2d. a Piece.

Hatters Bills 25/ per 1,000.

Paper Money 1d. per Pound, besides Paper and Cuts.

Party-Papers, Quadruple Journeymen's Wages.

Bills of Lading 6/ per Quire.

Apprentices Indentures 8d. a Pair, 6/ per Doz.

Bonds 4d. Single, 3/ per Doz. 5/ per Quire.

Bills of Sale 3d. — 2/3d. per Doz.

Powers of Attorney 4d. — 3/ per Doz.

Portage Bills 8d. each.

Journeyman's Wages

For composing Sheet Work, 6d. a 1000 Letters, to be reckoned by m's, and laid on its side between 2 Letters.

Small Jobs reckoned by the hour at 9d. per hour.

For composing an Advertisement, or any such small job, in Quarto, Great Primer or Double Pica, - 6d.

Folio Ditto - 1/

Blanks, 1 Side of Half a Sheet, in English or Pica, Pot or Pro Patria Size, 1/6d.

And other Jobs proportionately, according to Size of Paper and Letter.

Presswork, 12d. per Token, which is too much, if Pressman had constant Work, as compositors; but in America Numbers being generally small, they must often stand still, and often make ready.

For Jobs — An Advertisement, 60 No or 100. 6d. — and 6d. per 100 more.

If work makes less or more than even Tokens, all Numbers above 5 Quires to be reckoned a Token; all under, nothing; i. e. 4 Token and 5 Quires is but 4 Token; 4 Token and 6 Quires, 5 Token, &c."

Another entry in his Work Book for July 16, 1764, gives the following charge for a specific printing job:

"Thomas Ringold Esq.—Dr.

To Printing Remarks upon a Message sent by the Upper to the Lower House of Assembly of Maryland 500 copies making 4½ Sheets at 50/ (a) Sheet.

making 4½ Sheets at 50/ (a) Sheet	11- 5-0
To 5 Reams & 5 Quires of Paper for Do. at 14/	3-14-0
To folding and Stitching Do	2- 0-0
To Box for Ditto	7-6
	17-6-6"[31]

According to the above schedule, the 50 shillings a sheet in this account represented a labor cost of about 17 shillings, a gross profit to the printer of 33 shillings for each of the four and a half sheets. Add to this the 7 shillings representing the 10% profit on the cost of paper; he took from this job a gross profit of around £8. If office time, rent, lost time of workmen, deterioration of equipment, and other overhead charges reduce this amount to £6, his net gain on a typical pamphlet job was roughly 35%.[32]

The only charge in this account which could involve the binder is for folding and stitching. What was actually paid for binding during the colonial period is shown more clearly in other records. In 1662, Samuel Green charged 6 pence each for 200 copies of the *Indian New Testament*. This was a quarto of 33 sheets bound in leather. Two years later he received 2s. 6d. each for binding 200 copies of the whole Indian Bible, a quarto of 150 sheets, bound in full leather with clasps. John Ratcliffe, the Boston binder, received the same amount for those copies of the Bible bound in his establishment. He felt, that because of cost of materials, etc., that this was inadequate, and stated that he could not live comfortably on a rate less than 3s. 4d., or 3s. 6d. a book, "one Bible being as much as I can compleat in one day, and out of it (i. e., 2s. 6d. a copy) finde Thred, Glew, Pasteboard and Leather Claps, and all which I cannot supply myself for one shilling in this country."[33] In 1714, Elizabeth Short, the widow of Thomas Short, Connecticut's first printer, bound 2000 copies of the Saybrook Platform on eight sheet octavo printed by her husband in 1710 for 150s.[34] This is the first recorded instance of binding being done by a woman in America. It is a rather crude job in leather over birch boards, and the 6d. per copy was probably good enough pay for quantity production.

In 1731, Franklin paid his Journeyman-binder, Stephen Potts, 8 shillings for binding a Bible, 3s. 6d. for two other books, and six pence for two blank books.[35] In 1734, Franklin charged Thomas Penn £1. 10s. for binding "a great book of Birds."[36] The description suggests that this might have been Catesby's *Natural History of Carolina*, Vol. I of which was printed in London in 1731, which would explain the size of the charge. Franklin billed his customers in these cases for the amounts credited to Potts, with no profit for himself.[37] There would appear to be a great difference in the amounts paid to Ratcliffe in 1663 for binding the Indian Bible (2s. 6d.), and to Potts for a Bible in 1731 (8s.) but in addition to the probable size of the books, the former was an edition job, the latter a custom one. On edition work, Franklin's charge in 1731 was the same as that of Mrs. Short's seventeen years earlier — 6d. a copy for 1000 copies of Arscot's *Some Considerations,* a book of sixteen sheets, issued in two parts in 1732. In 1769, Hugh Gaine of New York informed William Johnson that binding the *Mohawk Book of Common Prayer,* an octavo of 26 half sheets, in plain leather would be 2 shillings currency a volume instead of the original estimate 1s. 6d. Those bound in morocco for which he must send to Boston, would cost an unspecified amount more.[38] In 1775, Valentine Nutter, a New York binder located opposite the Coffee House, charged Gaine 1s. 6d. a volume for 200 sets of *Chesterfield's Letters,* a duodecimo in four volumes at an average of 19 sheets each.[39] Timothy Green of New London received 5s. a copy for binding an edition of 505 copies of the Laws of Connecticut of 1784. This was a folio of 71 sheets.[40] The wages of John Stretch, bookbinder and journeyman for William Hunter of Williamsburg, are recorded as amounting to £98.45 from Jan. 14 to Dec. 31, 1751.[41]

The number of sheets in a volume and the format are important in comparing the cost of binding, as the binder received the books from the printer in sheets, flat, and before the actual process of binding began each sheet had to be folded by hand to accord with the format — once for a folio, twice for a quarto, three times for an octavo, and so on, and great care was needed to be certain that each resulting page followed the correct numerical sequence. This, and the succeeding operations of collating, pressing, and beating to make the signatures lie flat and the final collating and gathering before sewing the signatures — the first step in forwarding the volume was very time consuming.

The colonial printing office was always plagued with labor scarcity. While existing printing office records have often left the craftsmen actually engaged in binding in a singular anonymity, it is known that skilled hands in this trade were very scarce; and strong inferences may be drawn from the general labor situation and a few specific references that the binder in most cases was recruited from the ranks of journeymen printers and had to work in several capacities in the shop. From the beginning the printing office was very much a household establishment. Women and children in the family were a source of help too readily at hand to be overlooked. There are numerous instances of widows having taken over the operation of their husband's establishments. Among them are the widow of the Reverend Jose Glover in Cambridge; Dinah Nuthead and Anne Catherine Green in Maryland; Anne Timothy and Elizabeth Timothy in South Carolina; Ann Franklin, Sarah Updike Goddard and Mary Katherine Goddard of Rhode Island; Jane Aitken of Philadelphia and Clementina Rind in Williamsburg.[42] It is probable that women did more binding than the records show. From 1714 when Elizabeth Short bound the *Saybrook Platform* to the end of the century hardly any mention of women in this field has been found; but directories of the first years of the nineteenth century name six women binders in Philadelphia alone. Jane Aitken, who continued her father's printing establishment after his death in 1802, managed it in a thoroughly competent manner and executed some exceptionally fine bindings. It is evident that she must have had long experience in her father's printing office. That more women binders are not recorded throughout the eighteenth century is probably due partly to the custom of listing only the heads of families in the directories and partly to the general anonymity of binders.[43]

Aside from the family helpers, labor needs were supplied by the emigrant craftsmen, apprentices, and frequently by unskilled workers who were taught the various trades according to their abilities. Every working printing office had its quota of apprentices who were sometimes bound from infancy to help in any manner they could and learn the trade.[44] That in the course of their training they must have learned something of the various steps in bookbinding cannot he doubted, though no specific mention has been found of an apprentice binder in the records examined for this report. William Hunter's day book for August 28, 1750, lists the following item:

> *"Bookbinding Dr. to the Est. of Robt. Stevenson*
> *For a Servant Lad, Paul, and sundry Bookbinding Tools 17.5.3."*

This item, being charged to bookbinding, might indicate that the servant lad was to work at bookbinding.

Probably the most important labor source was the emigrant. He usually came to the colonies having a background of apprenticeship and occupation in his craft in England or Scotland and sometimes had his own tools. He frequently brought what distinction is to be found in both printing and binding during the colonial period.[45] Occasionally adults came to the trade under terms of indenture and were set to learn the various operations in the printing office. Some of these had considerable backgrounds in other fields and for one reason or another chose the indenture route for entry to the new country. George Webb, who was indentured to Keimer in Philadelphia, was an Oxford scholar. Franklin speaks of him at some length in his *Autobiography*.[46] In Williamsburg, Joseph Royle had working in his office an indentured servant, George Fisher, "by trade a bookbinder."[47] Although the apprentice system supplied a number of reasonably well trained new craftsmen, journeymen seem to have been very scarce throughout the colonial period. Not only they, but also their masters, were constantly on the move from one colony to another. Jonas Green, as journeyman and master, worked in three colonies; William Goddard in four; William Bradford and Benjamin Franklin each in two in addition to their employment in England; and William Parks in three English towns before he came to Annapolis and later settled in Williamsburg. The journeymen were even more inclined to change locations than the masters.[48] There was generally not enough work in the printing offices, nor was it constant enough, to justify the masters in training many apprentices. On the other hand, wages could not be paid on a full time basis which would have inclined more to take up the trade. Skilled journeymen, however, were able to pick up good jobs without much difficulty, especially if their training enabled them to work in several capacities. Stephen Potts, Franklin's binder, worked at press and possibly at other jobs for which there was need of a hand[49] and John Stretch, a journeyman printer for William Hunter, and probably also for Parks, seems to have done much of the binding.[50]

Runaways were a constant problem to the colonial printer. Nicholas Classen, a printer indentured to William Bradford, ran away and was advertised for in the *American Weekly Mercury* for June 13, 1728,[51] with a reward for his return. Hugh Gaine constantly advertised for journeymen and offered unflatteringly small rewards for the return of runaways. He described one as "pretty much pitted with the Small-Pox, wears his own hair and is much bloated by Drinking, to which he is most uncommonly addicted."[52] William Goddard advertised in his *Maryland Journal* in 1773 that he "wanted Immediately, one or two sober Journeymen Printers who can and will work."[53] Joseph Royle of Williamsburg inserted the following advertisement in the Maryland Gazette for May 2, 1765:

> *Williamsburg, April 23, 1765*
> *Run away from the Printing-Office, on Saturday Night, a Servant man named George Fisher, by trade a Book-Binder, between 25 and 26 years of age, about 5 feet 5 inches high, very thick, stoops much, and has a down look; he is a little Pock Pitted, has a Scar on one of his Temples, is much addicted to licquor, very talkative when drunk, and remarkably stupid. He had on, and carried away with him, several good white Linen Shirts, a Snuff color'd Cloth Coat, and a Suit of Light color'd Segathy, other good Wearing Apparel, a new Half-cut black Bob Wig, and a Set of Silver Buckles.*
> *Whoever apprehends the said Servant, and conveys him to the Printing Office, in Virginia, shall have Five Pounds Reward, and if taken out of the Colony, TEN POUNDS, besides what the Law allows.*
> *Joseph Royle.*

Evidently Fisher must have been apprehended and returned for the New York *Gazette* or the *Weekly Post Boy* for September 19, 1765, carried the following notice under the Williamsburg date line of August 2:

> *"Broke gaol, last Saturday night a servant man, named George Fisher, by trade a Book-Binder."* ... [54]

That individuals of such evident unreliability were found at all necessary to the printing operations of the colonies is a pointed commentary on the general labor situation.

NOTES

1. George Parker Winship, "Facts and Fancies" *The Colophon*, N.S. III No. 4,534.

2. Benjamin Franklin, *Autobiography* (N. York, Holt, 1916) 102; see also George Simpson Eddy, *Some Account Books kept by Benjamin Franklin* (New York, 1929) 38.

3. Hellmut Lehmann-Haupt ed., *Bookbinding in America, Three Essays* (Portland, Southworth-Anthoesen Press, 1941) ref. in Hannah Dustin French, *Early American Bookbinding by Hand, 1636-1820*, pp. 13-14.

4. Lawrence C. Wroth, *The Colonial Printer* (Portland, Southworth-Anthoesen Press), 1938, p. 195.

5. *Ibid*, p. 195

6. *Ibid*, p. 195

7. *Ibid*, p. 195-196

8. *Ibid*, p. 196

9. *Ibid*, p. 196

10. *Ibid*, p. 197

11. *Ibid*, p. 197; see also note 3, p. 325.

12. French, p. 27.

13. Wroth, *Colonial Printer*, pp. 169171.

14. *Ibid*, p. 198.

15. Ibid, p. 199. There are some interesting similar examples in the Maryland Hall of Records, at Annapolis.

16. Wroth, *William Parks, Printer and Journalist of England and Colonial America* (Richmond, Appeals Press, 1926) p. 39, note 12; see also, Wroth, *Colonial Printer*, p. 199.

17. French, *Early American Bookbinding by Hand*, pp. 18-30, contains an extended discussion of decorative development on early American bindings from which these paragraphs are drawn; see also Wroth, *Colonial Printer*, pp. 203-209.

18. Wroth, *Colonial Printer*, pp. 203-209 and French, *Early American Bookbinding by Hand*, pp. 18-30. The author's personal examination of a large number of ornamented early American books fully bears out the conclusions in works referred to.

19. French, p. 21, ref. in Thomas J. Holmes, *Proceedings of the American Antiquarium Society*, n. s. XXXVII, 39.

20. *Ibid*, p. 21.

21. *Ibid*, p. 21

22. A.C. Prime (Comp.) *The Arts and Crafts in Philadelphia, Maryland and South Carolina, Gleanings from Newspapers* (Walpole Society, Topsfield, Mass. 1929, 2 v.) I, 27-28.

23. French, p. 28.

24. Wroth, *Colonial Printer*, p. 2023.

25. *Minutes of the Commissioners appointed to Inspect the Public Records of the Province* [Maryland] June 4, 1729, p. 147.

26. Eddy, v. I, p.40.

27. French, p. 91, Ref. to Wroth, "Notes for Bibliophiles," New York *Herald Tribune*, Feb. 12, 1929.

28. See Joseph W. Rogers, "The Rise of American Edition Binding," in Lehmann-Haupt, ed., *Bookbinding in America*, p. 135.

29. Wroth, *Colonial Printer*, p. 161.

30. Wroth, *Colonial Printer*, p. 161-162.

31. Wroth, *Colonial Printer*, p. 181. This document is in the American Antiquarium Society, Worcester, Mass.

32. Wroth, *Colonial Printer*, p. 182.

33. *Ibid*, p. 198.

34. *Ibid,* p. 222.

35. *Ibid,* p. 200.

36. *Ibid,* p. 201.

37. *Ibid,* p. 201.

38. *Ibid,* p. 201.

39. *Ibid,* p. 201.

40. *Ibid,* p. 202.

41. The original day books kept by William Hunter and Joseph Royle are in the Alderman Library, University of Virginia, Charlottesville, Va. Colonial Williamsburg has photostat copies.

42. Wroth, *Colonial Printer,* p. 154-155.

43. French, p. 76-77.

44. Wroth, *Colonial Printer,* p. 156-157.

45. French, p. 27.

46. Franklin, *Autobiography,* p. 103.

47. *Maryland Gazette,* May 2, 1765 Advertisement for runaway.

48. Wroth, *Colonial Printer,* p. 103.

49. *Ibid,* p. 159.

50. *Hunter Day Book,* 1750-52, Entry for Dec. 11, 1751, Wages charged to Bookbinding.

51. Wroth, *Colonial Printer,* p. 160.

52. *Ibid,* p. 160.

53. *Ibid,* p. 160.

54. *The Arts and Crafts in New York* — 1726-1776 (New York, New York Historical Society, 1938)

This engraving is taken from Tome VIII of Diderot's Encyclopedia and shows a bookbinding establishment of the last half of the eighteenth century. Pictured are the following operations: Figure A — beating folded sections; B — stitching folded sections on the sewing frame; C — trimming the edges of a freshly sewn book on a plowing press; and D — pressing freshly bound books in a large standing press. The tools and methods shown here have been used in hand binding, almost unchanged, from ancient times to the present day.

This cut illustrates a typical book cover panel done in the so-called Cambridge Style, a commonplace colonial design.

Ice from Nature to Consumer
Tools and Methods

By BOB SIEGEL, JR.
April 1971

Among the first "icemen" were George Washington, who personally supervised ice harvesting at Mount Vernon in 1786, and Thomas Jefferson, who had two small icehouses at Monticello. Commercial ice harvesting began at Boston about 1800 with vessels carrying ice to southern ports and soon even to England, the East and West Indies, Rio de Janeiro and India. Previously, foods had been preserved by salting, spicing, pickling, smoking and dehydration (drying in the sun). As ice became easily available there was a constantly-increasing demand for its use in food preservation and also cooled drinks, ice cream, the brewing process, and hospitals. Expanding railroads carried ice to milder climates and returned with refrigerated fresh foods.

Ice froze to about six inches thick in the latitude of Ohio, and 30-40 inches in the latitude of Lake Superior. Two-inch ice could hold a man, four inches a horse, and five inches a horse and equipment. Farmers often travelled from one ice operation to another and did most of the hard work of harvesting. With a pair of ice creepers strapped to his

Figure 2

Figure 3

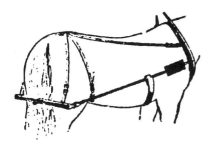

Figure 4

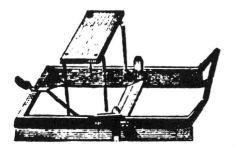

Figure 5

boots, a man turned an ice auger (Fig. 2) to bore a hole and test ice thickness with a measure iron.

Ice 14-16 inches thick was ideal for handling and storing. Snow had to be removed from the ice field as soon as possible, as it acted as an insulator and impeded freezing to the maximum depth. Often a snow occurred before the ice was thick enough to hold horses for scraping it off. Then the "wetting down" process was necessary. This started with a line of men using splitting chisels (Fig. 3) who punctured holes in the ice field usually at six-foot intervals. Rising water melted the snow, then froze, forming "snow ice." This unclear, and thus unmarketable "snow ice" also formed naturally with a successive snow, thaw (or rain) and freeze. If less than two inches thick it was usually shaved off by hand before delivery to customers. If thicker, horses fitted with ice horseshoes, plow rope and special harness (Fig.4), pulled an ice plane (Fig.5) over the field, shaving loose the "snow ice" (Fig.6). Deep cuts and later larger models of the ice plane (Fig.7) required the strength of two horses.

Success in this precarious business depended, to a considerable extent, in out-guessing the vagaries of the weather, in deciding when to scrape, plow and harvest. The ideal quick freeze, quick harvest, without rain or snow, rarely occurred. A sudden thaw or rain while ice was forming often flooded the ice field with dirty or sandy water from surrounding hills. Water on the ice field honeycombed and rotted the ice beneath.

The ice auger bored holes for pins inserted at two corners of the ice field. A line

Figure 6

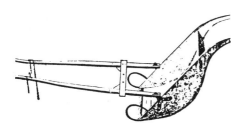

was stretched between the pins, and a man pushed the hand plow (Fig.8), cutting ½ inch deep along the line, guided by a board. The first crossline at right angles was similarly struck with the aid of a wooden square with ten foot arms. In the absence of a hand plow the line marker (Fig.9) was used.

Very early harvesting was done with only an ax and an ice saw. The horse-drawn ice plow (Fig.10) was invented in 1829, and after its perfection in follow-

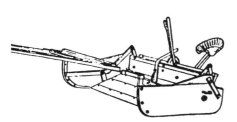

Figures 7. The clearing scraper and scoop scraper, later combined as the Boston scraper, removed snow from the ice field before harvesting.

Figures 8 and 9

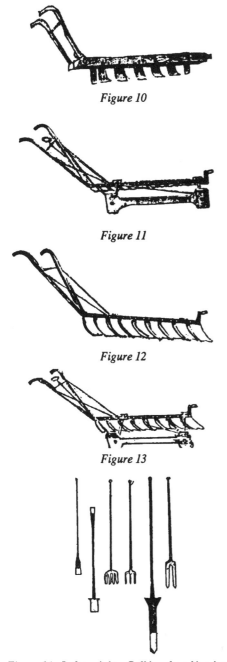

Figure 10

Figure 11

Figure 12

Figure 13

Figure 14   *Left to right.: Calking, breaking iron, four and three tined fork, splitting bar, and splitting fork.*

ing years, did the work of 50 men with the tiller-handled ice saw.

The marker plow (Fig.11), its swing guide riding in the previously cut groove 22 inches away, cut about three inches deep, 7/16 inches wide, and formed a checkerboard pattern on the ice field. Plows with successively longer, narrower teeth (Fig.12) cut two inches at a pass, until the ice was cut 2/3 of the way through. At least five inches at the bottom was left uncut to prevent accidental loss of men and equipment. Only one eight-inch plow, with swing guide attached (Fig.13) for the first cut, was used in small ice operations.

Calking bars (Fig.14) then calked ice chips into the ends of grooves at the separation lines where "floats" of 8-400 connected cakes were to be "barred off" (detached) from the ice field. All outside grooves having been calked, no water could enter the float and freeze. The "barring off" was performed by prying with a breaking bar (Fig.14). When ice was very thick or not sufficiently grooved, the heavier splitting bar (Fig.14) or splitting fork (Fig.14) was necessary. The tiller-handled ice saw (Fig.15) sawed some separation lines of floats where barring-off could not be done due to the absence of open water on the opposite side.

Floats, guided by men with 10-20 foot long ice hooks passed through the canals (wide, open waterways) toward the channel (the three—five foot wide waterway), which had been

GROOVING.

SAWING, CALKING AND BREAKING OFF.

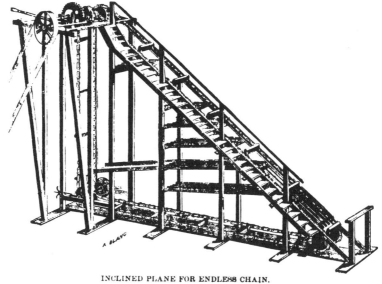

INCLINED PLANE FOR ENDLESS CHAIN.

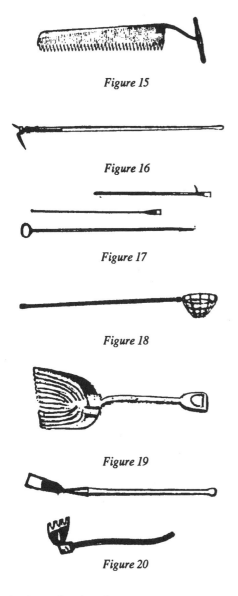

Figure 15

Figure 16

Figure 17

Figure 18

Figure 19

Figure 20

cut open with the tiller-handled ice saw, leading to the hoisting apparatus. Just before entering the channel, large floats were split into strips (single lines of cakes). Strips were towed down the channel by a canal grapple and rope or guided by six—eight foot long ice hooks (Fig.16).

In the channel the strip was split into single cakes with a tap by a needle bar or a three-tined fork bar or a splitting chisel or a canal chisel (same as splitting chisel but with a longer handle for use when men stood on a raised platform) or a hook chisel (combined canal chisel and ice hook) (Fig.17). Tools dropped into the water could sometimes be retrieved with an ice grapple. Ice chunks which clogged the channel were removed with a link chain scoop net (Fig.18) or a sieve shovel (Fig.19). Single cakes were guided into the elevator bucket by the elevator fork.

The steam-powered, endless-chain "elevator" (see previous page) (escalator in today's terminology), with buckets to hold cakes, was invented about 1850 and could raise 600 tons of ice per hour. Eight cakes 22" by 22" by 16" weighed just over a ton. Previously this "housing" was done by horse, pulleys, rope and ice gig (platform). A hoisting tongs would suffice in place of an ice gig.

Another method of housing was with horse, pulleys, and rope attached to a pole grapple or jack grapple. As the horse moved ahead on the ground, the grapple teeth pushed several ice cakes up an inclined plane. The grapple handle was held at the proper angle during ascent by a man walking alongside. Cakes were always elevated higher than the storage level and slid down "runs" (wood or metal slides) into the icehouse where they were packed in level tiers and covered with dry sawdust or planing mill shavings. Ventilating doors in the room allowed the escape of warm moist air. Good quality ice in a well-constructed icehouse could keep two to three years. Large icehouses held 10,000 to 90,000 tons. Little Random Lake (Wisconsin) had five 150 feet by 550 feet icehouses lining the entire south shore. Cakes were moved in, around, and out of icehouses on "runs" and were guided by short (3—5 foot long) ice hooks. If manoeuvered properly, an ice cake never stopped moving from the time it left the water until settled in its final location in the icehouse. In packing icehouses and (especially) vessels, the floor chisel or ice adz (Fig.20) shaved uneven spots off each tier of ice so the next layer would slide over easily. The drag tongs were handy in lifting and lowering cakes from elevated positions. Where ice was packed on edge, the ice cant hook "edged up" the cake. A bar chisel (packing chisel, socket chisel) adjusted cakes in packing so they were three inches apart, and also was used to cut around and loosen sides of cakes when preparing for delivery. When cakes were tightly frozen together, it was necessary to use a crescent saw or an icehouse saw.

The long-bladed separating chisel cut around the sides of cakes that had been packed on edge. The striking under bar "started up" a cake from the layer below, after it had been loosened on the sides. The house bar came into use only when ice was stored in 44" by 44" blocks, which first required grooving with the hand plow and then splitting with a pair of house bars (one in each hand), into four cakes. The summer bar combined the uses of the bar chisel and the striking underbar. Ice loss was about 10% in the ice houses and another 30-40% before reaching the customer.

An ice apron protected the ice delivery man's clothing. In the wagon, cakes of ice were split with an ice ax or sawed with an ice hand saw into 100, 50 and 25 pound pieces, suitable for the customer. Hand tongs lifted ice to the ice scale on the back of the wagon for weighing.

The trimmer bar or the ice breaker trimmed ice edges to fit in ice boxes. The ice chipper, chisel and awl (pick) chipped ice for use in ice cream and cool drinks. An ice crushing machine efficiently crushed ice.

Manufactured ice began on a small scale about 1880, equalled the production of natural ice by 1912, and took over the market completely in the 1920's and 1930's. Thus the natural ice industry, which for many years equalled the logging and lumbering business in manpower and importance passed quickly and is now virtually forgotten.

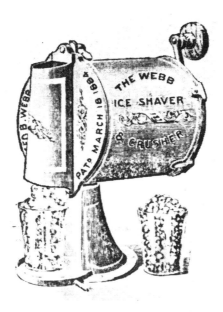

DELIVERING UP TOWN.

The Salem Waterworks

By LINDA LeMIEUX

September 1981

Fig.1. "A view of Salem in North Carolina — 1787" by Ludwig Gottfried von Redeken (Wachovia Historical Society)

"Among the many material mercies we have received we particularly mention that early in this year, after several years of longing and preparation . . . water was brought from a spring northwest of our town . . . Now in five places in the town water runs from a pipe, for drinking and for other necessary purposes."[1]

— Salem Memorabilia, 1778

Salem, North Carolina, was founded in 1766 by the Moravians, a Germanic people who had come to America from Central Europe. Their first colony, in Georgia, was abandoned for settlements in Pennsylvania in 1742, where Bethlehem became the central town. In 1753, the Moravians purchased 98,985 acres in Piedmont Carolina, which they named "Wachovia." The central town of Salem was the third Moravian settlement in the Wachovia tract, preceded by the towns of Bethabara (1753) and Bethania (1759). In their settlement of Salem, the Moravians continued a tradition of town planning and a highly organized system of community life in which the Church governed the social and economic as well as the spiritual affairs of its residents.

Humanists in their approach to community development, they sought ways to enhance the quality of life in their towns and cities.

Waterworks were relatively unheard of in 18th century America, where most cities and towns depended on wells, fountains and cisterns, or drew their water from springs and rivers. Nevertheless, Salem's waterworks had been planned since the town's inception. Its future water system was an important factor in the selection of the town site and the physical layout of Salem's streets, lots and buildings. Familiar with many of the 140 or more central water systems in operation in Germany by 1770, and the convenience of the pump-and-drive system which the Moravians built in 1754 at Bethlehem, Pennsylvania, Salem's settlers expected the same for their town.[2]

In March 1771, months before the organization of Salem when the majority of its 120 residents would move into town from Bethabara, community planners met " ... to speak of water in Salem, how we should provide ourselves with it."[3] Water was needed not only for drinking, cooking and fire protection, but to insure the economic success of Salem's many trades and industries. The dyer, brewer, potter, hatter — all needed good water, and nearly every craft depended on a reliable supply. The planners

Fig.2 White oak waterpipe and pipe fragment from the Salem waterworks, with iron ring used to connect the pipe sections. (Wachovia Historical Society)

estimated a waterworks to be less costly than wells and proceeded with plans for construction. By 1774 the Brethren (Salem residents) had agreed on the supply source for their system: two springs about a mile northwest of Salem Square, and the route of the bored log pipelines which would convey the water to the town. But it was four more years before they could turn their labor and finances from building homes and establishing trades toward the construction of their long awaited waterworks.

To make the pipes and superintend the installation of their water system, the Brethren chose the carpenter Christian Triebel, and the joiner, Johannes Krause:

"The Brn. Krause and Triebel are willing to undertake the bringing of the water in pipes to the town. The former is willing to superintend the laying of the pipes, the digging of the ditches for the pipes, and especially to select the route which the pipes shall follow; the latter, that is Br. Triebel, has agreed to cut and bore the pipes according to directions, and a contract will be made with him per yard."[4]

In January 1778, both men journeyed ten miles to Hope, the Moravians' English settlement on the Muddy Creek, to select and supervise the cutting of the white oak logs which would be drilled to make waterpipes back in Salem. Remarkably, Triebel, a man well into his sixties, and one journeyman were able to drill 96 feet of pipe per

Fig.3 One of the eleven ink drawings of Salem's second water system, drawn from memory by Mr. W.S. Pfohl of Winston-Salem in 1924.
(Archives of the Moravian Church in America, Southern Province)

Fig.4 Wrought-iron bits used to make waterpipes in Salem c. 1800-1850. The auger (left) has been broken from its shaft. Detachable reamer bits were attached to a common shaft. Extension plates could be attached to increase the diameter of two of these bits. (Wachovia Historical Society)

day, for which they received about 20 shillings.[5] Pipe mains measured about 10 to 12 feet with bores of 1½ to 4 inches; secondary pipes were somewhat shorter with smaller bores. (Fig. 2.)

To make each pipe, a log was solidly clamped onto a bench of convenient height and bored lengthwise through the center. First a small pilot hole about one inch in diameter, was drilled with a hand-operated auger. The bored pipe was then reamed out to size with a series of scoop-type bits of increasing diameter. Pod augers, the most common type of early American pipe drills, are the only type known to have been used in Salem. A variety of reams were used including a large bit with a hook at the nose to which a chain or rope could be attached allowing one man to pull the ream as another did the turning (Figs. 3-4).

While Triebel and his journeyman were busy hewing and boring the logs to size, Krause had plotted the system's course and set two non-Moravian German day laborers and several Single Brothers to work digging a 3-foot deep ditch from a reservoir near the supply springs to the town. More men followed laying and connecting the pipe sections with biconical iron rings, somewhat larger in diameter than the hole in the log; these "boxes" were placed around the hole and the pipes rammed together with large mallets. The rings tended to seal the system when the log pipes swelled after being placed under ground. To complete the project, Krause built two water outlets on Main Street and equipped the kitchens of Salem's Single Brothers House, Gemein House and Tavern with inside running water. The entire system, consisting of just over one mile of wooden pipe (322 rods) and accoutrements, was completed in only two months at a cost of £451:9:9 (a little over $2,000). To defray these expenses and support the operation of their waterworks, all residents over the age of sixteen were charged a small monthly water fee (originally 9d). This fund also paid the salary of the town water-master, who periodically inspected the pipelines and supervised necessary cleaning, repairs and extensions.

The mechanics of the Salem waterworks were quite simple, taking advantage of the gravity flow. Spring water was collected in a large reservoir above the highest point in the town. The pressure of the weight of the water distributed it through the pipelines. The pipes laid on the downgrade were bored to a wider diameter than those on the incline to achieve even greater water pressure. The location and type of water outlets were determined by the population and business needs in an area. The simplest of these was the standpipe: a short wooden pipe set upright on the pipeline, which was plugged at the top and fitted with a spigot. Water troughs, barrels and cisterns held water reserves. Most cisterns were designed to hold 3,000 or more gallons; these were lined with brick or wooden planks sealed with finely packed clay. Pumps with firehose connections were added to most cisterns beginning in 1805 (Fig. 5). Once the piped water reached an individual building it was conducted through a narrow copper pipe with a spigot. Water was stored in small interior cisterns and larger reservoirs outside the structure.

By 1800, Salem's water system was in poor condition because of rotten pipes and leaking joints. Rather than repeating the arduous task of ripping up and replacing the pipelines with newly drilled logs, the Brethren approached their master potter, Rudolf Christ, with the idea of substituting more permanent ceramic pipes for their waterworks.

Fig 5 Cistern with pump, second waterworks system (Old Salem, Inc.)

Their plan was prompted by a newspaper advertisement for "clay pipes for conducting water"[6] being manufactured at Charlestown on the Ohio River (present-day Wellsburg, West Virginia).

In March 1806, Christ requested that a sample be obtained "to get an insight into the matter."[7] Whether or not Christ used a Charlestown-made pipe as his model is not known, but the pottery began making waterpipes shortly thereafter. The first complete ceramic pipeline was laid in 1808, using "400 Stone burnt waterpipes" for which Christ was paid £40.[9]

Earthenware and stoneware water mains (Figs. 6 - 8) were made on the potter's wheel in two pieces; a "nose" section and "bell end"

Fig.6. Potter and his apprentice making ceramic waterpipes. Artwork: Jim Stanley.

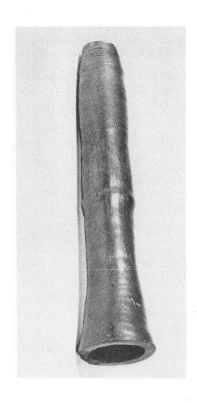

Fig.7 Slip-glazed stoneware waterpipe, attributed to Rudolf Christ, master potter in Salem 1789-1821. (Wachovia Historical Society)

were lapped together near the center of the pipe before being sent to the kiln to be fired. Each "nose" was turned with grooves to give a foot to the cement used to link the pipes. Smaller drain pipes were made in single section. A large plaster waterpipe mold (Fig.9.) in the Wachovia Historical Society collection suggests that Salem potters also attempted to mold their pipes, but none of this type survive.

The Brethren joined their ceramic waterpipes, nose end into bell end, with a mixture of gypsum and lime. This cement was apparently a much better sealing agent than the commonly used bonds of tar or linen. Unfortunately, the joints were all that held together when the ceramic pipelines were subjected to high water pressure. Years of experimentation brought little improvement, and the community was forced to rely primarily on their old wooden pipelines, supplemental wells and rainwater cisterns. Finally, in 1820, a waterworks committee decided to "stop the use (of) the earthen water pipes completely, except at those places . . . where the water has so much fall that the pipes will not have to suffer any pressure."[10]

The unsuitability of the ceramic waterpipes was not the only problem facing the waterworks committee by this time. Their primary concern was an insufficient and dwindling water supply. Salem's population had grown to about 500 persons, four times what it was when the congregation built its waterworks; families and trades had spread far beyond the town's original environs. Honest attempts to serve the entire community equally by means of pipeline extensions and additional water outlets had all but eliminated water pressure on the perimeters of the system. Water fees had to be reduced for residents in these areas, meaning less capital for system maintenance.

In 1823, the decision was made to build a new waterworks for Salem. New water sources were sounded and a variety of pipe materials were considered and priced. The Brethren hoped to install metallic pipelines, preferably modern cast-iron, but found the costs prohibitive. They agreed on a composite system using locally made log and ceramic pipes and a short line of cast-iron pipes, which they ordered from Pennsylvania.

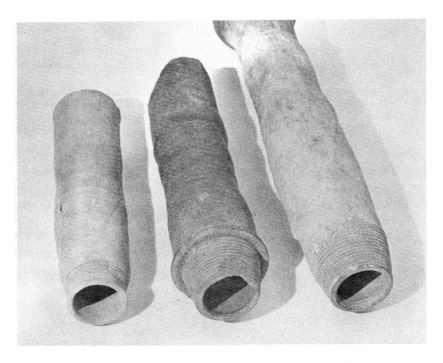

Fig.8 *Salem potters experimented with a variety of "nose" types to improve the fit between the pipe sections. (Wachovia Historical Society)*

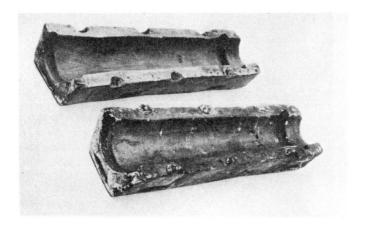

Fig.9 *Plaster waterpipe mold. No molded waterpipes survive. (Wachovia Historical Society)*

Fig. 10. (*Archives of the Moravian Church in America, Southern Province*)

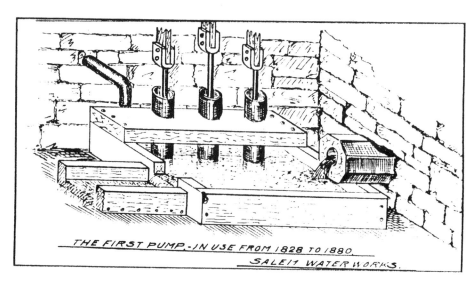

Fig. 11. (*Archives of the Moravian Church in America, Southern Province*)

Construction of Salem's second waterworks was completed in 1828. The new water source was a spring to the northeast, about 75 feet below the town. Here workmen erected an 18-foot overshot waterwheel and pumphouse equipped with a triple-piston pump. (Figs. 10-11.) The wheel was powered by water brought through a wooden trough from yet another spring two miles away. Revolutions of the wheel activated the pump, forcing water into a line of cast-iron pipe leading to a reservoir at the top of the hill. Gravity took over from there, dispersing fresh water through wooden and, then, ceramic pipes to the community. The Brethren were able to renovate and incorporate much of the first waterworks into the new system, including all of the town's standpipes, cisterns and water outlets in the buildings. Salem's second waterworks served the town until 1878, when a stock company, known as the Salem Water Supply Company, was formed. A cast-iron water system was installed at this time and most homes were equipped with running water.

In 1778, the 126 residents of Salem pooled their time and resources to build one of the first public waterworks in this country. Its construction was not the result of a discovery of new materials or mechanical improvements; towns and cities in Western Europe had been building waterworks using identical tools and methods for centuries. Salem's achievement was social rather than technological, rooted in the structure of the Moravian community. It was the disciplined yet community-caring nature of Salem which assured the financial support needed to maintain and extend the water system over its hundred-year history, and assured equal water service for everyone in the town.

NOTES

1. *Records of the Moravians in North Carolina*, translation by Dr. Adelaide L. Fries (Raleigh:North Carolina Historical Commission) Vol. III, p.1214.

2. Karen Zerbe Huetter. *The Bethlehem Waterworks*. (Bethlehem: Historic Bethlehem, Inc., 1976) pp.2, 11.

3. Elders Conference (Aeltesten Conferenz) minutes, 5 March 1771, unpublished transalation by Edmund Schwarze, on file at the Archives of the Moravian Church in America, Southern Province, Winston-Salem, N.C.

4. Records of the Moravians in North Carolina, Vol. III, p. 1180.

5. Aufseher Collegium minutes, 11 February 1778, unpublished translation by Erika Huber, on file at the Archives of the Moravian Church in America, Southern Province.

6. *Ibid.*, 18 March 1806.

7. *Ibid.*

8. *Ibid.*, 22 September 1806

9. Salem Diacony Ledger Book D 1808, on file at the Archives of the Moravian Church in America, Southern Province.

10. Aufseher Collegium minutes, 2 October 1820, unpublished translation by Erika Huber, on file at the Archives of the Moravian Church in America, Southern Province.

Building A New England Home 1831

By LAWRENCE B. ROMAINE
October 1954

Lawrence Romaine was a dealer in ephemera and the author of A Guide to American Trade Catalogs 1744 - 1900, the major reference on the subject.

I remember back in 1935 Homer Eaton Keyes, then Editor of *Antiques,* told me that there was no more tiresome reading in a world than a list of articles and prices. I wonder. It is possible that Mr. Keyes was thinking of the general run of human, and that our membership is unique in preferring history to "who done its?"

Another old calf bound ledger with delightful hand tooled edges sits upon my desk and leers at me. If I sell it, as I should in the course of business, it may be buried in an institution. And so I shall throw these notes at the door of the *Chronicle* and see what happens.

As so often happens, this veteran piece of calf contains everything but the owner's name—the one chap in whom we are most interested. The entries indicate that his house, factory and shop were in or near Enfield or Greenfield, Massachusetts. The schedules of work, neatly laid out by days and weeks for each helper, are clear and complete for 1831 and 1832. There is no question that he (and his employees) wove textiles. The schedule of work are entered in individual accounts of payments for hours and days and weeks and months of "breaking," "tentering," "satinetts," "finishing damaged satinetts," "finishing" yardage in quantity and various other fabrics. The payments for work record such names as Nelson Roberts, Elijah Gunn, Daniel Knapp, Experience Hosmer, Emaine Carter, Eunice Ames and Bathsheba Purington.

It is difficult, as usual, to figure out just what part of this man's life we are considering. There is no way of knowing how many ledgers he has already penned, how long his looms have been working and whether this house of which he writes was built after he had established his business. That the house was built in 1831-1832 is indicated by an entry of a payment to the Swift River Manufacturing Co., and not by an actual date heading the Expenses! Yet every other entry is carefully dated.

Under the date of Nov. 9th to Jan. 7th (1831-1832) there is a payment to one Thomas Jones for "36¾ days work in finishing the shop began the 4th Nov. & ended the 24 Dec. at one dollar a day & boarded.—$36.75." Shop??? A shop to house the looms? Or was this the country store which he also kept? While the looms were weaving and other Yankee hands were busy mending and creating, our friend also kept a store from which he sold "Small accounts not in the Ledger":—butter, whips, powder, satinets, sieves, shoes, shawls, hats, calico, towels, mittens, feathers, brass, tea, books, gimblets(?), clover seeds, flour, sheeting, "question books"(??), spelling books and even potatoes!

There is a long account of "wool bought for Thomas Jones" at Enfield, Mass., miles and miles of travel such as "to a horse 75 miles—4.50" and many entries of kettles of soap. As is always the case with these old accounts, there are facts that at the time they were written needed no explanation—but today, in 1953, require a good deal of thought, and often speculative imaginings. For instance, in this ledger it is quite impossible to say whether the writer is paying Thomas Jones for "making 4 kettles of soap" or whether Thomas Jones has paid *him* for the work.

The most exhaustive work on this section of New England, compiled by one of our members, Donald W. Howe of Ware, Massachusetts, does not mention any of the names in the ledger in his industrial account of Enfield. I am convinced that the accounts were kept in this area, and it is, of course, possible that our accountant operated his looms and store on an old country road not exactly in either Enfield or Greenfield. Perhaps he had no Post Office address.

However, the building of a homestead in New England in 1831 at a total cost of $893.32 is my text.

Expenses for Building House

Pine stuff for outside door		$1.75
285 ft. ½ inch Boards 400 c		1.14
Bill of lumber by J. E. Cathcart		18.17
for Shingles		4.73
for hill digger		.50
100 lbs. — 10* Nails	7½	7.50
4½" do	7½	.30
3000 Shingles	2$.-	6.00
6 lbs. Nails	7½	.45
Cash paid Sam'l		6.00
115 lbs. Nails	7½	9.62
Frank Naramore 1 man & oxen ½ day		1.00
670 feet Spruce clapboards		6.70
400 Spruce boards	$10.	4.00
for underpinning stone		7.00
20 Lime	55c	11.00
11 windows	3½ light	9.24
Pine boards		27.00
cutting & setting stone		14.12
30 lbs. Nails	7½	2.25
1 M Brads		.18
3 Fire Frames—		24.00
3 sheets sand paper		.06
Franklin Naramore laying wall — & his man — 2 days.		2.42
Franklin Naramore Oxen one day		.50

Theodore Parsons and David Carpenter one day		1.58
Daniel Hall		.42
Franklin Nar'e Oxen ¼ day.		.13
3728 shingles	$2	7.46
1500 shingles	$2	3.00
37 lbs. nails- 7½		2.75
22 lbs. nails 7½		1.65
288 feet lath 40c		1.15
1 day work by Captain Rice-		.83
1½ lbs. nails	7½	.12
40 lbs. nails	7½	3.00
2 sheets sand paper		.04
for lumber, stone, brick & labor-		100.59
6 sheets sand paper & 75 Lime		45.12
126 feet Lath 40c per—		.50
17½ lbs. nails	7½	1.33
½ lb. chalk— 6c		.03
8 lbs. nails	7½	.60
1 Door latch, Butts screws, etc. etc.-		.69
Latch, Butts, 10 doz. screws-		.91
8 lbs nails & 6 sheets sand Paper-		.72
2 doz. screws-		.12
for lumber-		1.60
4 lbs. nails 7½		.30
4 pts. oil-	7/ -	.59
2 qts. oil		.59
1 gal. & 2 qts. oil	7/ -	1.75
1 file 44c		.44
1 paper of brads		.25
½ dozen door handles	.25	1.50
14 lbs. of lead 1c		1.54
½ lbs. Umber — 16c & ½ do Lead-	9c	.14
9 rolls paper at 44c		3.60
1 roll paper—		.90
4418 feet Lath- 40c		17.67

Various bills for labor: —viz: — Mr. Blood —	$29.50	
Mr. Packard—	12.50	
Eben. White—	28.12	
Em. Cathcart—	3.50	
Isaac Wing—	21.33	
Len. Packard—	41.97	
Various others — transportation, hauling, etc.-		21.81
Total January, 1832.		$893.32

[Ed. Note: We have re-added these figures and find that they total $527.97. We don't know whether Mr. Romaine or the keeper of the Journal made mistakes in extension or addition, or whether there were some items omitted.]

As I check back on the above, trying to make words say what isn't there in ink, I wonder why he says "stuff" in the first entry. [Ed. note: "stuff" referred to the wood used.] I wonder how long the 285 ft. of ½ inch boards in the second entry was seasoned and who seasoned them for $1.14. I ask you, one and all, what was a "hill digger?" Has anyone ever seen a 3½ light window? I look at the fire frame in our front room, now embellishing the original opening ca. 1740, and wonder if it cost the James family, who lived here about 1830, $8.00 — or is it fancier than the one in the ledger, and was it $11.38? I look at our poor old barn roof and wish that I might take a horse and run down to the village and bring home 3000 shingles for $6.00. But also, being practical, I look at the front door latch and realize that with the butterfly strap hinges I wouldn't think of selling them for 60c — nor the dozens of smaller latches in the shop for 25c each! There are two sides to every equation. I wonder what that wall paper for 44c a roll might have looked like in 1831.

If Mr. Keyes was right and you yawn (as I have oft times warned you), drop me a line.

Old Time Fences

September 1949

(Condensed from an article in the *Rural New-Yorker*, September 4, 1948. Contributed by Mrs. Gillian W.B. Bailey).

One of the earliest types of fences was the brush fence. The pioneers would select the site for their barn in a thicket of small or medium sized trees and not too distant from the log house or cabin. The barnyard was enclosed in the brush fence which was made by starting at one corner and cutting the first tree so that it would fall parallel to the line of the desired fence. Then back from the butt of that tree the second tree would be felled in such a manner that it would overlap on the first tree. This process was continued until one side of the desired enclosure was completed. The other sides were constructed in the same manner. All the trees were cut four or five feet from the ground and the stumps left standing to anchor and strengthen the fence. It is not hard to visualize that, with those trees overlapping each other, an impenetrable mass was built up several feet wide and many feet high.

A second type was the stump fence. The logs were either used or burned, and the stumps themselves incorporated into a fence. The stumps were pulled by ox teams and hauled to the borders where they were to be built into weird yet picturesque stump fences. The roots were cut off one side of the stumps, which sides were then laid on the ground in a straight row, the tops of the stumps all facing in the same direction, while the roots on the other side of the stumps were left to extend high into the air. The roots which had been cut off were then used for chinking to fill any gaps where necessary.

The next fence to come into general use was the rail fence, which was usually made of chestnut. The rails split by Abraham Lincoln for fences were black walnut, but they wouldn't be used today. An expert at making this type of fence used a very simple device for laying it out:

With a stake set at the farther end of the line, he would keep his eye constantly on the stake as he walked toward it on the center line of the zigzag pattern. After so many paces a helper would hand him a good sized stone or a block of wood. With right arm extended full length he would drop the stone or block. Where it struck the ground the helper would place it in proper position. Then the operator would move forward again the correct number of paces and this time the object would be dropped with the left arm extended. This process was continued the length of the line, first with the right arm and then with the left and when he had finished the 12-foot rails would go on the stone or block foundation perfectly without a miss.

The buck fence was also made with rails which were wired or otherwise fastened to posts or stakes. In more elaborate types the posts were mortised and the ends of the rails were fitted into the mortises.

Stone fences or walls were used extensively in New England and parts of New York when land was being cleared, and many have been repaired or restored making attractive and practical enclosures.*

Slab fences utilized the end products of the saw mills. Extra heavy posts were set and the slabs were pinned to the posts with wooden pins. Cut, wrought iron spikes were later used for this purpose. As saw mills and hauling methods became more efficient, board fences became the vogue. These were made of six-inch boards nailed a few inches apart to uprights, also of six-inch boards, cut to the desired length.

These fences were usually six feet high, and the panels twelve feet long. Holes were bored in the posts at the proper height into which rather heavy right angle iron hooks were driven. On these hooks the board panels were hung.

The panels overlapped on the posts a few inches so that opposite ends of two panels hung on each post and by a single hook. After the panels were placed, the hooks were driven in a bit farther until they held the panel tightly.

This fence was probably the easiest to move from place to place around the farm. The holes for the hooks were usually bored all the way through the posts. When necessary to move the fence, a hammer and a punch would be used to drive the hooks back a fraction of an inch, after which the panels were easily lifted from the hooks and moved to the desired location.

*In Leicester, Mass., at Stone Wall Farm there is one stone wall 20 feet across. During the panic of 1873 Mrs. Edward Flint, then owner of the farm, hired the unemployed of the town to clear her fields of stone and build stone walls. This extra wide one was made with two parallel walls, then stones were tossed in between and the whole leveled off. When Col. Samuel Winslow lived there he won a wager that two teams of horses could be driven on top of the wall and pass each other comfortably. *Ed.*

Hay Burners

by JOSEPHINE H. PEIRCE
September 1949

Today cavalry men call their horses hay-burners, but when pioneers from eastern states settled on the treeless prairies, a hay-burner was the source of heat.

At first these settlers lived in sod houses and met the fuel problem by burning buffalo and cow chips, called by the users "Prairie Coal."

However these soon became scarce in some sections; there was no wood for any purpose; homesteaders had no money to spend for coal; so any substitute was used.

Woody weeds such as sunflowers were advocated and sunflowers seeds were advertised in 1871, with the claim that one acre would produce twelve cords and furnish enough fuel for the winter.

However, the fuel most universally used was the one most easily found. Heavy slough grass was plentiful and contained heat units equal to that of light wood. The hay was twisted in stovewood lengths known as "cats." Children were early taught to do their stint, and large piles of "cats" were nearby the house, with so many piled inside there was little room for inmates or visitors.

An ingenious device for twisting the hay was invented by James S. Foster of Yankton, Dakota Territory, in 1876 which consisted of two upright pieces of wood carrying a crank and roller. Hooks caught the long prairie hay and wound it until it was a solid stick of the proper size. The sticks were then cut into proper lengths like stove wood.

Several stoves for burning hay were invented and placed on the market. One was known as a drum, but the one in most general use was the "Wash Boiler Type." It was constructed of sheet iron or steel and shaped like a wash boiler, the only difference that the hay-burner was twice the depth of a regular wash boiler. Into this the hay was firmly packed, then the front lids and cross bar were removed from the ordinary fourholed cook stove and the hay burner inverted over it. Lighted, the draft and damper were regulated to where the hay would be consumed slowly or rapidly as the need demanded, and sufficient heat would be developed to cook on the back of the stove; or light baking could be done in the oven. The burner, well filled and tightly packed, would hold the fire for two hours or more.

A unique type of stove used magazines for the hay. The stove was built like an early two-hole cook stove with an elevated oven at the rear, the fire box in front. Two cylinders about 25 inches long and 6 inches in diameter were filled with weeds, hay or straw and fitted into place under the oven, where two openings went directly into the firebox. The hay was set on fire in the box and as it burned, a spring at the other end of the cylinder pressed the hay into the firebox as fast as it was consumed. Several pairs

Two-cylinder Hay-Burner. M.L. Wood's Patent, Jan. 16, 1877. The stove was purchased in 1878 by E.H. Bush from J.S. Harmon, pioneer hardware merchant at Tecumseh, Nebraska. It is now owned by the Nebraska State Historical Society, and is on display in its museum.

of these magazines were kept filled and on hand. Then when one pair burned out, a ratchet was turned winding up the springs, and another pair were inserted.

While hay made a hot fire, it was unsatisfactory as it required such constant attention and there was great fire hazard. Changing drums or magazines or opening the stove to insert new "cats" gave the embers an opportunity to get out and start a blaze.

Corn was used after its production became established. Cobs were used in the kitchen and stalks were used in the fireplace or heater. The stalks were often pressed, with a hand-power device, into compact bundles that were bound with patent steel bands. Advertising stated that a bundle would burn from 20 to 40 minutes. The machine cost twenty dollars. The cobs made a hot fire, and when corn was cheap it was burned on the cobs in large quantities. The price at this time was about ten cents a bushel.

Another type of hay burner was called the "Prairie Pioneer's Pet." It was the type built by the Mennonites who came from around the Black Sea and settled in Manitoba, Minnesota, Kansas and Dakota 1874-75.

It was a brick stove (or stone or clay) varying in size with the size of the house. However the average was five feet long and two and one-half feet wide. It took about 600 bricks to make one and the general idea was that it have six "stories." 1) the ash box; 2) the fire box; 3) the oven; 4) a smoke passage; 5) hot air chamber; 6) smoke passage to chimney or drum in an upper room.

It was built in the middle of the house, so that the oven door opened into the kitchen; the hot air chamber into the living room, or in the event there were two rooms to heat, there would be two doors, perhaps one into a bedroom.

For this type of hay-burner the hay did not have to be prepared, but was thrust into the fuel chamber with a fork much as one would thrust fodder into a rack. For about 20 minutes, the hay was forked in and pressed down, then the drafts were arranged, and it would not need fuel again for eight to twelve hours, depending upon the weather.

AUTHORITIES

Everett Dick, PhD., *The Sod House Frontier, 1854-1890* (Lincoln, Nebraska, 1937)

Nebraska Farmer, Vol. 2, No. 6

U.S. Patent Lists, 1876-1882.

From Bushkill to Easton on a Raft in 1896

By FRANK LeBAR
June 1963

The bulk of this paper consists of my memories of a trip down the Delaware River on a raft made when I was a boy twelve years old. The footnotes at the end, which attempt to place this story within the larger perspective of early rafting days in Pennsylvania, have been added by my son, Frank M. LeBar, Ph.D., Research Associate in Anthropology at Yale University.

I was fortunate in being born in the historical village of Shawnee-on-Delaware, Monroe County, Pennsylvania; Shawnee being located in the beautiful Delaware Valley with the Delaware River flowing peacefully by Great Shawnee Island, Depue Island, Tocks Island, these islands forming benikills together with the flat bottom land known for its fertility. My grandfather, Judge J. Depue LeBar, moved to Shawnee from Pahaquarry Township, New Jersey, about 1850. He opened a general store in Shawnee in 1854, this store remaining in my family until the death of my father following the great flood of 1903. My father, Hiram LeBar, farmed about 200 acres on Great Shawnee Island and adjacent land in Pennsylvania, this farm comprising the present golf course and grounds of Shawnee Inn and Country Club.

My grandfather began at age sixteen (i.e., about 1830) to run on rafts to Philadelphia and in his later years was considered one of the most expert steersmen on the river (Matthews, p. 1084).[1] He was a member of the crew on the pioneer trip of the steamboat *Kittatinny* running from the Delaware Water Gap to Port Jervis in the spring of 1879. When he was eighteen or twenty years old (i.e., about 1867) my father accompanied my grandfather on rafts to Easton, learning to steer and learning the channel of the river. After a successful trip they would walk home. Father always said they took short cuts whenever possible.

When I was a boy in the early 1890's, my father was carrying on an extensive lumbering business in addition to farming and storekeeping, running rafts annually to Portland and Easton.[2] He would buy ties during the winter months, they were put in piles and rafted in the spring. During the summer Father would buy a lumber tract and then lumber it off during the winter. He lumbered in New Jersey, one mile above Shawnee, where one hundred and sixty acres of land went with an old stonehouse on property which belonged to my grandfather, then to my father. One hundred acres was timber. We had plenty of snow in those days and men would cut the logs, snake (i.e., drag) the logs with a team of horses to a sluice, then slide them down the mountain to near the river where they were put in piles ready to raft down to Easton in the spring of the year. Father always had four lumbermen hired during the winter, and (in addition) the two men who boarded with us and worked by the month, drove team and helped pile the logs. These lumbermen always had their breakfast and supper at our house, their dinner was packed in baskets or pails by the work girls. Father also lumbered near and around Bushkill, piling the logs on the river bank near Peters' boat landing.

In the spring of 1896 Father told my brother and me he was going to give us a ride on a log raft from Bushkill to Easton. I was twelve and my brother, John, fourteen. Were we excited. He didn't tell us the exact day but one evening after finishing supper he said: "You boys go upstairs and get to bed. I will call you in the morning." It seemed to me we had only slept a short time when Father opened the door, told us what clothes to put on (the warmest we had), and when we came down the men were eating. My mother was there and we had to eat a good solid breakfast. We left the house at six o'clock in a three-seated wagon for Bushkill. I remember it was a clear day but chilly. Well, we had a large wooden shoe box from the store. They nailed that fast to the logs in the center of the raft, then two or three blankets and a large box with a lunch that Mother fixed for us. At seven o'clock the men untied the ropes, which were fastened to trees, rolled them up, put them on the raft, and my Dad called "Ready to leave." It was a four oar raft, two oars front and two rear. There was one man to each oar, the steersman (my father) standing at the rear of the raft, Pennsylvania side; this was the position all steersmen occupied. Father called "hold" (i.e., rest oars) and we started to float with the current. We were fortunate as we had a high river and this meant we would make wonderfully good time. We sighted two rafts quite a way up the river, floating down. We didn't see any ahead of us. We were gone but a short time when I noticed my father getting very active. As our raft was about 120 feet long with a width of some 25 feet, he had a responsibility on his hands. We were entering the narrow channel below Bushkill with very swift water and some dangerous rocks in the channel. As a heavy log raft took at least two feet of water, it was necessary to avoid any rock that didn't have three feet of water pouring over it.

In order for the front oarsmen to obey his orders Father used his arms as well as his voice; for the Pennsylvania side he held his right arm out and for Jersey his left arm. To signal for "hold" he held his arms straight up.

The raft had gained speed and we were shooting along at a rapid pace, the rocks towering up on both sides of us, and large boulders sticking out of the water, some with water pouring over them. I marvel as I write this how they managed to get those tremendous big rafts through those rifts without an accident — but in no time we were opposite Wallpack Bend and entering Sambo and Mary Rift. We shot through that and then Father called to my brother and me to "watch for Van-Camps' nose." Sure enough, on the right side of the raft we passed a large rock sticking out of the water in the shape of a man's nose. Before we knew it we were passing Heller-Dimmick ferry and as we neared Tock's Island we looked back up the river and counted seven rafts.

We went through Walter's Eddy and LeBar Eddy and in a short time were down to Walker's Ferry at Shawnee. As we were entering to go down back of Muskrat Island I looked across and could see our house and barn. Many a day we youngsters stood up by "the post," as we called it, right near our home, where we had a wonderful view of the Delaware River, and we would watch and count the rafts going down the river — right where our raft was at that time. I wish I could describe in words the feeling it gave me — one of the highlights of the trip.

We soon sighted the N.Y.S.& W. (Railroad) bridge and I noticed my father was getting very active again. The piers under the bridge were quite close together for a large raft; also the rift right below the bridge caused a current and there was a strong wind blowing. Father had to take all these things into consideration. I noticed the oarsmen were watching him very closely. Well, we went sailing through under that bridge and then the rift, which as I later learned was a dangerous one as the current tended to throw the raft to Jersey and the oarsmen had to really pull to keep the raft from hitting the rocks along the shore. We were really going and to this day I get a thrill thinking about it. Then through the Delaware Water Gap on a raft. My, those mountains seemed high to me. Well, there wasn't a dull moment.

We were entering the rift just below the Gap and in no time sighted the old Portland covered bridge — again a narrow place to pilot a raft between the stone piers. A swift flowing rift just below Portland, and then in one-half hour we were nearing Foul Rift — about one mile below Belvidere — the most treacherous rift of any in the Delaware, with the greatest drop. More rafts were "stove up" (i.e., broken up) in Foul Rift than any place on the river. Years ago the government spent considerable money to make the river navigable for boats and rafts, and as you pass through Foul Rift you can still see dynamite holes in the rocks along along the shore.

As we sighted the rift, Father told us to stay on top of the store box and keep our feet up on the box. Well, even though I was twelve years old, I noticed the men at the oars were watching my Father very closely. He was looking straight ahead, giving orders with his hands and arms. The roar of the water was such that the men couldn't hear his voice. I can see that rift just as it looked that day. The river being very high, it seemed as though we just shot along. The roar of the water was indescribable. Half way through, where there is a sudden drop in the river, the front of the raft dove under causing about eighteen inches of water to come over the logs. I knew then why Father nailed that store box to the logs and why he told us to keep our feet on top of the box.

We were through the rift in no time and all was peaceful and calm again. In about one-half hour we entered Saundt's Eddy, an eddy caused by very deep water and no drop or fall in the river for a distance of five miles. For the first time the raft didn't make over three miles an hour. Half way through the eddy Father called "Pennsylvania," and the men pulled the raft to shore. They threw out the ropes, tied up the raft and Father said: "Now you boys get out on shore, run around and get some exercise. The men have one hour rest and we will go up and get our dinner." It was a farmhouse and they served meals to the raftsmen. It was one o'clock, meaning that the men had stood at those oars for six hours and most of the time busy. In a short time more rafts were landing back of us and father seemed to know all the steersmen. What a time they had visiting and telling about things that had happened on the trip. Some of the rafts had been two and three days on the way, coming from way up in New York State on the East and West branches of the Delaware.[3] My brother and I got lots of attention as we were the only boys on any of the rafts.

We left at two o'clock and Father said we would reach Easton at five o'clock and get the six o'clock Pennsylvania train home. When we finally pulled out in the current again I don't know how many rafts were tied up for dinner and more coming. Some of the rafts didn't stop so we now had rafts ahead of us and in the rear also. When we got to Easton we could see both Jersey and Pennsylvania shores were lined with rafts. We pulled in at the first opening, and by the time Father finished his arrangements for selling the raft it was train time. My uncle met us at the Delaware Water Gap station with the three seated wagon but I never woke up until we reached home.

Rafts in my father's day were constructed as follows: They used white, also black and white, river birch poles about 3 inches in diameter and about 20 feet long to hold the logs together. A horseshoe-shaped iron was clamped over the sapling or pole and driven into each log. The oars were immense in size, and hung very similar to a boat oar. The oars had to be perfectly balanced, thus making them easy to handle. When a raft was sold and dismantled at the Easton lumber yards, they put the irons in nail kegs with the owner's name on the keg. Father would send a team of horses to Easton and have a heavy load of irons to bring home, ready to use again.

Notes

1. The story of rafting on the Delaware would appear to be somewhat less fully recorded than that on the Susquehanna and Allegheny river areas. For the former, reliable information is found in C.T. Curtis, *Rafting on the Delaware* (W. Heidt Jr, ed. Revised edition, DeWitt Historical Society, Ithaca, N. Y., 1956). This booklet is especially good on the details of rafting logs, sawn timber, stone and tanbark in the latter 19th century on the upper Delaware around Callicoon aod Narrowsburg. Two papers in the proceedings of the Bucks County Historical Society by T.S. Kenderdine (Vol.4, 239-252, 1917) and J. Pine (Vol. 6, 467-524, 1932) furnish additional valuable information. A. Mathews, *History of Wayne Pike and Monroe Counties,* Philadelphia 1886, contains some secondary information. A series of papers, pamphlets and books by J. D. Tonkin, J. C. French, M. J. Colcord, J. H. Chatham and others, record the story for the central and western parts of the state (cf. J. D. Tonkin, *The Last Raft,* Harrisburg, 1940; J. H. Walker, ed., *Rafting Days in Pennsylvania,* Altoona, Times-Tribune Co., 1922). Henry C. Mercer, in his *Ancient Carpenters' Tools* (3rd ed., Doylestown, 1960), drew heavily on the personal communications of French and other informants of Potter County, so that much of his very excellent treatment of rafting tools and techniques pertains to the Susquehanna rather than the Delaware.

Rafting the Delaware traditionally commences with one Daniel Skinner who is supposed to have run the first raft, from Cochecton to Philadelphia in 1764 (although the date appears also as 1746 and 1796). By 1828 Hazard's Register (Vol. 3, p. 384) was recording as many as 1,000 rafts in the spring floods of that year, estimating 50 million feet of mostly sawed timber. Rafting reached its height on the Delaware, by all accounts, in the 1840's and by 1855-1860 had begun to decline. However, up to May 31, 1875, some 3,140 rafts were counted for that year at the Lackawaxen Dam across the Delaware (Curtis, p.32). Rafts continued to be run as late as 1907. Rafting on the Susquehanna and Allegheny rivers appears to have started later than that on the Delaware, and to have reached its height somewhat later, i.e., around 1850 to 1860. The largest raft ever run on the Delaware, according to Curtis (p. 30), was 85½ feet wide and 215 feet long, containing 120,000 feet of lumber.

Most of the timber rafted on the Delaware came from the upper reaches of the river, in Delaware and Sullivan counties in New York and Wayne County and the Lackawaxen River area of Pennsylvania. White pine, bouyant and readily marketable as ships' masts and spars, was the chief export timber at first. The magnificent stands of pine stretching all across northern Pennsylvania and remarked by early naturalists and explorers of the area, furnished the initial impetus to rafting on the headwaters of the Delaware, Susquehanna and Allegheny. By the 1870's, however, hemlock had supplanted pine; fully ninety percent of the lumber shipped to market in log from Wayne County at this period was hemlock (Matthews, p. 327).

J. C. French (in Walker, p. 84) estimates that 50 billion feet of white pine, and an equal amount of hemlock, were lumbered from Pennsylvania forests during two centuries of intensive exploitation.

2. The importance of lumbering and rafting to the economy of the nineteenth century farm should not be overlooked. Nearly every farmer did some lumbering and many ran at least one raft during the season. Speaking of the situation as of about 1850, T. S. Kenderdine (p.241) says: "Farming . . . was only a side line; the hay and grass they raised to partly feed their working cattle; for getting out logs and running sawmills was their profitable business. Their crops in, the farmer, his boys, and his hired men, hied them off to the woods . . ." It was the timber trade that brought prosperity to many an early farm. Curtis (pp.16-17) states that by 1835 there were 208 sawmills in Sullivan County (N. Y.) alone.

3. Curtis (p. 30) notes that under good conditions the trip from Callicoon (Sullivan Co., N. Y.) to Dingmans — a distance of 70 miles — could be made in a day's run of 10 hours. From Callicoon to Easton required 2½ days; and from Callicoon to Trenton 3½ days.

The use of horseshoe-shaped irons for fastening lash poles to logs, as herein described, is a late variation of the earlier, and more prevalent, bow and pin method. As pointed out by Curtis (p. 20), the irons required no boring and therefore involved less labor. The bow and pin method is ably described by T. S. Kenderdine (pp. 243-44) from personal experience of Delaware River rafts of the period 1840-50:

"Logs were sledded to some point suitable for raft building, a level stretch at an eddy, when, before a rise of the river, they could be lashed together in shape for floating to market. The appliances for this were slender saplings cut into what were called lash-poles . . . They were laid across the logs when arranged for floating, and fastened to the logs with wooden staples called bows, made of ash. . . Lash-poles were from 2 to 4 inches in diameter . . . and generally made from water-birch or iron-wood, or whatever wood best held its size. The bows were of ash, split out 16 inches long and 1¼ inches wide, and half that thickness, and then steamed and bent. The logs were in units of length generally 16 and 18 feet, the joints being 'broken' by lengths of a half more or double, that the float might be stiffened for going through rough water at the falls or rapids. The lash-poles being regularly spaced, holes were bored on each side of the top of the log with an auger with an iron shank over 5 feet long with a crank near the top like that on a brace-stock. With this the holes were quickly bored, the bows inserted and fastened with wedges, and soon a raft from 100 to 200 feet long and from 16 to 36 feet wide was ready for the oars. These were monsters . . . The shafts were 30 feet long, tapering up from a hand-span to 8 inches across, with blades 15 inches wide and as many feet long, and were hung on stout headblocks with projecting pins, and nicely balanced."

The construction of a "bow and pin" log raft is shown in detail in the accompanying sketch, from J. D. Tonkin (facing p. 64). As Tonkin points out, the ash and oak bow and pin system, when wet, expanded forming a rigid platform. The system of "breaking" joints to enable a raft to navigate rifts and rough water without injury is well-illustrated in Tonkin's sketch. The long raft auger, as mentioned by Kenderdine, is illustrated in H. C. Mercer (p. 45) together with the comment that a raftsman balanced on a single log found it easier to stand rather than stoop at his work, hence the extraordinary length of the shaft.

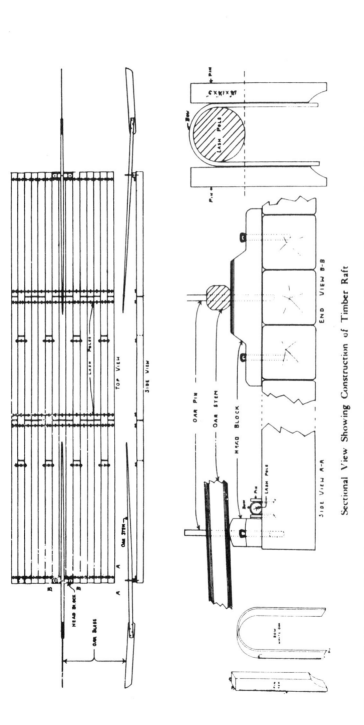

Sectional View Showing Construction of Timber Raft

Shown is a "half-raft" or "piece" consisting of three "platforms" coupled end to end. Larger rafts were made up by joining two or more "pieces." The top view shows lash poles pinned in place at either end of each platform. A cross sectional view shows the method of pinning, by inserting a white oak bow in holes drilled in a log on either side of the lash pole and drawing the bow down tight with square ash pins driven alongside the ends into the holes. Stiffening of the overall structure was obtained by the use of three or four "hinge sticks" which, as shown, extended some distance into the adjoining platform; these were likewise secured with bow and pins using extra short lash poles. The side view shows the heavy oars nicely balanced and secured to head blocks with stout oar pins or thole pins, usually of white oak. Shown in the end view (B-B) is the method of securing head block to logs by pinning with the center tholepin and two wooden side pins. The oar stem was slotted at its larger end to a depth of 2 to 3 feet; the blade was inserted into this slot and pinned in place as shown.

—204—

The Maine Woodsmen

By JOSEPH COBURN SMITH
December 1972, June 1973, and September 1973

This is a portion of a Master of Arts thesis in the field of labor relations the author wrote at Harvard in 1926. It is based on his experience in the woods and on research. He shows the "state of the art" as it existed just as the logging industry was being mechanized.

The author's speculations about tools are quite astute. His comments on mechanical methods of cutting trees state that one had to have a very flexible tool, as machines do not think. He differentiates between a tool and a machine. The tool he was looking for was, of course, the chain saw, which he mentions, but only in passing.

In the area of transportation, the ox has been replaced by the horse and the horse is in the process of being replaced by the tractor. The caterpillar tread has not been completely accepted. He mentions the drag line, but the balloon and helicopter, of course, were still 20 years over the horizon. CHRONICLE EDITOR

The Tools of the Maine Woodsmen

Most callings have their symbols: the blacksmith and his anvil, the artist and his brush, the priest and his crucifix. Just so, we invariably associate the woodsman and his axe. Ever since the caveman made the momentous discovery that by binding his sharp flint onto the end of a stick he could make a deeper dent in a piece of wood, the axe has been the traditional instrument for cutting down trees.

Archaeologists tell us that the axe was discovered independently by nearly every primitive race, but evolved slightly differently in each case. Thus, the blade of the ancient axe used in Northern Europe was broader than that of Southern Europe because of the difference between the softwood trees of the North and the hardwood trees of the South.

In the same way, America has evolved a distinctive model of the axe. Strictly, of course, the first American axe was the tomahawk, but usually this term is used to distinguish the axe with which we are familiar as compared with the axes found in other lands. One of the features of the American axe is the peculiar crook to the helve, or handle. Recent Egyptian excavations have disclosed ancient adzes with handles of the same pattern, but the shape of the typical American axe helve was probably evolved simply by finding the model which gave the best balance and "feel." In the early lumbering days, each man whittled out his own helves to suit himself and from these countless experiments there has come to be a standard design. Another characteristic is the curved edge, which enables the force of the impact to be concentrated on a small point and so results in a deeper cut with each blow. These and other refinements characterize the tool which has been, and still is, the mainstay of the woodsman, the farmer, and the pioneer.

The use of the axe in the specialized field of logging, however, called forth a radical change in design. This is the two-edged or double-bitted axe. It is not a modern invention. We find that in ancient Rome "the double axe was a usual tool and also in a sacred form. Ceremonial copies which could not be hafted were used in various northern centers, apparently as standard weights."[1] The history of the woodsman's double-bitted axe is rather vague but, so far as can be discovered, the first one was made by William Mann, a Pennsylvania lumberman, somewhere around 1850. In that state, the tool came into use rapidly, but its spread into other regions was slow. One old woodsman writes that his uncle tried to introduce this type of axe into Michigan in 1865, but it was looked upon with suspicion and did not find favor until a pattern called the "Michigan" was put on the market. The original pattern was called the "William Mann" for many years and now goes under the name of "Western." In the Maine woods the double-bitted axe was not generally used until about 1900 and even now is not by any means in universal use.

The chief advantage of this type of axe is that one edge can be kept sharp and used only for straight chopping, while the other can be used for lopping off limbs, cutting through knots, and any other rough work which dulls the blade quickly. This axe is apt to be longer and heavier than the old type. The handle is straight, since the curved kind would be correct for only one of the two blades.

Although saws of different types were in common use when the logging industry of Maine was developing, they could not be used to advantage in the felling of trees. To be sure, there were cross-cut saws, that is, blades several feet in length and with wooden bars on either end as handles, but these had the usual "V"-shape teeth and were very slow cutters. A woodsman could chop a tree down in much less time than he could saw it. The axe was even used to cut the trunk into log lengths.

In 1878, Jerome C. Dietrich, a Canadian saw manufacturer, patented the "lance tooth" type of saw, which was a tremendous improvement. These teeth were in pairs, quite long and beveled on opposite sides, thus making two cuts or scores an eighth of an inch or so apart. Following each pair was a "raker" or "cleaner" tooth. This was not sharpened to a point, but had an edge like a miniature plane and was slightly shorter than the points of the cutting teeth. Its function was to plane off the tiny strip between the two scores made by the cutting teeth and also to rake out the sawdust or little shavings thus produced. Innumerable designs have since been tried and are now on the market, but all are of this same principle. A cross-cut saw with these teeth can bite through a tree quite quickly, and its advantages were soon seen by the lumbering companies.

In the lumber sections of the South and West, the use of the saw for felling trees had been established for many years before it was adopted in Maine. About 1900, the Simonds Saw Company realized that this area was undeveloped as a market for their product and sent a demonstrator into Maine and Newfoundland to show the effectiveness of the cross-cut saw as compared to the axe. One of the important duties of the salesman was to instruct one or two men in each camp how to sharpen the saw.

The new tool, however, was not received with any enthusiasm by the Maine woodsmen. The inexperienced saw filers could not keep the saws in proper condition, and the typical skilled axeman disliked to learn the use of a new tool. The timber owners and operators, on the other hand, were quick to see that by sawing a tree down there would be more usable wood in each log because it was possible to cut closer to the ground. Moreover, a log sawed off squarely at the base saves the twelve inches more or less which is wasted in chopping a notch deep enough to fell the tree (there is additional wastage in cutting up the trunk into log lengths with an axe). Hence, permits for lumbering and similar contracts began to specify that the trees should be cut by the saw. The woodsmen were thus forced to adopt the saw. As they began to learn the knack and as the saw filers learned how to keep the tools sharp, they found that this was the easiest and quickest way to cut a tree.

The method now in universal use, therefore, is as follows: After clearing away some of the surrounding brush, the man chops a notch two or three inches deep into the trunk on the side towards which the tree is to fall. The more skillful the woodsman, the more accurately he can aim the tree in this way. The traditional stunt is to drive a stake into the ground by felling a tree exactly on top of it. This ability is often necessary to prevent the tree from lodging in the branches of another, or to keep young saplings from getting needlessly broken down. This notching takes only a few blows of the axe. Then two men take the saw and start cutting from the other side of the tree trunk. As the saw cut gets about three-quarters of the way through the diameter, the tree begins to sway and crack. The men withdraw the saw and stand aside or perhaps push on the trunk until it slowly topples over and comes to the ground with a resounding crash, amid a spray of snow and flying twigs. Then the axes come into use again for trimming off the branches, and the saw once more for cutting off the top and cutting the log into standard lengths.

The problem of felling the trees has often stimulated the minds of inventors. One device, said to have been tried in Europe, cut the tree by means of a platinum wire, heated to a white heat by an electric current, which would burn its way through the trunk. It is doubtful whether this would eat through the wood as quickly as a saw, but even if it could, its use would seem to be limited to forests which happened to be conveniently close to power houses or transmission lines.

Power-driven saws, such as the air-driven drag saws called "steam dagos," have been used successfully in parts of the country where the forest is level and open for cutting the trees into logs. There is an "endless chain saw," in which the links are similar to saw teeth and press continually around a steel form. This is also used for "bucking up" logs which are so large as to take considerable time by hand. Neither of these devices, however, has been adapted to the problem of felling the trees.

There have been tree-cutting machines patented, but they do not seem to have met the test of actual practice. A typical device of this kind consists of a sort of push cart on which is mounted a gasoline engine which drives a horizontal circular saw. The idea is to shove the apparatus against the trunk and thus saw it through, then trundle it over to the next tree and repeat the process. Providing that the mechanical problems of the machine could be perfected, it is a question whether such an apparatus would be economical in the Maine woods where the floor of the forest is covered with hummocks and boulders, dense thickets, and two to four feet of soft snow.

This brings us to the chief obstacle to the mechanical revolution of the logging industry. Machines can be made to do anything that man can do—except use judgment. The first problem in inventing a machine is to standardize the raw material. An example is the pea canning industry. It was easy to make a device for pulling pods off the vines, but it was impossible to make a machine that would choose only the full pods. Only when Burbank evolved a pea plant which would ripen almost all the pods at the same time could they be harvested by machinery. Just so in the woods. It is a comparatively simple problem to work out a machine which would cut through a trunk in a fraction of the time now necessary. The problem which has baffled all inventors is how to make such a machine so flexible as to be adaptable to all the varied conditions of the woods. A man can work in cramped quarters, can adjust himself to the kind and size of the tree, can move about easily up and down hill, through the underbrush, over the snow, and can adapt himself to all the different factors according to his judgment. Until these factors are standardized, there seems to be not even a remote chance of any mechanical logging.

This, however, does not mean that there is no chance of any development in the operation of felling trees. There is always the possibility, if not the probability, of improvement in tools. A tool is quite distinct from a machine: a machine is automatic; a tool is entirely dependent upon the hand of the operator. We have seen how the usefulness of the axe was increased by the double-bit and then how a change in the design of the cross-cut saw made that tool a still better way to cut down a tree. What the next change will be is impossible to predict, but it is safe to say that the tools of the woodsman will become further and further perfected.

In another branch of the logging industry, much more use has been made of modern mechanical science. This is in the field of transportation of the logs. When civilization deals out the area of land to be put to different uses, the forests get only what is left after the cities and farms have taken their pick. The woods are naturally the least accessible parts of the country. Yet the products of the forests must be taken to the city. That is the problem.

In Maine the logical method was to use the rivers. The state abounds in waterways, but the four river systems which are most important to the logging industry are: the St. John River, which drains the northern part of Maine and flows into Canada; the Penobscot River, which enters the sea at Bangor; the Kennebec River, which drains the Moosehead and Dead River sections; and the Androscoggin River, which is near the New Hampshire border and drains the Rangeley Lakes. Ever since the early Colonial days these rivers have been floating millions of logs from the backwoods to civilization every spring. It is also true that civilization, in the shape of pulp and paper mills, has been moving towards the backwoods as, for example, the tremendous establishments at Millinocket, Madison and Rumford.

But the waterway offers only half the solution. One cannot, as a rule, chop down a tree and have it fall into the river which is to carry it to the mill. Furthermore, to move the logs from the interior of the forest to the water by manpower would be a problem comparable to the building of the pyramids. Hence, there was a need for some kind of power which had great strength, although not necessarily great speed. They did not need to search farther than the Maine farms to find what they wanted—the ox. This patient beast was used in teams of four, six, or even eight.

Even their strength, however, would not be enough to drag the logs unaided. The winter snow was utilized to eliminate much of the friction. Also, they used what was called a "go-devil" or "bob-sled," possibly because of its bobbing motion as it is drawn over the humps and hollows. This was very simple, merely a pair of wooden runners with a heavy crossbar or "bunk." Usually these sleds (as well as the yokes for the oxen) were made on the spot with only those versatile tools of the pioneer—the axe and the auger. The only part brought into the woods was the iron pin which attached the tongue to the sled.

To load the logs onto the bob-sled was also somewhat of a task because in those days only the great trunks of the white pine, from two to six feet in diameter, were deemed worth handling. To accomplish this, a small log, or "skid," was placed with one end on the sled and the other at the side of the big log, thus forming an incline. Next, a chain was fastened and wound around the big log and the other end attached to a yoke of oxen. When all was ready they pulled on the chain which, as it unwound, rolled the end of the log up the skid onto the crossbar of the bob-sled. The other end remained on the ground and had the bark peeled off on the underside so it would drag easier. The number of logs to the load varied according to their size. After the ends were securely chained to the sled, the oxen were hitched up and after much striving and creaking, the load would slowly get under way. One enthusiastic woodsman of those days describes the pleasurable excitement of these operations as similar to those experienced at a ship's launching.

As time went on and the lumbering operations were carried on farther and farther from the banks of the streams, the problem of hauling became more serious. Therefore, what is known as the "wagon-sled" was invented by Jonas Colby, a lumberman of Moose river. This might be described as two bob-sleds hitched together so that both ends of the logs are carried, instead of one end dragging. The gauges of the runners vary from 4½ to 14 feet, according to the region. The broader sleds will hold more logs and travel easier, but it is much more difficult to build a road for them. The usual wagon-sled in the Maine woods is 5½ feet wide.

About this time a change was taking place in the character of the farm animals. Hitherto oxen had done all the heavy work and the horses (mostly the Morgan strain) were used only for light work and driving purposes. About 1870, the Percheron and Clydesdale breeds (or "western horses," as they were called) began to come into Maine and these heavy draft horses were powerful enough to be used for winter work in the woods. By means of a wagon-sled, a load of logs may be hauled with much less friction and so horses were plenty strong enough. It is said that a man with a team of horses and a wagon-sled can do as much as with eight oxen and a bob-sled.[2] The combination, therefore, was a big advance step in the problem of hauling logs. In many cases where the distance is not great, a bob-sled and a team of horses is still used, but for hauls of more than a mile or so, the wagon-sled is the method that generally prevails today.

With the introduction of the wagon-sled, a new technique of road-making has evolved. Glare ice is about as effective in reducing friction as are ball bearings. Accordingly, the woods roads which are used for the long hauls of the wagon-sleds are carefully tended. After each fall of snow the hollows are filled in and the humps smoothed over. Then the "rutter" goes over the road. This is an apparatus with chisel-like pieces of iron the same width and distance apart as the sled runners and which make two grooves in the snow. The finishing touches are put on by a sprinkler—a water tank mounted on a sled. This is usually done at night so it will be ice in the morning and the procedure leaves a hard, perfectly smooth trace over which the wagon sleds laden with logs will glide with a minimum of friction. The smooth road thus produced is a great contrast to the same trail in summer, where one clambers over boulders, tree roots, and small bushes. It shows the importance of the winter snows in the logging operations.

Not always does the woodsman take such pains to make the road slippery. Where there is much of an incline, he must use all his ingenuity to slow up the sleds, or else the heavy load of logs will start sliding downhill faster than the horses can run, with disastrous results. Sand or straw strewn on the tracks is sufficient to retard the sleds if the grade is not too steep. Sometimes the teamster will wind a chain around the runners in order to make them drag as much as possible. But where these measures are not sufficient, a "snub line" must be used. This is a hawser some three inches thick and as long as the slope. At the top of the hill a tree has been felled so as to leave a stump four or five feet high. The teamster hitches the snub line to his load and passes it around the stump a number of times. Then he starts slowly down the hill. The friction of the folds of the hawser passing around the stump, together with the weight of the length of the line being dragged up the hill, is enough to retard his progress sufficiently. Sometimes two men are stationed at the stump to play out the slack and check it against the stump with a lever if the load gets to going too fast.

Both methods, however, are full of danger and in this connection many tales are told of gruesome accidents as well as miraculous escapes. Sometimes because of a flaw in the rope or a rock or sharp piece of wood cutting a few of the strands, the snub line will break. The teamster, who must always be on the alert for this event, must then yell to his team and whip them into a run, pell-mell down the hill, hoping to keep ahead of the weight of the load. If one of the horses stumbles or the sled leaves the track, the driver must jump for his life while the poor beasts are mangled and crushed under the pile of logs. Almost anything may happen and whether the event proves to have a tragic or happy ending, it will serve as the basis for many an evening's yarn in the bunkhouse.

When logging in some of the mountainous regions of Maine, gravity has been used to carry the logs to the water. Up on Squaw Mountain, for instance, a sluice was built nearly a mile in length which carried the logs down into Moosehead Lake. In a lumbering operation on Mount Katahdin in 1905-06, the logs were rolled down the side of the mountain to the level whence they were hauled to the stream on sleds. Frequently along the upper Kennebec, where the level of the river terrace is many feet above the present water line, one sees a strip of the steep bank torn up and clear of bushes where the logs have been rolled into the river.

The problem of using mechanical power has long commanded the attention of inventors, and as early as 1889 four steam log haulers were tried in Michigan.[3] They were not successful, however, and the credit for the first workable device belongs to O.A. Lombard of Waterville, Maine. In appearance, the Lombard Log Hauler resembles a small locomotive with headlight, smokestack, boiler, cab and whistle. Instead of the driving wheels, there are two endless chain treads, 12 inches wide and 14 feet long, which will give good traction over the snow. Under the front end are two heavy runners which steer the machine. It weighs 14 tons. The first one was tried in the winter of 1900-01 and was so successful that this type of log hauler was put on the market. Many of them are still in use.

The same firm constructed an electric trolley log hauler which was tried on Alder Stream, near Eustis, Maine. The current was supplied from a cascade about 200 feet from the road. It was not considered a success and was only tried one winter. Needless to say, it was not an arrangement which could be put into very wide use.

In 1911 the Lombard Traction Engine Company developed a gasoline tractor. This has a big six-cylinder engine rated at 100 horsepower. These machines have been manufactured continuously since that time, displacing the steam haulers which were less powerful, four tons heavier, and more expensive. The makers claim that one of the gasoline tractors will take the place of between one and two hundred horses and can travel four or five times as fast. During this last winter in a demonstration, one of these tractors hauled eighteen sleds containing nearly 90 cords of frozen pulp wood, which is the equivalent of about 220 tons.

There are other kinds of tractors being used in the Maine woods, but all have the "caterpillar tread", which was the invention of Mr. Lombard. It is this feature which makes it possible to develop motive power over the snow. One interesting model uses a standard Fordson and is equipped with a special tread of this kind. It is not so powerful as the bigger makes, but the cost is only a small fraction. It will haul about half as much again as a team of horses and do this three or four times as fast. Probably there will be an increase in the use of these small tractors on operations that are too small to make it worthwhile to employ one of the bigger and more expensive types.

The greater speed of tractors makes them useful where the haul is too long for horses. The latter cannot be expected to travel much more than 20 miles a day, while a tractor can go several times as far. In any operation that has a haul of more than five miles and enough material to keep the machine busy for the season, it will pay to use a gasoline tractor. In Maine, the operations are small compared to other regions, but according to the information supplied by the various manufacturers there have been well over 200 machines in use this past winter.

Besides the problem of getting the logs to water, there is the task of carrying the supplies to the camps. In the early lumbering days, when the highway system consisted of a very few stage routes connecting the towns near the coast, the operations had to be conducted sometimes as much as 200 miles from any road, to say nothing of any railway. In such cases, it was out of the question to attempt to keep bringing in supplies. A small crew would be sent ahead in the summer and fall to clear out the woods roads, build the camps, and cut what hay they could find in the nearby river meadows. Not until the first snows came could the main crew start out. They would load up with their winter's supply of staple foods, as much hay and grain as they would need, and start out with the oxen on a slow laborious journey into the interior. Sometimes they would have to wait until the lakes and streams froze hard enough to carry their teams. Often when the same camps were to be used a second year, the oxen would be left to forage for themselves all summer on the marsh grass, and the men would take the chance of finding them again the next fall. In this case, the men and their supplies would come up the waterway by bateaux.

This remarkable type of craft deserves a few words. What the canoe is to the Indians or the gondola to the Venetians, the bateau is to the woodsmen and river drivers. As the name indicates, it is the product of the Canadian voyageur and was evolved to meet the task of carrying great loads upon the swift and rocky streams of North America. In form it is something like a very large dory, only with much narrower, tapering ends which overhang the water for several feet. It is large enough to carry huge loads; small enough to be maneuvered through boiling rapids; heavy enough to withstand the buffetings upon the rocks; and light enough to be carried around falls or between lakes on the shoulders of men. The bateau made it possible for Benedict Arnold and his men to journey up the Kennebec, over the Boundary Mountains, and down the Chaudiere on his ill-fated expedition to Quebec. It was the bateau that played the part of both passenger and freight carrier on the vast network of water highways which were for so long the only means of reaching the timber of the north woods.

Today with the various railroads and many highways, there is no lumber camp which is not in communication with the outside world. There is at least one fine automobile road forty miles long which was built by one of the paper companies and kept passable all winter for the sole purpose of serving its camps. The last five to twenty miles into a camp is usually along a tote road through the woods, but this distance is not too great for the men to come on and off the job at will and for the supplies to go in every day. The ordinary means of transportation is the "tote team" which makes daily trips to the nearest railroad station. One camp which is 56 miles from the railroad uses a Lombard tractor to portage in the supplies. Perhaps before long the horse will give way entirely to some automotive device. The "snowmobile," which is an ingenious adaptation of a small automobile into a vehicle with runners in front and a caterpillar tread supplying the traction, is in wide use as stage and mail carrier in the rural districts of Maine in the winter and may be used on the woods roads sometimes. There is another new snow vehicle which appears to have promise. This is a Fordson tractor which runs on two pointed revolving cylinders about three feet in diameter. These have spiral grooves which, as they revolve, carry the machine over the snow on the same principle as a screw bores through wood. The machine is particularly adapted to going over drifts of soft snow. But before any machines displace the tote teams, they must be flexible enough to go over the rough places, be reliable enough to prevent the possibility of interruption of supplies, and be cheap enough to be more economical than the tote team. For the present, a team of horses is nearly always the best method of getting the supplies.

There are several minor tools of the woodsman which should be mentioned. The peavy is said to have been invented and first manufactured by a Bangor man of that name. This cant-dog, as it is called, is a stout wooden handle some five feet long equipped with an iron point and a peculiar swinging arm or hook which makes it possible to grab a log and roll it over. It is used in loading and unloading the logs from the sleds and handling them generally. In many places, a wedge is used to assist in aiming the fall of the tree or to prevent the weight of the tree from binding the saw in the cut. Where four foot pulp wood is being cut, a shorthandled hook is used to get a grip on the pieces and toss them around. Also, each sawyer must have a measuring stick to show the exact four-foot length. After these bolts of wood are stacked in a pile, they are marked with a stamper, which is similar to a heavy hammer with the raised initials of the company on the head. A blow with this on the end of a log will imprint these initials. Often, however, the mark is painted on the ends of the logs instead of being stamped.

It is in the West and South that the nearest approach to a genuine mechanical revolution of the logging industry has taken place. Here the actual felling is still done by the saw and the axe, but the operations of transporting the logs is often entirely by steampower. The logging railroads perform the function of the Maine waterways. From the main line of one of these railways temporary spurs are projected to the edge of the uncut timber. Here will be installed one of the several methods of steam skidding. These operate by means of wire cables and grapples attached to the logs. Some kinds pull the logs along the ground, while the "sky line" system of overhead cable ways picks up one end of the log or bunch of logs and drags it to the edge of the railroad. Here a steam loader, similar to a derrick, puts them onto the special flat cars which carry them to the saw mill. The whole scene with its puffing engines, tracks, cables, and machinery of different kinds is a far cry from the winter woods and jingling bells of the Maine lumber camp.

Perhaps as trite a remark as can be made is the statement that this is an age of machinery. Yet, it is true. The grain harvester and binder have made "bringing in the sheaves" merely a picturesque phrase. The multiple spindled automatic loom has relegated the spinning wheel to the antique shop. Even in the realm of domestic duties we see the sewing machine, the electric iron, the washing machine, and the vacuum cleaner. The question naturally arises: when will this tide of mechanical invention reach the logging industry? Perhaps we may envisage a future where the forests consist of crops of trees of uniform size, equally spaced on level ground, and where the lumberjack is replaced by a mechanic who drives some kind of gigantic mowing machine. But aside from such fantastic speculations, we can foresee no mechanical revolution. The tools will doubtless become further refined and mechanical power will be more and more utilized in the problems of transportation, but the fundamental operation, the felling of the trees, will remain for some time hence purely a handicraft—the craft of the axe and the cross-cut saw.

The Work of the Woodsmen

"They could erect camps, make axe handles and sleighs, and many of them were blacksmiths, sawyers and carpenters capable of undertaking almost any kind of work."[4] So one man describes the typical Maine woodsman of his youth. In those days it was necessary for the lumber worker to be somewhat a jack-of-all-trades. In fact, as we have seen, he sometimes even had to be a cook. About the only specialized job was that of the teamster who drove and cared for the oxen. Since that time, the operation has been divided up into somewhat more specialized jobs, but nothing approaching the minute division of labor which is so characteristic of most modern industry.

The first division is between the direct producers, who actually handle the logs, and the indirect producers, who take certain special duties off the shoulders of the workers. The functions of these indirect producers will be described first.

The camp boss has the responsibility for the general supervision of the work, for keeping the production up to schedule, and for planning the direction of the cutting. Where a single operation includes several camps, there is also a "walking boss" in charge of them all. There is the clerk who keeps the records of the daily production, the payroll and men's charge accounts, the invoices of supplies and provisions, and tends the camp store. The scaler has the duty of measuring the number of board feet (or cords, if pulp wood) in the logs as they are piled up. His salary is paid half by the company and half by the timber owner since each has an opposite interest in the amount of the recorded cut (unless the operator is also going to sell the logs according to his reckoning). The saw filer spends the day in sharpening the crosscut saws. The duties of the cook and cookees are self-evident. Most larger camps have a blacksmith who is kept busy shoeing the horses, mending the sleds, making new runners, and repairing broken tools. One teamster drives the tote team which brings in the supplies from the nearest railway depot.

Among those who are directly engaged in cutting and hauling the logs, there is also a certain amount of division of labor, but this depends upon the stage of the winter and also upon whether long logs or four-foot bolts are being hauled. All summer a few men are kept to watch the camps and to cut the hay in the river meadows. In the fall and early winter all the crew is engaged in building the camps and "swamping out" the new logging roads. Then comes the two or three months of cutting until the snow is too deep for convenience, after which the men devote their time to hauling the logs from the "yards," where they have been collected after being cut, to the "landing" on the ice of the stream which will eventually carry them away. After that, some of the men stay to follow the logs down on the "drive." Thus, a man who has stayed on the job all winter will have done several quite different kinds of work. In other words, the specialization of jobs is not hard and fast.

Where long logs are being cut, the usual unit is a gang of the following six men: two choppers, a sled tender, a teamster with a pair of horses, a yardman, and a swamper. Each of these has a particular function. The choppers fell the tree and cut it into log lengths. The sled tender helps the teamster load the three or four logs onto the bob-sled. The teamster then drives to the yard where the yardman helps him unload. The swamper is engaged in constantly extending the road to the newly cut trees. Sometimes the area being cut is not more than half a mile from the stream and in that case no yard is needed, since the logs are hauled directly to the landing and the yardman becomes a landingman. Besides helping the teamster unload, it is the duty of the landingman to chop the company's mark in the bark of the log or paint it on both ends. This is so that the log may be distinguished and sorted out from the logs of the other companies when they all float down the river in the spring.

Where the trees are a number of miles from the stream, the men stop cutting in February or March and spend the rest of the winter in hauling the logs from the yards to the landing. Here is where the wagon sleds come into use and the need for a smooth, slippery road creates a new kind of special job—the road tenders or "road monkeys," who use the sprinkler and rutter which have been described. Where the mechanical log haulers are used, we find them operated by specialists who are rather different from the traditional woodsmen.

The cutting of four-foot pulp logs requres a slightly different division of labor. Here a typical gang will consist of five men and a horse. Two men will fell the trees. One man and the horse will "twitch" the log (haul it along the ground) a rod or two to a small yard. Here a sort of trestle or saw horse has been built onto which the log is rolled and then sawed into four-foot lengths by two "buckers." These bolts are then piled up so that the scaler can estimate the number of cords and discount any that may be partly rotten. Each camp will have one or two men who go around to the new piles and mark the ends of the logs, either by stamping or by painting the company's initials or special mark. As soon as there is enough snow to make going good, they start hauling the wood to the landing on the bank of the stream, just as with the long logs, except that being much smaller and easier to handle, one-horse sleds are often used.

It is only within recent years that this "four foot stuff" has been cut, but that it is one of the most significant developments in the logging methods of the Maine woods will be seen when we realize that roughly seven-eighths of all the pulp woods, which means about 70% of all the timber cut, is in the shape of these short logs. Why is it better to cut up the logs in the woods by hand than to wait until they get to a mill where a circular saw could buzz through them in a fraction of the time? There are several reasons. It is possible to drive these short logs down streams so small that full-sized logs would jam at every bend. Much less water is needed to float them. Then, too, only the sound wood is sent to the mill. Long logs which are half rotted would be sent along for the sake of the good part, but if cut into short lengths, the bad part is left in the forest and does not have to be handled all the way to the mill. This point has been more important since the bud worm made such ravages in the spruce forests of Maine in 1920-21. Another point is that a short log operation is less interrupted than a long log operation by changes in the labor force. Although the gang of five men and a horse which has been described is an efficient unit, yet it is possible for even one man alone to carry on production and fell the trees, saw into lengths with a buck saw, and pile at the stump. In other words, the production of short logs is more flexible and so much less is lost through the high labor turnover.

However, the most important and significant reason is that it requires a less experienced type of woodsman to handle these four-foot bolts. Partly this is because smaller trees are cut for pulp than would be profitable for lumber purposes. No big pines are taken, only the spruce. Then, too, after the tree is felled, it is cut up and all subsequent operations handle only the short lengths. This is much simpler and does not require the knack of handling the long logs and knowing how they will act, which is one of the important elements in the skill of the best woodsmen. The difference in the quality of the workers in the two kinds of camps is fairly evident. In one typical short log camp, for example, about half the workers were boys under 20, while in a typical camp where full-sized logs were being cut, there were only a few of these less experienced boys.

The question arises: Is this method of simplification of the job the cause or the result of the scarcity of skilled woodsmen? Probably it is the result, because for many years there has been a shortage of the best woodsmen. Doubtless this was one factor which, perhaps more or less unconsciously, influenced the logging operators in deciding to adopt the short log method.

Although this may be a step towards what in other industries has been called "the degradation of the craft," by so simplifying the method that special skill is no longer necessary, after all it is only a short step, and the logging industry is still a long way from the point where totally inexperienced men will be acceptable labor.

We have pointed out how there is a considerable amount of division of labor and simplification of the work, so that the woodsman does not have to be the versatile craftsman described at the opening of the chapter. To just this extent, it might be said that logging has developed from a primitive towards a modern industry. Yet we have only to compare the work of the woodsman to the work in some highly developed industry to see how few of the possibilities of division of labor have been developed.

The meat packing industry offers a good example of how far this principle may be carried without much machinery being employed. Here the work which was once performed by one butcher is divided up among 30 or more men, each of whom does only one small operation upon the carcasses as they pass by on the conveyor. One man pulls off the tail, another pounds off a certain area of hide, another cuts off a more valuable part of the hide, another splits the backbone, and so on. All this makes for cheaper costs of production, but the human costs involved in the repetition of a single operation day after day, year after year, are forming one of the serious problems for sociologists.

For this reason, from the standpoint of the well-being of the woodsman of today and tomorrow, it is probably fortunate that the logging industry is very slow to follow the trend of industry in general towards extreme specialization and minute division of labor.

The Life of the Maine Woodsmen in 1926

The occupation of logging is somewhat peculiar in that any consideration of this industry must take into account the living conditions as well as the work itself. In fact, this point is so important that most men, if they are choosing whether to go into the woods or not, will base their decisions largely upon whether they want to spend the winter in the lumber camp or in the town, instead of deciding whether they want to do this kind of work or not. This is because the woodsman's life is such an integral part of his job. In most industries, such is not the case. The bricklayer, the filing clerk or the locomotive engineer is not restricted to any particular mode of life. His standard of living is the same as that of any other worker who has about the same income. The woodsman, on the other hand, obviously cannot live in the same way as a city or village worker who may happen to get the same rate of wages. He must live in the lumber camp, and camp life is at best inconvenient, and at worst full of hardship. In other words, his wages must pay him for living in the camps, as well as for the actual work of felling trees. For these reasons, in order to understand the woodsman, it is necesary to describe the conditions in the Maine lumber camps. We can have a more accurate picture if we first get an idea of the woodsman's life a hundred years or so ago (ca. 1826) and compare it with the typical present day conditions.

In the old days, the camps were much more primitive. Each establishment would have but two buildings—the men's camp and the ox hovel. These were built of logs notched at their ends to fit closely and with moss stuffed in the chinks. There were no windows, and only a small opening for a door. Both camps were made in the "lean-to" shape. The hovel was sometimes higher in order to have space for storing hay under the roof, or else a shed was built in front for that purpose. It was necessary for the health of the oxen to have a floor, and this was made of small poles laid closely together and smoothed on with an axe. In the men's camp, however, the floor consisted of dirt, the only preparation being to remove the turf, which was very inflammable when dry.

The roof was composed of long "splits" of cedar or spruce laid on in shingle fashion and held in place by the weight of poles, boughs and eventually snow. A big hole was left over the middle of the front side to let out the smoke of the camp fire. Sometimes a crude fireplace of stones or a chimney of green wood was built, but usually the fire was made on the ground underneath the hole with merely two stakes to prop up the back logs. Naturally, the danger from fire was quite great, and instances have been recorded where whole crews perished while asleep. Ordinarily, however, the fire would be discovered and put out with snow before it gained much headway. In any case, the roof, not being fastened down, afforded an easy way of escape.

The interior of the camp was furnished in a meager fashion. The bed consisted of an area of the ground with the low back wall of the camp as a headboard and a log for a footboard, with fir boughs piled between as a mattress. The coverings were old quilts and blankets, often sewed into one piece wide enough to cover the whole crew as they slept side by side with their feet towards the fire. Along the foot of the bed was the "deacon seat"—a long bench made by mounting a half round log on stakes. Sometimes that was all the furniture they had. The deacon seat would also serve as a table for the one frying pan which was plate and platter for the whole crew. Later, however, it is recorded that "the luxury of a temporary table is now pretty generally enjoyed, with plates, knives and forks, and tin dippers."[5] A few homemade stools completed the equipment.

The daily life of the woodsman in those days was largely one of toil from daylight until dark, with a short respite for the noon meal. His only relaxation came after supper before turning in. Perhaps the truest picture of his leisure time can be given by quoting the following description of a typical evening, bearing in mind that the writer, in his sentimental regard for his youthful experiences, may have made the scene somewhat more idyllic than was always the case:

"Could you take a peep into our snug camp some evening, you might see one of our number seated perhaps on a stool in the corner, with a huge jackknife in his hand, up to his knees in whittlings, while he is endeavoring to give shape and proportions to the stick he is cutting to supply the place of a broken axe-handle. The teamster might be seen driving a heated 'staple' with jingling ring into a new yoke, which is to supply the place of one 'Old Turk' split while attempting, with his mates and associates, to remove an immense pine log from its bed during the day.

"There sits another fellow staring into vacuity, while between his lips, profusely covered with a heavy beard, the growth of a quarter of a year, sticks a stub-stemmed pipe . . . Beside the deacon seat, a little removed, sits the cook with a large pan between his knees, with shirtsleeves furled and in dough to his elbows, kneading a batch of bread to bake for breakfast . . . Yonder, at the further end of the camp, in close proximity to the fire, sits a lean, lank little man with thin lips, ample forehead, and eyes no larger than a rifle bullet, piercing as the sun, poring over the dingy pages of an old weekly, perhaps for the tenth time. Songs, cards, or stories possess but little attraction for him. Intellectually inclined, but miserably provided for, still the old newspaper is a more congenial companion for him.

"Behind the deacon seat, lounging upon the boughy bed, you may see half a dozen sturdy fellows—the bone and sinew of the crew—telling yarns or giving expression to their feelings in a song, while the whole interior of the camp is lighted with a blazing hardwood fire, which casts upward its rays through the capacious smoke hole, gilding the overhanging branches of the trees. All within indicates health, content and cheerfulness."[6]

Perhaps the woodsman of a century ago was fairly healthy, contented and cheerful in such an environment, but that kind of a camp and that kind of food would no more satisfy the modern woodsman than the average living conditions of today will probably satisfy the woodsman of tomorrow. It all depends upon how the life in the woods compares with the life which the man would have to live otherwise. In general, the standard of living is steadily rising. The luxury of yesterday is the convenience of today and the necessity of tomorrow. The standards of camp life will always lag a certain distance behind those of town life, but they are improving at just the same rate. While the living conditions in the usual lumber camp of today are far from being the best that is possible, yet, by looking back, we can see the improvement.

Instead of a small crew of a dozen or so, the typical camp of today includes thirty or forty men, and the buildings are, accordingly, larger and more numerous. Sometimes the logs are built up "crib fashion" with notched ends, but often they are placed upright, a method which is cheaper since smaller and shorter logs can be used. The builders are no longer entirely dependent upon the forest for their materials as it is usually not too costly to haul in rough boards for the roof (which is then covered with tar paper), the floor and the doors. A few small windows and a skylight are put in for light and ventilation.

The number of buildings varies, but every establishment has at least four: the men's camp, the cook camp, the hovel and the office. There may be also a blacksmith shop, a cabin for the saw filer, a storehouse, a shed for the sleds, or other small structures.

The men's camp, or bunkhouse, contains double-decked wooden bunks built along both sides. Two men sleep above and two below. The mattresses are usually fir boughs and the bedding consists of "woods puffs," which are quilts covered on both sides with heavy (and usually dirty) blankets. At the foot of the bunks in front of the stoves is the traditional deacon seat. The two stoves—a vast improvement upon the old-time campfire—are placed in the middle of the camp and separated by a stack of firewood. Poles or wires for drying clothing are strung overhead. At one end is a sink and basin for washing. Often there is a grindstone for sharpening axes.

The cook camp serves as kitchen and dining room for the men, as well as living quarters for the cook. Usually it is only a few feet from the end of the men's camp, and the gap is covered over, forming a small shed, called the "dingle," used for storage. The kitchen occupies about half the space. There are two ranges with firewood piled between, while along the wall are the sink, a long built-in table, shelves, hooks, and other equipment of the cook. The other side of the camp is taken up with the oilcloth-covered tables and the benches.

The hovel is smaller in proportion than it used to be when oxen were used, because whereas in those days six or eight oxen were needed for every sled, now a pair of horses will suffice. The hovel contains stalls and storage space for hay and grain.

The office camp is much smaller. Besides performing the function of office, it serves as the camp store and living quarters for the boss, clerk, scaler, and any visitors who may be present. The "wangun," as the store supplies are generally called, include tobacco, writing materials, mittens, boots, and other clothing or supplies which the men may want. The furnishings of the office usually include a desk and counter, a small safe, the telephone (a homemade line connecting with the other camps in the section and the railroad station), a stove, bunks, and some stools or crude chairs. Sometimes homemade beds with real springs take the place of bunks, and even sheets and pillow cases have been known.

The food of those early lumber camps sounds most unpalatable to us, especially since the fare was seldom varied. Salt pork, bread and tea were the chief items, with sometimes beans, salt fish, and an occasional deer. The writer's great-grandmother is said to have cooked up whole barrels of fish hash and applesauce for her boys to take into the camp, and this must have furnished a welcome change of diet. The food was abundant, however, and what it lacked in variety it made up in quantity. The following quotation indicates that at least some of the woodsmen in 1814 were not stinted as to their daily rations: "We are but eight hands all told and don't need much. A barrel of rum, another of molasses, three barrels of pork and six of flour, and that meal bag over there full of tea, with the corn I have already got for the oxen, will do to begin with."[7]

Obviously, the preparation of such meals was no great task, and so in many small camps there was no cook, the men taking turns in doing the work, or else dividing up the duties. At other times, the cook was appointed not for any particular culinary skill, but because he was "some one of the hands who has not so much efficiency in getting timber."[8] On this subject we have an account of the experiences of a boy only eleven years of age: "To me was assigned the duty of cooking. Although not very proficient, I managed well enough as our fare was limited to pork, beans, bread, molasses, tea and dried apples—not a well diversified diet, but as good as could be obtained under the circumstances. The men were satisfied mainly, perhaps, because the time did not afford a higher standard by which they might measure the short-comings of their own lot. Nor did their health suffer for lack of luxuries, edible or otherwise. Sickness was rare."[9]

The food of today is, as a rule, much better than in the early days. The change has been brought about mainly by the opening up of railroads and highways to within a fairly accessible distance from the camps. In the old days, a crew would have to carry its provisions for the whole winter when it went into the woods and to remain as completely shut off from civilization as an arctic expedition. Nowadays, no camps are so remote that they cannot "tote" in fresh supplies at least once a week, and most of them have daily trips. Furthermore, the atmosphere serves as a perfect refrigeration device. As a result, the variety and quality of the provisions is limited only by the economy of the operating company or jobber. The supplies ledger of one camp, for example, showed items of more than sixty different kinds of feed, which is a great contrast with the old time tri-daily diet of salt pork and beans. One paper company maintains a large truck farm situated in the heart of the woods near Moosehead Lake for the sole purpose of supplying its camps.

The quality of the fare, of course, depends to a large extent upon the skill of the cook. This important personage is better paid than the other men and is well worth it because a good cook is one of the most important factors in keeping a high morale in the camp, in attracting the better woodsmen, and in keeping a low labor turnover. Also, from the operators' standpoint, a good cook must be economical with his provisions as well as able to prepare appetizing meals. The cook has one or two assistants called "cookees."

As soon as the meal is ready, the cook lets out a halloo and the men come in from the bunkhouse to take their places at the table. At each place is a tin dipper, tin plate, knife, fork and spoon, all somewhat stained and battered. On the table are pans of some hot meat or stew, potatoes, plates of thick slices of bread (often surprisingly good), or hot biscuits, and pitchers of strong coffee, well steeped tea, or cold water. There are always dishes of prunes, applesauce, or pickled beets within reach. For dessert there are cookies and pies. The two cookees hover over the tables and replenish the platters as needed. There are no restrictions on quantity. The mealtime is curiously quiet; there is almost no talk, each man tending strictly to business. When he has eaten his fill, he picks up his utensils, walks over to the sink, places each in its respective pile and goes out.

That is a description of a typical dinner or supper. Some camps will have a more elaborate meal and some less, but that is a fair sample. The meat may be beef, veal or pork and in the form of steaks or stew, and the other items may vary somewhat. Breakfast consists mainly of beans (baked in a hole in the ground), and oatmeal. Often where the working place is too far from the camp for the men to come back for the midday meal, the cookees will carry out a lunch of hot coffee, beans and meat sandwiches. In general, the food consists largely of the protein and the "fuel" materials, which are required by the combination of the strenuous work and the cold weather. A scientific study of the subject disclosed the fact that the average woodsman consumed a greater number of calories per day than almost any other type of heavy worker and nearly three times as many as the average sedentary worker.[10] On the whole, we may draw the conclusion that, unless the cook is unusually poor, the food in the Maine lumber camps is as good as, if not better than, can be had in the average workers' boarding house of the cities.

The leisure time of the woodsman of today is spent in much the same manner as that described in the old time camp. The two chief diversions are playing cards and swapping yarns. The daily mail makes it possible to have a few magazines and newspapers which circulate around the camp. Some evenings will be given over to various rough games and stunts. Once or twice a year some Catholic priest will visit the camp and hold Mass. In late years, the State Y.M.C.A. has been sending a secretary around with a portable moving picture apparatus, which forms a welcome change. There is also at least one paper company which has been doing the same thing. There is no work on Sundays and the men spend the day washing their clothes, cutting fresh boughs for their beds, hunting (in season), or loafing. On the whole, it is a pretty monotonous life and perhaps it is this point more than any other that makes the work of the lumber camps less attractive to many men than a job in the city.

Probably it is this scarcity of the usual recreations during the winter that has made the spring "coming out party" of the woodsmen so often the wild drunken spree that gives so many people their only idea of these men. The drink problem is as old as the industry. A century ago, hogsheads of rum were considered indispensable and a bill of supplies in 1827 contained an item of 180 gallons of rum for fifteen or twenty men. However, it is stated that "of late [ten years later], very few respectable lumberers take any spirits with them, and the logging business is consequently carried on with much more method, economy, and profit."[11] As time went on, the daily ration of rum, like the grog rations on shipboard, became a thing of the past. This, nevertheless, did not prevent the men from taking a bottle or two to hide in their bunks and especially did not mean an end of the traditional orgy when coming out of the woods. Since 1858 Maine has been a prohibition state, but the spectacle of the thousands of men coming out from the camps every spring with a roll of bills or check representing their winter's wages offered such a glittering prospect to unscrupulous men that ways were found to make the liquor traffic "safe" in Bangor, Greenville Junction, and other lumber towns. Agents would board the trains and hand out free drinks in order to start the ball rolling and then sell the men all kinds of goods from shoddy suits to ridiculous trinkets. When the men had been paid with a check instead of cash, merchants would cash these only at an exorbitant rate of discount, relying upon the ignorance or befuddled state of the woodsmen. Other agents would give out free booze in order to induce the men to go to their particular resorts. The net result was that usually the man would wake up in a day or two and find himself stripped of his roll of bills and even his new clothes. He would not be particularly resentful. He had had his fun and now was ready to go home or take the first job offered him.

This situation had long been deplored, but nothing had been done about it until the State Y.M.C.A. made a careful study which resulted in the erection of the Y.M.C.A. Hotel in 1911 in Greenville Junction, financed largely by one of the big paper companies. This is the first town that a large proportion of woodsmen strike when coming out of the woods and was the scene of the same kind of exploitation as at Bangor. This hotel has fifty rooms and offers the woodsmen the luxury of a single room with a clean, white spring bed, and access to showers or tub baths (a tremendous attraction). There are many kinds of wholesome recreation—music, bowling, pool, magazines—as well as a good restaurant. The idea was to give the men a chance to celebrate without a debauch so that they would be less easy marks when they went on to their homes or other jobs. Another service was a banking arrangement so that the men could deposit their money and send for it later or have it when they came back. By doing this, they could run the gauntlet of the agents and saloons without losing all their winter's wages. Naturally, this service was looked upon with great suspicion at first, but after a few bold spirits had tried it and spread the word that they really did get their money back safely, more and more men began to take advantage of the service. On the top floor of the Y.M.C.A. building is a well-equipped hospital which treats many injured and sick woodsmen. Until this was built, they had to go eighty-five miles farther to Bangor in order to get any surgical treatment.

National prohibition has made a great difference. Enforcement by federal officers has made the open saloons too risky, and the scarcity of liquor has made the trade too expensive to be carried on in anything like the former scale. Of course, the men coming out of the woods do get drunk often enough, but the grand spree has ceased to be an accepted institution.

This description of the life of the woodsmen would not be complete without mentioning at least one of the instances where the living conditions are far above the average. The American Thread Company has model camps in several places where it cuts the birch wood for its spool mills. One of the special features in these camps is that instead of the customary bunks filled with straw or boughs, they provide double-decked iron cots with real springs, mattresses, white blankets to sleep between, and pillows. Another innovation is a little cabin which provides a bathtub and a place to do laundry work. Every bunkhouse has a Victrola. More attention than usual is paid to other matters of comfort and well being of the men.

This example is significant in that it shows the direction in which the living conditions in the camps are developing. As time goes on, the probability is that the life of the woodsmen will be less "roughing it" and more and more like the life of the industrial workers in the cities.

FOOTNOTES

1. W.M.F. Pietrie, *History of Tools*, Annual Report of the Smithsonian Institution, 1918, p.565.
2. State of Maine, Bureau of Labor Statistics, 1899, p.71.
3. R.C. Bryant, *Logging*, p.172.
4. Isaac Stephenson, *Recollections of a Long Life, 1829-1915*, p.81.
5. J.S. Springer, *Forest Life and Forest Trees*, p.72. This quaint and interesting book was published in 1856 and describes the author's experience as a young man in the Maine lumber camps. The quotation, therefore, may be assumed to describe conditions somewhere around 1815.
6. Springer, op.cit., pp.129,130.
7. North American Review, April, 1844, p.333. The incident quoted had happened "some thirty years ago."
8. North American Review, April, 1837, p.350
9. Isaac Stephenson, *Recollections of a Long Life—1829-1915*, p.43.
10. Woods and Mansfield, *Studies of the Food of Maine Lumbermen*, U.S. Dept. of Agriculture. Experiment Station Bulletin No. 149, 1904.
11. *North American Review*, April 1837, p.352.

Winslow Homer's Lumberman Print

The following two letters from the September 1973 issue concern the Winslow Homer print of a lumberman which appeared as the cover picture (and also as a separate print) in *The Chronicle*, March, 1973.

I cannot refrain from commenting on the Winslow Homer woodcut on the cover of the March 1973 *Chronicle*. Apparently you are aware of some of the errors when you say, "It would be interesting to know whether Homer sketched this scene from life or whether he was drawing on his imagination plus some studio posing....Homer usually worked up his compositions from on-the-spot sketching, but one is a little suspicious of this one."

First, the workman in the rear would take all morning to get through the log if he tried to work through the log with such a narrow notch. In the first place, a woodman would ordinarily use a saw on a log of this size. I have never found it practical to use an ax on such cuts if the tree was more than six or eight inches in diameter.

I doubt if the awkwardly placed woodman could have made the cut shown in the standing tree. In the first place, the cut is wrong. In the second place, I never saw a man cut down a tree while standing on snowshoes, nor did I ever stand on snowshoes while doing the job myself. Snowshoes were sometimes used to get into the woods, but to the best of my knowledge were seldom if ever used while a man was working with an ax. It is possible that such could have occurred in some other part of the country, but it certainly was not common here in the East.

Let me tell you how we used to cut trees here in New Hampshire. First, one trimmed out the small bushes or branches that would handicap the swinging of the ax or the footing of the woodsman. Then one stamped down or cleared away the snow around the tree. The thoroughness of this task depended upon how fussy the owner was about the height of the stump. It was sign of wastefulness and laziness to leave a lot of high stumps, even if the snow was high. Those stumps represented a few more board feet of good lumber and a few more Yankee dollars.

Having cleared the area for work, the woodsman selected a line of fall which would be clear of any other obstructing trees and in line with the rough trail along which the logs would be later "snaked" to a yarding area. Having selected the line of fall, the notch would be made so that it would face directly towards this line. The tree would eventually fall almost directly at right angles to the line formed at the base of the notch. This notch needed to be made to a depth of only three or four inches into the tree.

The woodsman then took up a position on the other side of the tree and started to work with his one-man or two-man crosscut saw. As soon as the saw was well into the tree, wedges would be driven into the sawcut with a sledge hammer or the poll of the ax. As the saw went deeper and deeper into the tree, the wedges were driven farther in. When the saw got to within an inch or two of the notch, the wedges would really begin to bite to lean the tree towards the direction of the fall.

The woodsman in Homer's woodcut would certainly waste a lot of time making a cut such as shown here. It would, of course, be all wasted motion since the base of the cut appears to have been made with a saw and not with an ax. It would be extremely difficult to use an ax cross grained in this manner.

I have a strange feeling that I may be rising to the bait that you have set for the EAIA readers. You must expect a lot of responses on this one. Anyway, it is fun to be able to recall some of my earlier experiences in the woods. I like Homer's woodcut, but his woodcutting is for the birds. I really think he should haved stayed with seascapes.

RICHARD A. MARTIN

I enjoyed your article in the March *Chronicle* but I have to confess I've been bothered by Winslow Homer's woodcut. It seems questionable to me for a number of reasons.

The lumberman's position with respect to the tree would severely reduce the impact of the axe on the wood. For greatest impact the tree should be directly in front of him, or, perhaps, slightly to the left. In this position, the axehead has the greatest velocity and the woodsman is positioned to take best advantage of his shoulder muscles.

To be realistic, the lumberman's snowshoe should be on the left side of the trunk and his body should be rotated 30 degrees counterclockwise.

Correct technique has been used in creating a wedge and leaving the "hinge" the full width of the tree. This encourages the tree to fall in the direction of the wedge and creates maximum resistance to falling at right angles to the wedge—on top of the woodsman. However, instead of continuing to chop in the wedge, he is now commencing to attack the side of the hinge.

It is not believable that a woodsman of this man's stature could chop the top angle of the wedge as portrayed. In order to chop this 60 degree angle his axe would have to be way over his head and descending at a 60 degree angle. He would more likely chop a lower angle or make his cut farther down the tree.

One final matter—I wonder if a lumberman would fell a tree while standing on snowshoes. While I have had no experience with snowshoes they would seem to impede his ability to make a rapid retreat if the tree fell in an unanticipated manner. Unless the snow was very deep—which is not true in this case—he would have greater maneuverability without them.

It would be interesting to know whether these objections would occur to a professional axman—if there are any of them still around.

ALAN FRANE

Cutting and Hewing Timber

By Frank G. Bawden

September 1976

My interest in hewn timbers and their hewing goes back to my boyhood days on a farm where I was intrigued by some huge hewn beams in an old barn on the place. I was so interested in these beams that I asked an old carpenter who worked around the place many questions about how anyone could chop such nice square beams from round logs. One day he brought a broadaxe to work and explained to me how in the old days men squared logs with just such an axe. I am sure that was some time before 1920. This little teaching lay dormant in my mind until twenty years ago, when the desire to hew timbers really came to me. This desire was brought about shortly after I took up residence in Newtown, Connecticut.

One day while I was working in our woods cutting firewood a neighbor came along, and we talked a little and he watched a bit. He remarked that I was pretty handy with an axe. Well, it turned out that this neighbor had made a living years back hewing

Figure 1. The starting of the saw kerf after the boxing is completed. The author is in the background.

Figure 2. Saw kerf well along and just about time to drive the felling wedge home.

—228—

Figure 3. The tree felled, with the crosscut saw and one wedge still lying on the stump undisturbed by the felling action. The second wedge not having been set deeply fell to the ground due to its own weight as the tree toppled.

Figure 4. Tree lying on the ground supported by the stump, with space cleared for good working conditions.

railroad ties. This did it. We struck up a real friendship, and he taught me his method of hewing. The method he taught me seems universal. About ten years later my interest in early tools led me to buy Mercer's *Ancient Carpenters' Tools,* where hewing is described almost exactly as I had been taught. Some time later I found that Moxon's *Mechanick Exercises* describes it in a similar manner.

At our spring meeting at Sturbridge when I was about to demonstrate hewing, a member (Douglass C. Reed) from Hagerstown, Maryland, approached me and informed me that he was doing hewing for some primitive restoration work. After some conversation about his work and mine, he joined me in the demonstration. Anyone watching us work would not have realized that we had never met before and would have thought we had been taught by the same teacher. The only difference was that Doug worked with the log on his left, as described by Mercer for right-handed workers, while I worked with the log on my right, as I am left-handed. This, of course, required Doug to use a right-handed axe and me to use a left-handed one. I have always been ambidextrous when using any kind of axe. My right eye is my master eye, and with the log on my right I sight from a point directly above the hewing line. This to me simplifies cutting a straight and plumb cut as the axe is dropped to the line.

I will now endeavor to describe the various techniques I use in cutting and hewing timber. I first select a straight tree of the species and of the size desired, as free from limbs as possible. Bear in mind that there was plenty of virgin timber when hewing was the way of working lumber, so the selection was a small problem. After the tree is picked out, determine the most convenient place to drop it. At a point as close to the ground as is convenient, use a felling axe to box into it (that is, notch it), about one-third of the way through). This is done on the side on which the tree should fall. The boxing helps in felling the tree where desired, and when properly done, it prevents much damaging splitting of the trunk as the tree tumbles over.

The next step is to saw into the tree on the opposite side from the boxing. This kerf is started about two inches above the bottom of the notch on the opposite side. Continue sawing until a slight movement is noticed. Either the kerf will tend to close up and bind the saw or it will start to open slightly. This is the time to start a wedge into the saw kerf directly oposite the center of the boxing.

Some stubborn trees require more than one wedge. If so, place the next wedge right beside the first one. I use a sixteen-pound sledge hammer to drive these wedges. Locally, this is called a "persuader." From here, saw a bit, then snug up a little on the wedge or

Figure 5. Timber length determined, the log is being cut to length. Note saw wedge set to prevent binding and saddles in place to receive the log.

Figure 6. The saw has cut completely through. The saw wedge slid along the cut in the log as it dropped to the saddle, leaving the butt resting on the stump.

wedges. Repeat this, keeping a close lookout for movement in the kerf until the wedge begins to advance with much less effort. At this point, stand to one side, drive the wedge home, and step number one is accomplished. Once the wedge is started, the saw is never withdrawn from the kerf. When the tree tumbles, the saw will be left lying on the sawing portion of the stump (Figure 3), undamaged, as the action is all to one side and above the saw. Common sense warns one that only the most experienced woodsmen should try to fell a tree away from the side with the most limbs, though with proper boxing, sawing and wedging, a tree can be dropped anywhere desired.

The tree is now lying horizontally on the ground. One must clear away any brush and interfering objects. Assuming an eight-by-eight-inch timber is desired, experience has taught that a log twelve inches in diameter will require a minimum amount of hewing. From a point on the log where it is about twelve inches in diameter, measure down toward the butt the required length for the timber needed, adding a few inches for final cuts by the carpenter at the building site. If you are lucky, a couple of lengths

Figure 7. The log in the saddles and dogged, a plumb line is now made at the center of each end.

Figure 8. End lines have been connected by snapping chalked lines from end to end and log is now ready for scoring.

of firewood can be cut from the swell of the trunk where it grew out of the ground. This will eliminate a lot of hard work involved in removing the excess wood by hewing.

Before sawing either end to the desired length, cut two pieces from the toppings about eight or ten inches in diameter and five or six feet long (Figures 5 and 6). These pieces are notched in the center with a felling axe and placed under the log to be hewed, about three feet in from each end. They are to form saddles on which the log can rest while being hewed. Often the tree is held above the ground by resting it on the stump at the butt and the limbs at the top. When the piece is sawed to length, it can be dropped into the notches in the saddles. If by chance it is not supported above ground, a cant hook is used to roll it on to the saddles. Once the log is seated on the saddles, dog it to them to prevent rolling and shifting. When the dogs are properly placed, the chisel-like portions of them will be driven parallel to the grain in both the saddles and the log.

The log is now in position for squaring. Start with a plumb bob (Figure 7), making a plumb line at the center of each end. Then, using a rule, measure outward from these lines the distance needed to obtain the size timber required. In this case it would be four inches each side of the center line. There are now three plumb lines at each end at four-inch spacing. Next, connect the outer lines by snapping a chalked line from end on top of the log (Figure 10). These are the hewing lines.

When hewing, I stand on the log and, with a felling axe, score the sides at about six-inch intervals (Figure 9), scoring as plumb and to a depth as close as possible to the hewing line. When both sides have been scored, a broadaxe is used to remove the wood between the scoring and to obtain a good face for the finished timber (Figure 10). When

Figure 9. Scoring the log. Note position of hands on axe handle. The left hand slides along the handle as the axe is dropped to the cut.

much wood is to be removed, a second scoring and pass of the broadaxe may be necessary to obtain the size timber required.

The broadaxe I prefer is wide in the bit, thirteen and one-half inches, and, with its twenty-inch handle, weighs nine pounds. The reason I prefer a wide-bitted axe is that when hewing, about one-fourth

Figure 10. Left leg against the log, right eye in position for good sighting to direct the axe to the hewing line.

Figure 11. Portion of axe gliding over previous slice and cutting to the line.

Figure 12. First two sides of timber completed.

Figure 13. Timber in position for final squaring. Top is level, with end lines plumbed and hewing lines being snapped.

—233—

Figure 15. The timber is covered with canvas to retard drying as much as possible in order to keep air checking to a minimum. Keep covered as long as time allows - at least one month - and throw bucket of water on it every day or two.

Figure 14. Timber completed. The end was left in the round only because this will be used as a display in a local Bicentennial exhibit.

Figure 16. Grinding the broadaxe.

of the bit is gliding over the portion that was cut with the previous slice (Figure 11). This is a great help in keeping a straight and plumb cut, as once started right it is simple to follow through the length of the log. Occasionally a log turns up with grain running in every direction. On these it is convenient to be able to pick up a right-handed axe and work in reverse until past the obstinate grain.

When both sides of the log have been hewed to these lines, remove the dogs and clean up the unfinished portions where they were fastened, roll the timber over on the flat side, and repeat the procedure. When completed, snap a chalked line along the center of the timber, checking it for size and any additional trimming that may be required.

The broadaxe, if not kept sharp and in good condition, can be a most hazardous and unruly tool, yet if kept sharpened and with a good solid handle that fits the user well, it can be a joy to use.

Now a few words on sharpening the axes I use. With the broadaxe, the entire back must be stoned to a perfectly flat, smooth, polished surface. For this I use a natural Arkansas stone, one inch by two by eight inches, with one side coarse and one side fine, using a thin household oil for a sharpening oil. A rusted or pitted axe just cannot be sharpened as every minute imperfection in the back will appear as a nick in the finished cutting edge. Once the back is conditioned by using a wet grindstone (Figure 16), I grind the front or face by holding the axe at right angle to the stone and working back and forth across the bit until the width of the cutting edge is ground to about a twenty-to-twenty-five-degree angle, to a reasonably sharp but rough chisel-like edge. For final sharpening, the same oilstone is used that was used on the back. This can produce a razor sharp edge.

Sharpening the felling axe is a simpler task as these are ground from both sides to a knife edge at a point in the center of the bit, and any rust or pits will be ground away in the process. As these axes vary greatly in thickness and taper generally from the eye toward the cutting edge, I find that grinding the thicker axes back about three quarters of an inch to an inch and the thinner ones about one-half to three-quarters of inch and letting the taper of the axe take care of the rest, works well. Too much of a bevel nicks too easily and tends to stick in the log after each blow; not enough makes a poorly cutting axe. The fine sharpening is accomplished with the same oilstone used on the broadaxe.

I don't pretend to be an expert at the work I have described, but just want to pass along the knowledge I have gained through working with men who have passed the information along to me.

Tools used in cutting and hewing timber. These are lying on a pile of chips. Starting from the bottom and from left to right, a pair of log dogs, a two-man crosscut saw, a plumb bob with a chalk line attached, a felling axe, a left-hand broadaxe, a pair of splitting and felling wedges, a right-hand broadaxe, a sawing wedge, a cant hook, and a sledge hammer or persuader. All photographs by the author and his wife.

Hewing and Squaring a Beam

by DOUGLASS C. REED

December 1976

The hewing or squaring of beams and wall logs was a task most men of the backwoods and rural areas knew how to do well. When America was first being settled, hewing logs with a broadaxe was a way of obtaining these necessary beams. Even when water powered sawmills came into general use, it was far more common for a man to square his own beams by hand than to take them to a sawmill.

The reason that water powered sawmills did not replace the practice of squaring beams by hand was because it was often simpler to cut the tree, hew the log to its proper size where it was felled, and take the finished beam directly to where it was intended for use. Otherwise the trees had to be felled and the logs dragged to where they could be loaded on a wagon. Once loaded, they were transported to a mill, cut, and once again slowly transported to where they were to be used. It has been said that water powered sawmills were so slow that when a log was positioned and the sawing begun, the sawyer took a long break and was back before one log was completely cut through.

Unfortunately, as with most of the old building methods, the knowledge of hewing of beams has nearly vanished. It's rare to find any man today who has actually squared logs by hand and who is still capable of doing so. It is even more difficult to find complete agreement concerning the methods used to hew the logs. It seems that each man had his own ways of hewing a log. Although most of the methods were similar, it all boiled down to two types of axe and to whatever way worked best for a particular man using those axes.

My information about hewing logs has come from the elderly people of the Catoctin Mountains in Maryland. Until recently they were isolated from the "modern" world. Most of the men learned how to hew at an early age by following their fathers around the woods and sharpening the tools while they watched their fathers work. As these boys grew older, they learned to hew beams and railroad ties, often leaving school in the early grades in order to help support their families.

The hewing of a beam is simply the process used to flatten a side of a round log into a more useful shape. I learned the skill of hewing from C. Paul Lewis, who at the time was 69 years old. When he was eleven, Mr. Lewis was supporting his family by hewing railroad ties. His method is similar to all the others described to me, but I chose to use his example because he was the only man still physically able whom I could find to teach me how to hew a log into a beam.

Figure 1. Position the log to be hewn with any twist up. Mark the underneath side of the log on either side of where it rests on the saddle block. Roll the log out and cut a saddle notch.

Figure 2. Cut this notch almost concave. If the log balances on just the two sides of the notch, it will be much more steady.

The necessary tools are few and easy to maintain, though they do require skill and experience. A poleaxe, also known as a felling axe, is necessary for scoring a log. Another tool is the broadaxe. This is used to hew a flat smooth surface on a log.

By tradition the poleaxe has a straight handle measuring 28 inches in length from the end of the handle to the top of the axe head. The average weight of the poleaxe is from four to four and a half pounds. The broadaxe handle is 24 inches and curved. With the head of the axe lying flat on a table, the end of the handle is about three to four inches off the table. This crooked handle keeps the man's knuckles from scraping the log when hewing. The preferred weight of the broadaxe is from six to six and one half pounds.

In selecting trees for this type of work, only those free of knots, windshakes, and catfaces are selected. Of course, the straighter and greener the tree, the less work involved. The tree should be felled so that it falls to the side with the least amount of ground cover. If this is done, the axe, when swung, will not snag in any brush. It is easier to hew the log where it is felled and there is obviously much less weight after the log has been squared and it will be lighter to move afterwards.

Figure 3. Roll the log out so that the slightly concave saddle notch rests squarely on the saddle block. This method of saddling a log makes the use of log iron dogs unnecessary.

Figure 4. Go to the small end of the log, measure up for a size desired, and cut two small nicks in the end of the log to aid in holding the chalk line.

To prevent the axe from digging into the ground and for ease of working and safety, the log should be placed on two small logs called "saddle blocks." One small log or block is placed under each end of the log to be hewn. Place the log with any hump or twist up. At the large end, mark the size of the tree on either side of the saddle block. Roll the log over and cut a flat, almost concave, area between these two marks on the log to be squared. Then turn the log into the notch so the flat area rests on the one saddle block. If this is done correctly, the hump should be up and the log should not roll. Do not cut an additional flat area over the second saddle block. It is not necessary and it is too difficult to get both notches to rest evenly on the two saddle blocks. This method of saddling the log precludes the use of logging dogs. Iron dogs are hazardous to the axe if it should strike the iron and chip the edge.

It should be mentioned at this time that the bark should not be removed from the log. The axe man loses an invaluable aid in the hewing process. This will be explained later.

Now the log must be marked as a guide to hewing. To determine where to place the line, go to the small end of the log and measure the size of the beam required. By measuring the smaller end of the log first, a uniform size throughout the beam will be

Figure 5. This log would yield a 6"-wide beam. The eye and experienced judgment make levels and plumb bobs unnecessary.

Figure 6. These small notches will serve to hold the chalk line in place when the line is snapped. With this method one does not need a helper.

assured. If the butt or large end of the log were measured first, there possibly would not be enough room at the smaller end to get the same measurement.

Make two small notches on the top side of the log on the small end that will indicate where the sides will be cut. Make the same measurement on the larger end of the log and again two small notches on the top side of the log. These small notches will serve to hold the chalk line in place when the line is snapped.

Figure 7. The string was pulled up from the log and allowed to snap, thus leaving a visible line to hew by.

Figure 8. Scoring is done by standing on top of the log and striking vertical kerf marks with the poleaxe two to three inches apart.

The granddaddy of the chalk box was a small forked twig, a tin cup, a string, and a colored liquid obtained from berries. An example would be the purple liquid of the pokeberry. The forked twig was in the cup of juice with the string placed between the forks. The string was pulled through the juice the necessary length. Immediately after this, it was stretched between two notches on the same side at either end of the log. Held by two fingers, the string was pulled straight up from the log and allowed to snap back. This procedure left a clearly visible line on top of the log for use as a guide in hewing.

The next step in squaring a log is known as "scoring." Stand on top of the log and strike vertical line kerfs with the poleaxe, two to three inches apart. This will chip off most of the rounded side, leaving score marks and a very rough side. Try not to cut deeper than the line, and make sure that care is taken to strike vertically with the guide line on top of the log and straight to the ground.

This takes a great deal of experience since the eye is the only instrument used to determine if the side being scored is flat and vertical. In the old days when this type of work was common, a man never used a square to guide his hewing. The fewer score marks that remained on the log was an indication of the man's skill. Today we have some awkward notion that handhewn beams should look as though a madman had

Figure 9. The broadaxe is used to smooth up the rough side left from scoring.

Figure 10. The vertical kerf cuts allow the hewer to shave off large sections with very little effort compared to hewing a log that

Figure 11. Very few score marks indicated a man's skill with the axe. The white marks are the result of a pair of nicks in the broadaxe. These will not be noticeable once the log dries.

butchered them. In fact, beams of the highest-quality workmanship did not even require the use of an adze to finish them for exposed use inside a house.

Once an entire side of the log has been scored, hew that side smooth. This is done with the broadaxe by working forward along the side of the log. Start at the end so that the rest of the log is in front. Hew the rough side as smooth as possible while trying to keep that side vertical from the guide line to the ground.

It is during this process that the bark is of some aid. As previously mentioned, the bark should be left on the log. This helps to hold the broadaxe to the line that is being cut and prevents the slipping and ricocheting of the axe as the wood is being hewn.

The hewing process just described is relatively simple. All it requires is a determination and time to learn. The easiest way to learn to use an axe is by cutting and splitting firewood. It was by cutting our own firewood one winter that I learned the pure pleasure of striking with an axe that mark that I aimed for. This exercise also helps with the fuel bill.

Needless to say, the log house builders hewed only what was necessary. Wall logs in log houses were hewn on only two sides and were rarely squared. Rafters were hewn either on only one side (pole rafter) or on all four sides, depending on the amount of work the builder intended to do on his house. Most barn beams were squared, as were most second-floor joists. First-floor joists were hewn on one side if no basement was under the house. They were hewn on two sides if a basement was planned.

As a point of interest, it should be mentioned that wall logs in a log house were hewn with the intention of eventually covering them over. The outside of the house was usually sheathed with wooden siding. The inside walls were either plastered or faced with boards. Builders generally did not have the time or the money to fool around with such unnecessary work as hewing wall logs to obtain a "neat" effect.

It is true that some log houses never were covered, but for the most part these houses have deteriorated beyond repair. Those that were never plastered or faced on the inside were invariably whitewashed to the point of being almost impossible to clean sufficiently to suit modern man's tastes.

Those houses that are in excellent condition were without a doubt sheathed soon after the house was built. Even I have to admit that a newly exposed log wall looks nice, but my better judgment tells me this is wrong unless great precautions are taken to check future deterioration.

The author supplied the following additional information:

This information came down to me in an unusual old-style apprenticeship of myself under C. Paul Lewis. He not only supplied me with the oral tradition that was in fact passed to him by his father, but Mr. Lewis was still physically able to show the use of the axes to me in 1973.

Since then I have hewn thousands of face feet of logs. I have tried every method I have heard of, and I truly like the one described herein. I hope this will suit your taste and be of interest to the members.

This article is a companion piece to the article by Frank C. Bawden in the last issue. The editor solicited both articles when he saw the authors demonstrate their knowledge at the meeting in Old Sturbridge Village.

The Crooked Knife

By NEWTON C. BRAINARD

September 1962

Ed.Note: These three articles on the crooked knife are interesting both for their explanation of the tool and for the rather uncharacteristic controversy that the subject aroused, a reminder that ideological values often lurk just below the surface of historical research.

Crooked knife—this one has a blade 4 inches long and the handle 7 inches. This is a bit larger than most of them.

The crooked knife seems to have been overlooked by the writers on woodworking tools. It is neither described nor listed. It is a tool of the woodsman of the north, most commonly found in eastern Canada and probably in northern Maine. A one-handed drawknife, it serves to shape the items which the woodsman needs in order to survive. It will strip the bark from a canoe pole and shape the subtle curves of an ax helve. The knife which is pictured was found in the window of a junk shop in Hartford. On the thumb piece was written in pencil the name of its former owner, from Rocky Point, New Brunswick.

The blades of these knives are commonly made from a worn-out file which has been reshaped, ground and tempered by the local blacksmith. On the best examples, the outer part of the blade is curved so that it is possible to hollow out concave surfaces. Probably this is the characteristic which accounts for the name of the tool, although it might relate to the angle between the handle and the blade. As the picture shows, the blade is usually bound to the handle by a thin strip of ash, which is applied when green and which shrinks to make a firm binding. There is no standard shape either for the blade or the handle. Each knife represents the taste of its owner. A clever whittler often spends many of his idle hours in the logging camp shaping the handle and carving it with fancy designs. A fraternal member often puts the society's emblems on the handle. Sometimes the name of the owner is there too.

The crooked knife is different from most cutting tools, in that it is used by cutting toward the user. This sounds dangerous but it is not. Pulling it toward the chest, one would have difficulty in bringing the blade edge in contact with the body. The knife is grasped by the right hand with the palm up, the blade of the knife pointing to the left of the user and the thumb against the flat side of the extension of the handle. The blade is flat on the bottom and all of the bevel is on the top. If working with a long stick, such as a canoe pole, the stick is tucked under the right armpit as the work proceeds, the thumb against the extension of the blade. The angle between the thumb piece and body of the handle varies considerably. Some like this are almost straight and others lean back almost at a forty-five degree angle.

The crooked knife is not a belt knife, nor is it carried on the person. In a rough scabbard, it remains in the kit bag in camp until there is a job to be done. As time goes on, successive sharpenings wear down the blade, like an old pocket knife. As a result the blade is often short and narrow. With an ax and a crooked knife, the woodsman can build himself a home and make almost everything that is needed to furnish it.

Notes on the American Indian Crooked Knife

By JOHN WITTHOFT

March 1963

I am startled to discover that the well-known carving knife of our northern Indians has been dropped into the Yankee tool-box, and that this distinctive tool can be discussed at length without any mention of its cultural context. Newton C. Brainard, "The Crooked Knife," *Chronicle* of the Early American Industries Association 1962, has welcomed the crooked knife into the fable of the pioneer woodsman. I suppose this is inevitable, for as fast as we have destroyed the Indian and his culture, we place him in a fantasy-world of sentimentality and adventure, the only realm in which we can rationalize a very bad conscience. Now even his craft tools have become ours.

The crooked knife is a universal and basic tool of Indian peoples of the northern woodlands. Many tribal and regional varieties are known, but they have not been adequately studied. Otis T. Mason's, "The Man's Knife among the North American Indians," 1897 *Report of the U.S. National Museum*, pp.725-45, Washington, 1899, is a preliminary monograph on the crooked knives of North America and Siberia, but is only an introduction to these fundamental tools. The Micmac specimen which Mr. Brainard illustrates is typical of this tribal style, but is only one out of many hundreds of specific types and forms of this exotic tool. As far as we know, the crooked knife has never had usage in a white community. It is a part of a totally different set of craft-skills and motor-habits, and like the rest of Indian culture, it is an emblem of contempt and degradation in the white mind. Like the specific tools of the Gypsy, the Negro, and the tinker, it is an implement that symbolizes a mean existence, and no Voyageur, no white trapper, no farmer would lower himself to use this powerful precision-carving knife.

Distinctive tribal and areal styles of the crooked knife are apparent, but have been poorly characterized. Micmac specimens are normally plain with a narrow thumb-grip. Malecite specimens commonly have a broadened thumb-grip, and are frequently carved and decorated in a fanciful manner. Some even have mirrors set into the handle. Mohawk examples are most like Malecite. More northern types, from the Montaigne, Nascapi, and Cree have shorter handles, narrow thumb-grip, and are generally bound with spruce-root or leather or wire rather than with ash basket- splint. Ojibwa examples are much like Malecite knives, but the Mississauga examples I have seen have a long, acutely pointed handle without a thumb groove. Seneca and Cayuga specimens have a simple cylindrical handle without a thumb groove, and the blade is bent into an S-shaped curve. Steel blades from Seneca and Susquehannock archeological sites of the seventeenth century look more like eastern Canadian types, however. Delaware and Cherokee knives of this type are unknown, but these peoples have adopted the scorper or arc-bladed knife of the cooper as a hollowing tool. They use a pocket knife for most wood carving, whereas the northern peoples do all carving with the crooked knife.

Steel blades for the crooked knife, made in Sheffield, are retailed by the Hudson Bay Company, and have been stocked in northern trading posts for at least the past century and a half. They are made in England only for this trade. However, one seldom sees a knife with a factory-made blade, and the people prefer file-steel as a material for these knives. Most northwestern crooked knifes and Eskimo crooked knives also include home-made, rather than manufactured blades.

Among the Eskimo and northwestern tribes, many other varieties of this knife occur, carrying this basic tool through a continuous distribution from Maine to Bering's Strait. In Asia, crooked knives of obviously related forms carry the continuous distribution of this knife across the spruce forest zones of Siberia. Although the knives are made of old world steel, the knife itself, its style, its motor habits, its applications, have nothing to do with Europe or with the white man. It is entirely a product of the northern forests of America and Siberia, made from introduced irons when these were available, as an innovation upon native carving knives with narrow cutting bits shaped from beaver teeth. Ethnological specimens of beaver tooth knives are well- known from the MacKenzie Valley, and archeological prototypes with antler handles are known from New York, Vermont, and Ohio. (William A. Ritchie, *The Pre-Iroquoian Occupations of New York*, Rochester Museum Memoir No.1, Rochester, N.Y., 1944, pp.115, 117, 360: pl. 54, 55, 56, 62, 65, 68) At least one example with a meteoritic iron copy of the beaver tooth is known from Ohio. Archeological specimens from the area of the central

Eskimo appear to include small crooked knife blades of both meteoritic iron and of Norse iron. Pre-Columbian archeological material from St. Lawrence Island and adjacent Alaska includes what appear to be crooked knife blades of Siberian or Chinese iron. These ancient indirect trade contacts had practically no effect on the events of American Indian culture history, but they did supply small amounts of iron for the making of purely American knife forms. Although the history of the crooked knife in the American northland is obscure, there can be little doubt that it is entirely a product of Siberian-American soil, and that at least a few steel-bladed examples existed prior to any direct contact with Europeans. The crooked knife is a beautifully designed craft tool that the white man has never adopted or used. It is the tool of an entirely different sort of man, a man who has met little in us except our scorn and contempt. It is no wonder we understand him and his tools so poorly.

The Crooked Knife

By WENTWORTH P. BLODGETT

December 1964

My interest and desire to do further research into the use of the crooked knife by the white man was aroused by the E.A.I.A. member, Waldon B. Hersey, who informed me that he had one which belonged to his Grandfather, and possibly goes back before that—certainly in his Father's time and his own, and was used by the whites. Both grandfathers of Mr. and Mrs. Hersey were partners in the lumbering business and ran the first log drives down the East Branch of the Penobscot river and were associated with the early days of that section of Maine.

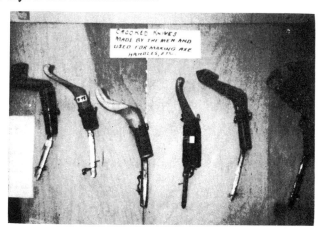

Figure 1. Crooked knives displayed in the LUMBERMAN'S MUSEUM, Patten, Maine.

There seems to be no doubt that the Indians figured out the form and the usefulness of the crooked knife. However, the white man who came in contact with it saw its advantages and soon adopted it to his own use. The knives became fairly common in the lumbering districts of Northern Maine and were used extensively to finish off axe helves, paddles, cant dog handles, pick poles, and other such articles after they had been roughed out with an axe. They were also used in carving gum boxes—little wooden boxes usually in the shape of a book—that were filled with spruce gum and given to their loved ones when the loggers and trappers came out of the woods.

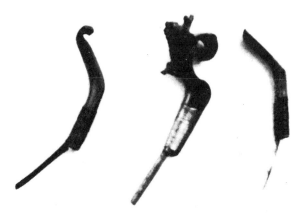

Figure 2. Crooked knives used by lumbermen.

The crooked knives shown in Figure 1 are from the collection of the Lumberman's Museum at Patten, Maine. The card attached to the display reads: "Crooked Knives made by the men and used for making axe handles, etc." In Figure 2, the plainest knife on the right was owned and carried to the logging camps before the turn of the century by Mr. Whitney of Bingham, Maine, whose wood lore is extensive as he has spent his entire life as a logger and timber cruiser. He informed me that when he went to the woods in the early years a great many of the loggers had these knives. He stated that some of the knives were really fancy and showed much skill in carving, as does the knife in the center of the picture. Of course, men being as they are, some had exotic female forms on the grips! The two fancier knives in Figure 2 were used by ancestors of a man I encountered in camp this summer.

The knife blades I have seen are about 4 inches long, more or less basiled on the top side, slightly dished upwards and usually made from an old file. The tang of the blade is fitted in a slot in the handle and usually bound with wire although I have seen one bound with rawhide.

From the evidence I have gathered from these and other sources there is no question but that the crooked knives were adopted by the white man in the northern regions.

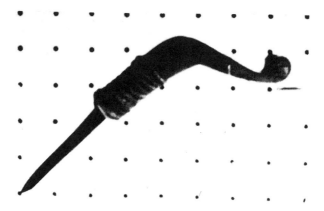

Crooked knife in author's collection.

How to Check the Set of an Adze Blade

By Arthur E. Woodley

March 1980

At the time of writing, Mr. Woodley, of Crib Point, Victoria, near Melbourne (Australia) was a veteran shipwright and a collector of shipwright's tools. He was in charge of the shipwright's shop of the Melbourne Maritime Museum.

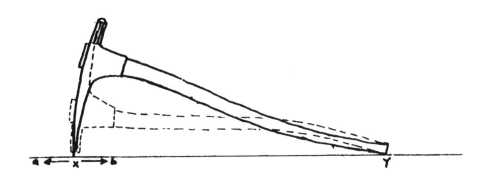

To check the "set" of an adze blade (as shown me by an old shipwright), place the adze pin down on a surface and place a mark at the end of the handle (Y); then make another mark at the centre of the pin (X). Turn the adze over, placing the end of the handle again at (Y). The cutting edge of a properly set blade should be at mark (X). In other words, the span between centre of pin and end of handle should be the same as from the cutting edge to the end of handle.

If this test should show the blade to be offset forwards toward (a), it is considered to be "on-the-wood," which gives the cutting edge a tendency to dig into the cut. Conversely, if the edge is set back towards (b), the set is "off-the-wood," causing the cutting edge to deflect off the wood and making a lean cut.

In using a badly set adze, one has to compensate for the offset with the grip of the handle, which soon becomes tiring, particularly when working to a line where accuracy counts, say, on a spar. It is obvious that the curve of an adze blade is part of the arc of the swing action in use, and if the cutting edge is out of line it will not function properly.

Adjustments to the set are made by shaving or packing the handle within the socket. The head end of the handle was usually supplied proud of the adze head for such adjustments.

It seems to me that most of the old shipwrights with whom I worked (many of them from shipyards in Britain) preferred to have their adzes set just very slightly "off-the-wood," particularly when doing a good class of work. Some of them even had two or even three adzes in their kits, tuned up for different classes of work. I clearly remember how fussy they were about maintaining that keen cutting edge, and God help the man they caught using their adze!

The Charcoal Burner

by JARED VAN WAGENEN

September 1947

In the years of which I write, charcoal was one of the standard commodities. It was the sole fuel of the absolutely indispensable blacksmith—then a personage of very much more importance in the community economy than now. The tinsmith (in New England sometimes called "tin knocker") used it to heat his soldering coppers. In a few cases there were iron manufactories which demanded it in large amount. Even today charcoal finds considerable use in the world, but an ample supply is available as a by-product of the so-called "acid-factories" where wood is distilled in iron retorts.

Several old men have written me concerning the method of burning it, but the most exact data is from a correspondent in Oswego County.

Wood for charcoal burning was commonly cut about four feet long. Any kind of wood might be used, but elm was considered especially desirable. The man who laid the kiln began with a pile of light, dry kindling wood in the center. Around this the wood was set up on end, leaning toward the center, and this was continued until a round pile was made about twenty feet in diameter. On top of this lower pile, another pile was constructed, the wood being set on end as before. A "pit" commonly contained 25 to 30 cords of four-foot wood. When complete the pile was thickly covered with earth and sods, and in general shape might have suggested an old-fashioned straw bee hive. Openings were left in the earth covering to give some draught at the beginning. When all was ready, the pile was fired by thrusting burning poles into it, and as soon as it was thoroughly on fire, all openings were closed with earth. To burn a "pit" satisfactorily required no small degree of skill, judgment and vigilant care. If the fire threatened to break out at any point, the spot must be immediately covered with fresh earth. The danger of the slow smouldering combustion turning into a conflagration was so great that the pit must be watched day and night for nearly two weeks. So the burner built a little cabin where he might be sheltered and possibly snatch cat naps as he watched the fire. Sometimes two pits were laid close together and the cabin built between so that a man might watch two fires instead of one.

My informant writes me that a cord of four-foot wood commonly made about thirty bushels of charcoal. It would seem, theoretically at least, that it should make more, but these were the figures given me. When burned, the charcoal was drawn several miles to Oswego City and sold for 14 cents per bushel. In other words, after all his labor and care, he received say $4.20 per cord of wood—an indication of how our fathers toiled for a little money.

Perhaps I may be permitted to tell a charcoal story in which my great-grandfather figures as the victim, but I believe the same disaster has happened to other men. More than a century ago, my mother's grandfather—one Alexander McNeill—was a Scotch-Irish pioneer in the town of Carlisle, Schoharie County. He once finished burning a "coal-pit", and at evening the wagon— topped with a big slatted charcoal rigging, like a barley rack— was loaded in order that a very early morning start might be made for Albany—some 40 miles distant. He laid down to a peaceful sleep and at the first streak of dawn arose and harnessed his team for the trip. When he came where his load stood, he found the irons of the wagon and nothing more. Evidently somewhere in the coal there had been a tiny speck of fire and when the air reached it, the inevitable happened.

Gathering Sawdust

By Jason Almus Russell

March 1950

My early home was in the midst of a New England lumber country. There, second- and third-growth pine forests alternated with oak, yellow birch, hard maple, and ash in a natural crop rotation. Almost every winter father hired a portable sawmill set up on some particular woodlot which had arrived at maturity. All too soon the removal of the great white pines, the feathery hemlocks, and tufted spruces, opened up new vistas across the valley to the gently rising highlands beyond, leaving in their place a wilderness of brush, slash and broken saplings, Such careless deforestation constituted a constant fire hazard for many years.

Prominent in the midst of this desolation a great yellow pyramid usually with a broken cone at the top, rose higher and higher with an ever broadening base. A blower (run by steam generated in an iron boiler, heated by a plentiful supply of dry resinous slabs) blew out a steady "stream" of sawdust to the ever-growing pile. This dust perfumed the air for a long distance with its pleasantly pungent pitchy odor.

After the pile had dried thoroughly on the surface, we made ready to procure our fall supply. Indeed, the local term "gather" was employed for no sensible reason except that of native idiom.

Attached to the East barn, the sheep-"linter" had long served as our sawdust repository. Once a year at the end of the summer, we proceeded to repair the shingles, nail boards over the cracks in the weatherbeaten boarded sides, tack pieces of tin cans over the frequent knotholes, and "plant-in" the broken flooring.

Our hired man went to considerable pains in nailing the side-boards on the wagon and wedging burlap bagging in between the cracks of the flooring. A goodly supply of tarred rope, bags, shovels, hoes and baskets completed our equipment.

Now the wagon bumped precariously over the rough cart-road to the distant sawdust pile. Once there, father drove the cart close to the slope of the pile for easier loading. The hired man and I shovelled the dry dust into the wagon while father filled bag after bag with which to ballast the top of the load. All day long we made trips back and forth from the homestead until the "linter" overflowed. Meanwhile our shirts and trousers were filled with the gritty particles producing a skin irritation, and we coughed incessantly from breathing the fine dust.

In the meantime, "neighbor children" who had ridden out with us in the empty wagon, amused one another by climbing up to the top of the pile, then running swiftly down the steepest part of the cone.

Dry sawdust was useful for many purposes. In the late fall father banked the base of the house and barn with it, taking care to cover this deep layer of home-discovered insulation with pine, hemlock or spruce boughs, primarily weighting it down to prevent the light "dust" from blowing away, but also for artistic reasons.

In addition, sawdust was the material used in the hollow shell-like partitions which insulated the icehouse. Sawdust also covered its earthen floor, and filled the chinks between the layers of ice. With dry sawdust we bedded down the cattle before spreading a protecting layer of meadow hay on top of it. In wet sawdust we buried wet scions from the various apple and pear trees against the spring grafting. Grandmother even sifted the dust, using the fine powder to fill rag dolls and pin cushions. These were but a few of sawdust's multifold uses.

Today, the casual motorist, the hunter, the country lad—all see nothing romantic in the sawdust heap, disfiguring the landscape, amidst the slash of a ruined woodlot. My father and grandfather however valued such a pile gleaming golden in the sun for its many uses around the farmstead in that thrifty period when nothing was allowed to go to waste.

Such was an early American Home Industry.

Another Tool Classification

By HARRY BAER

September 1964

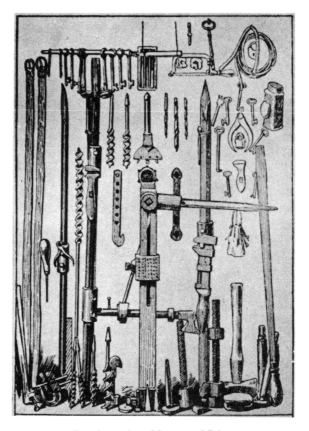

Burglar tools — Museum of Crime.

We are all normally accustomed to thinking of tools as implements wholly beneficial to man's welfare. Unfortunately, this is not always true. Tools used for a positive purpose have been, and can be, used for a negative purpose. It may, therefore, be of some interest to collectors to re-examine their more peculiar looking tools with a view towards fitting them into another possible category of use. By doing so they may, perhaps, understand the puzzling design of some tools and may even be able to classify those which are distinctly "whatsit" in character. We are, in fact, suggesting that there may be those who are the innocent possessors of old-time burglars' tools.

Although no dates are available as to the period when various types of burglars' tools were used, we have reason to believe that some of the "offbeat" tools pictured in the engraving go back to approximately 1850 when combination safe locks were first introduced. We say this because combination locks presented problems which the criminal, in the beginning, had to overcome by plain force. Lack of a key to duplicate and a keyhole to "pick" did not thwart him. Except for the "Raffles" type of criminal,

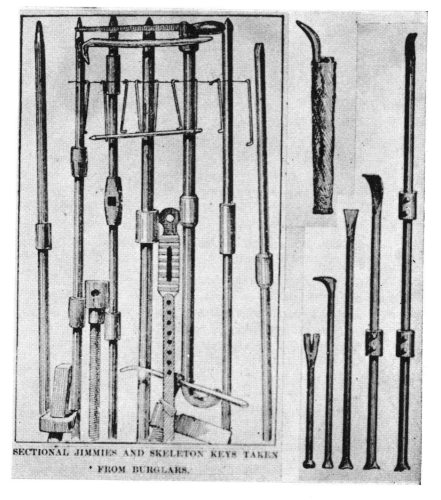

SECTIONAL JIMMIES AND SKELETON KEYS TAKEN
FROM BURGLARS.

Left, Jimmies and Skeleton Keys. Right, Sectional Jimmies. At the top is the carrying case.

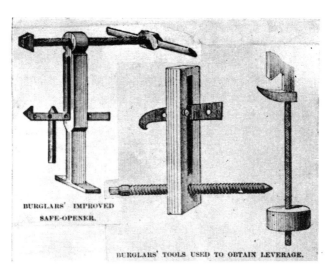

Burglar' Drag, an improved safe-opener.

who, as fiction would have us believe, could open any combination lock, the practical, hardworking old-time crook did not care to spend too much time twirling a combination lock.

The safe crackers, bank robbers and burglars of yesteryear often displayed remarkable mechanical ability not only in adapting and modifying time-tested tools for their illegal activities, but also in devising ingenious special purpose implements. Although the ancient crook had to procure his tools by devious methods, almost any blacksmith was available to make tools if it was within his capacity to construct them — with no questions asked. As a rule, however, the making of such implements was confined to mechanics who were actually involved with the criminals.

A basic problem confronting the burglar then, as now, was the need to construct his tools with an eye towards portability. While he could overcome this aspect of design to a reasonable degree, the heavy weights he had to carry must certainly have made him a very tired man at the end of a night's work.

One type of tool peculiar only to the burglar's trade was the sectional "jimmie," a most important device for forcing entry by powerful mechanical leverage. Here a coupling or couplings joined the pieces which were assembled as needed on the job. Wrecking the bank's or merchant's safe required first the use of a drill to make a small opening, after which the safe door was sprung with the jimmie until the door hinges broke. Steel wedges pounded into the spaces around the safe door, widening the spaces until the jimmie could be inserted, accomplished the same purpose as a drilled hole, but created more noise. Silence was indeed golden.

If the jimmie and the wedge were ineffectual, one of several types of apparatus was brought into play. The jackscrew or "drag," simple as it looked, was extremely

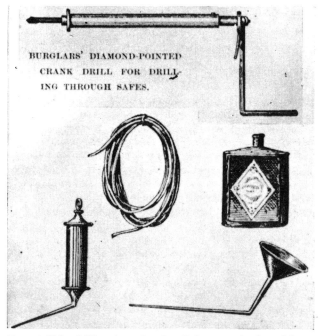

BURGLARS' DIAMOND-POINTED CRANK DRILL FOR DRILL-ING THROUGH SAFES.

Top, drill for drilling through safes. Below, powder funnel, fuse, hand pump to blow the powder into the safe crevices.

powerful and very quiet to operate. In this case, after the hole was drilled, the bend end or filed shoulder was hooked into place on the inside ledge of the hole and the screw turned by a long handle. It was a power that few old-time safes could withstand.

The more reckless of the safe robbers, of course, used explosives, but the patient, careful craftsmen preferred the noiseless method of wrecking the safe with tools — noiseless, that is, except for his grunts of exertion.

While the tools illustrated are mere curiosities of a bygone age, collectors possess-ing such implements should avoid carrying them about in a bank on a dark night.

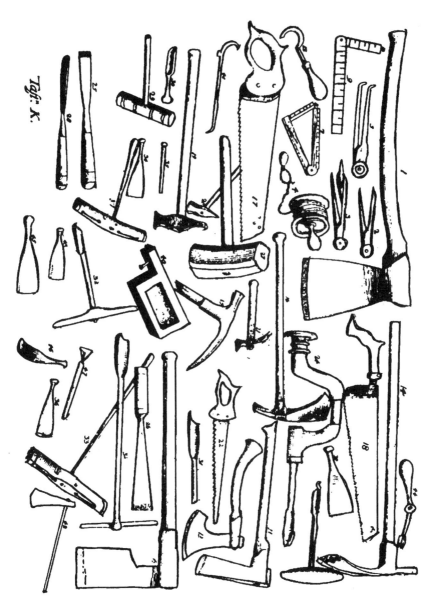

PLATE FROM RALAMB'S BOOK ON SWEDISH SHIPBUILDING 1691

Tools and Equipment of the Early Settlers in the New World

By W. L. GOODMAN

September 1976

The author, W. L. Goodman, is one of the outstanding living authorities on woodworking tools and is the author of the well known books British Plane Makers from 1700 *(New York, 1968) and* The History of Woodworking Tools *(London, 1964) and many articles. This essay was the Early American Industries Association's contribution to the Bicentennial.*

For some years I have been enjoying a regular correspondence with Benno M. Forman of the Henry Francis du Pont Winterthur Museum, Winterthur, Delaware. Of late he has been sending me a series of inventories of the estates of woodworking tradesmen of the seventeenth and eighteenth centuries, which he has been studying in connection with the research for his forthcoming book on early American furniture of this period in the museum, mainly for comments on the tools mentioned. These lists were, of course, all new to me and give fascinating insights into the type and range of the tools and equipment in use in this period of rapid development of the craft of woodworking. I felt I must do something to get this fresh information properly organized.

By a fortunate coincidence, Elliot M. Sayward came up at this time with a request for something for *The Chronicle*. He suggested as a theme, in connection with the Bicentenary anniversary of the Declaration of Independence, what is known of some of the carpenters and joiners who emigrated to the New World in this early period, and the kinds of tools they were likely to have brought with them. It is not often that such an opportunity occurs to kill two birds with one stone and I could not refuse the challenge. The present article is an attempt to answer it. I hope members find it as interesting to read as it was for me to write. Mr. Forman has very kindly given me permission to use his material, for which I am humbly grateful, and I have also included some research of my own on this side to fill in the background. In all the lists I have, for the sake of the record, retained as far as possible the original spelling and arrangement. In some cases, the prices are given, but it must be remembered that as a rule these are useful only for comparative purposes in the same list or in others of about the same date from the same locality. The notes deal only with the more important or problematical entries, but the ordinary tools have been left in to complete the record.

In Volume 5 of the *Bristol Apprentice Register* there is an entry on f. 23 dated 4th December, 1616. It is in Latin, which was the custom at the time, but a rough translation runs as follows:

"John Morgan, son of Edward Morgan, of the City of Bristol, sailor, dec[eased], puts himself apprentice to Francis Eaton of Bristol, Carpinter and Dorothy, his wife, for a term of Seven Years, at the end of which term he will have 4s 6d for the fee for the freedom and double apparel etc., and one set of tools appertaining to the art of a Carpenter aforesaid." In the margin there is a note in English: *"The mr [master] is at New England"* and below the entry, another: *"The mr covenanteth at thend of said terme to convey to thapprentice & his heires forever 25 acres of land lyinge in New England in America & alsoe to give vnto him 15 bushells of Wheate he servinge him truly the terme of his apprenticeship."*

Nine years earlier (Vol. 4, fol. 151v. 1617), Francis Eaton, carpenter, and Sara, his wife, had enrolled William Rocke of Clevedon, Somerset, for nine years, paying for his freedom and promising him in English "one toole of everie sort belonging to his trade." Three years later, Francis Eaton and Sara sailed in the *Mayflower*. He is reported to have acted as the ship's carpenter. But this, if true, would have been a temporary arrangement, as a ship of that size would have carried a ship carpenter as a regular member of the crew. Also, a house-carpenter like Eaton would not have had the proper tools for the job. But working at his own trade on his arrival, he would have been and apparently was a very useful member of the pioneer community. Unfortunately, his wife Sara did not survive the first winter, and in 1622 he married again; his second wife's name is not known. Possibly she was the Dorothy mentioned in the *Apprentice Register*. Later, in 1626, he married for the third time, Christiana Penn (a surname common in Bristol), then eighteen years of age, by whom he had four children. There is no other mention of Francis in the Bristol records, but in 1587 a John Eyton of Purseahed, Somerset (now Portishead at the mouth of the Avon, about six miles from Bristol) was put to John Batten, Junior, for many years one of the leading Bristol house-carpenters, and his wife Maria, for seven years and one year covenant, which would bring him to 1595-96. According to the American records, Francis Eaton was born in 1595. Assuming that John was his father, Francis could have served his apprenticeship to him without being enrolled, which was quite usual for father and son. In 1617, he would have been 22 and able to take apprentices himself. Nothing more is known of John Morgan or William Rocke, but Francis Eaton died in 1633 at Plymouth, Massachusetts, "of an infectious fevoure." It was interesting to find that one of the Pilgrim Fathers was a Bristol carpenter and in all likelihood had a North Somerset background.

Some idea of the kind of tools which a house-carpenter like Eaton would have had about this time can be obtained from another entry in the *Bristol Register* for 10th February, 1611 (Vol. 4, fol. 37v).

"Thomas Thomas, son of David Thomas of Pentether in the Co. of Glamorgan puts himself to Thomas Phelpes of the Citty of Bristoll Howscarpenter and Maria his wife for 8 years and one year Covenante, having 40s and 4s 6d for the liberty of Bristoll and

double apparel, and one axe one Addys two Chezilles one Mallat two borriers one Squire & one Compasse one Joynter one foreplane one Rabbett plane one bowltle plane & one Cadgment plane and one smoothing plane at the end of the yeare Covenant."

The clerk had started to write the usual formula, but fortunately for posterity the master insisted on naming the tools (but forgot to mention saws). These were only the basic tools, of course. The apprentice would build up his kit on these as he went along. The "borriers" were augers; from the German *Bohrer,* the Bristol dialect being then, and still is to some extent, almost pure Saxon. For a carpenter the plane group is well represented and includes a rabbet plane and two moulding planes, as well as the bench planes; jointer, fore or roughing plane, and smoother. Before the introduction of the Italian names for moulded profiles some time later in the seventeenth century, English masons (cf. Moxon, p. 267) and woodworkers used two types of moulding: the "boltel" or "bowtell," for projecting mouldings such as beads or ovolos, and the "casement" or cove, later known as the scotia.

In connection with moulding planes, a list of the tools promised to his apprentice John Wysse in 1551 by Thomas Woode, an aptly-named Norwich carpenter, is of interest:

"...a ioyncter a chesill a former iij inboweng planes a handsaw a hatchet two wymbles a handplane a stapler hoke a parser."

"Inboweng planes" was a late-medieval term for the precursors of our modern "hollows and rounds," which as late as 1899 were still described in an Alex. Mathieson of Glasgow catalogue as "casement planes." "Inbowing" was probably regional and most common in East Anglia and the east Midlands (see references in Salzman's *Building in England to 1540*), but appears to have died out and the very meaning of the word forgotten (see *Oxford English Dictionary*). Yet it must have been taken to America by one of the early settlers, as Benno Forman has found it in an agreement for work done on a ship being built in William Greenough's yard in 1685 by John White, joiner, of Boston, New England, and Arthur Tanner, mariner:

"...To Inbow all rails that shall be placed on said ship with a fffife raile. . ."

It also occurs in a contract from Portsmouth, New Hampshire, dated April 6, 1698, which specifies that *"the carpenter is ordered to imboe all the Summers, girts and beames. ..and plane the parts of the framing members...which is to be seen after the Finishing of the house."* A picture of the room in question was taken before its destruction in 1920 and shows that *". . .the lower margin of the chimney girt and the interior margins of the framing members is a planed quarter round"* (what an English joiner would call a "square or common ovolo.) In addition to this, the diary of Joshua Hempstead, a farmer/joiner of New London, Connecticut, has an entry dated November 14, 1717: *"I worked at Jno Coits' Emboing Waals for ld per foot and John to help me. I did 48 foot."* No doubt a good day's work. It seems quite likely that these men came originally from Norfolk or the Cambridge area in England and were still using the term over a century later than Thomas Woode. Incidentally, his "Stapler Hoke" is probably a form of Draw-bore pin, known to Moxon as the "hook pin."

Another kit of carpenter's tools from Norwich is that promised by John Fellbrygge, carpenter, to his apprentice Robert Clarke in 1560, also containing some interesting regional tool names:

"...a mortas wymble a fote wymble too draft wymbles too little wymbles a former a chesyll a belt a hatchet a Joynter a fore lopper a hand plane a little hand sawe a passer a wrest a fyle a payer of compas & a sqwyer or 7s 4d in consyderacion of the same toolys at the choyse of the said Robert..."

The problem here is whether all these different "wymbles" were augers or what we now call bitstocks or braces. My own view is that they could have been either or both. After all, they did the same job in more or less the same way and the name was simply transferred from one to the other. The "mortas" and "draft wymbles" were probably fairly large augers for starting the mortices in the framework and boring the holes for the "draft hooks" and trenails to hold it together. The "fote-wymbles" for longer holes at the foot of the posts and the "too little wymbles" were very likely bitstocks, complete with their own bits, for pilot holes for the augers and giving the nail a start. All these are shown, both in use and lying around the site, on the well-known picture of the building of the Ark in the Bedford *Book of Hours*.

The "belt" is the East Anglian term for a large, general-purpose carpenter's axe; the "hatchet" is the smaller side-axe. The "fore lopper" among the planes is a direct transliteration of the Dutch *voorloper*. This is not surprising, as Holland is only on the other side of the North Sea from Norfolk. The "passer" was a small auger or gimlet. Later, in the form "perser," as in Moxon, it was used for the brace itself.

A typical list of carpenter's tools, together with those of other tradesmen, is shown in the inventory of George Barell of Boston, Massachusetts, dated 1643. It has been suggested that he was an ironmonger, but Savage states that he was a cooper. With a name like that, he could hardly have been anything else.

Tools in the shop

23 Chisels	11:0	4 compasses, 3 prickers	6:0
10 fyles, 2 wrests	0:9	15 Augers	1:3
10 Saws	1:13:0	A turne & 15 turning chisels	
Four Copers axes	1:16	and other things in that house	1: 5:0
4 Carpenters axes & two hand		7 Squires, 3 tileing axes,	
hatchets	10:0	1 Carps ads	8:0
4 adses, 4 havels	0:10	9 wedges, 1 Crowe yron,	
12 shaves	:12	4 smale screws, a lead	
four hammers, 3 pairs of pinsors	7:0	panne, 2 holdfasts, 1 vice	12:0
41 perser bitts, 2 Phrowes	7:0	89 trusse hoopes	1:2
24 plaines	1:11:0	For Cowper ware in the yard	
2 Joynters, 2 Crookes	14:0	& a grindstone	2: 6:0

The "havels" were "howel adzes" for cutting the curved recesses around the inside ends of the staves to take the heads, and, generally speaking, what coopers could not do with an adze they did with some sort of "shave." The "perser" as a bitstock is not mentioned, but there are plenty of "bitts" to use with one. This is an unusual way to spell "frow," but coopers used them a good deal for cleaving timber and splitting hoops. The large cooper's "Joynter" was supported at one end by a stand of some kind, here described as a "Crooke." The use of "turne" for "lathe" follows the French *tour.* The word can, in fact, be traced as far back as the Greek. "Wedges" would also be required for riving staves and heavy timbers. The "trusse hoopes," a rare but valuable entry, are the extra-strong hoops of standard sizes for setting up the barrels. Considering the date, it is a pity the "24 plaines" were not given in more detail, but we must not complain.

Another typical list of mainly carpenter's tools is given in the inventory of William Adams, Jr., a farmer of Ipswich, in the probate records of Essex County, Massachusetts, dated 1658:

one handsaw	4s	a hollow shave	1s
one Broad & narrow chessell		fowre planes	5s 6d
& goudge	4s	a sqare	2s 6d
two Addes	4s 6d	a payre of pinsers	5d
one croscut saw	5s	one beetle & 2 small wedges	4s
fowre Augeers	7s	fowre axes, broad and narrow	16s
five cheesells & goudges		a small hand vice	5s
for turneing	5s 6d	a plane, alls & gimlet	1s 3d
a percer stock & bitts	4s	a grinstone	8s

We now have the "percer stock & bitts" in the clear, and Adams includes the "beetle" or large ringed mallet he used with his wedges. Other special tools for cleaving timber are shown in the lists of Thomas Baynley, a turner and carpenter of Concord, Massachusetts, dated 1643, including a "clapboard axe" and "3 wedges" and that of Thomas Bloggett of Charlestown in the same year, who had "a lathe frowe" priced at "1d."

Up to the time of writing at least two Bristol-trained joiners are known to have emigrated to New England by 1634. In 1621, Jenkyn Davis, the son of a shoemaker of Cadwally (now Kidwelly), Carmarthen in South Wales, was apprenticed to John Faylande, a Bristol joiner, and Elizabeth his wife for eight years. In 1630, a year after finishing his time, Jenkyn took as his own first apprentice his brother George. In 1634 John Humfrey of Sandwich, Kent, who later became deputy governor of the Massachusetts Bay Company, emigrated to Lynn, Essex County, near Boston, taking with him as a "servant" Jenkyn Davis, a joiner, who was sworn as a freeman in Lynn on 9th March 1637. He is the earliest-known joiner and furniture-maker in this part of the colony. Presumably his brother George served the rest of his time with him, being made a freeman at Lynn in 1647. By his wife Sarah, who is not mentioned in the Bristol record (suggesting that he married her after settling in America) Jenkyn Davis had a son John, born in 1641. A man of piety and in good esteem with his neighbors, Jenkyn Davis served his term as constable in 1650 and died in 1662. In his will he left his joiner's tools, valued at £8, to his son, not yet twenty-one years of age. At the time of John's death in 1702 his estate was valued at £102.12s, including joiner's tools worth £5.10s and "borde timber for the Joyner's trade £2."

It is not known what Jenkyn Davis would have taken out with him in the way of tools in 1634, as the custom of promising specified tools to apprentices completing their term of service had been discontinued early in the seventeenth century and had never been very common with joiners in Bristol. But as a comparison, in 1591 John Runhalle, a Norwich joiner, took William Whall as apprentice for seven years and promised:

"...to teach the said Willm Whall the younger his seid occupacion as far as his wit and capacitie will serve and to finde him meate drinke loding & aparell in sickness and in health & in thend of the seid terme to pay & delyver to the said Willm Whall the younger the summe of Twentie shillings of lawful money of Ingland & also afore playne a joynter a passer a hammer a handsawe a paring chisell & Six carving toles meete & convenient for a workman & two sutes of aparell aswell lynning as woolen vidz one for the holy days & one for the working dayes."

Whether Runhalle had some doubts about the mental abilities of his pupil or was diffident about his own, is not clear, but he treated him very generously. The paring chisel is notable as a typical joiner's tool and like all joiners of this period, he was provided with a set of six carving tools. In fact, some joiners were known in Elizabethan times as "carvers," as all Scandinavian joiners are to this day.

A few years later in Bristol (*Apprentice Register,* Vol. 3, fol. 32, 1594), a joiner's apprentice was dealt with even more generously:

"John Sparke son of John Sparke of the Citty of Bristoll puts himself apprentice to Humfrie Bryne of the same Joyner and his wife for Seven Years paying him at the end of the years 4s 6d for the liberty of Bristoll & double aparell and two years Convenant at the end of the years aforsaid and paying him at the end of the Covenant years 53s 4d and a Rule a compasse a hatchet a hansawe afore plane a Joynter a smothen plane two moulden planes a groven plane a paren chysell a mortisse chessell a wymble a Rabbet plane and six graven Tooles and a Stryking Plane."

The 53s 4d for the two extra (covenant) years after the usual term of apprenticeship would be about one-third of a full craftsman's annual wage at the time. It would have been a useful sum to set him up on his own and marry a wife, which was very often their first move.

It would be nice to know what the "Rule" looked like. The full complement of bench planes was supplemented by two moulding planes, probably a hollow and round like the carpenter's "imbowing planes" or even an ogee and ovolo, which are sometimes found on paneling and furniture of this period: a rabbet plane and a grooving plane, possibly a fixed plow. The paring chisel was also included with a special mortice chisel and the usual set of carving tools. The "Stryking" plane with the latter was almost certainly a small carver's router plane for leveling the background of a carved panel, a process similar to striking off a measure of corn. Moxon mentions a "striker" as a tiler's tool on page 248.

A further stage in this development is shown by the earliest known list of tools of a New England joiner, George Coall or Cole, of Lynn, Massachusetts. He was a former apprentice of John Davis, son of the Jenkyn Davis from Bristol mentioned above. John Davis attested this inventory in September 1675, when George Cole made a will before joining an expedition into the Narragansett Country. The tools are as follows:

3 saues	8s	3 plaine irons & 3 bits	1s 6d
2 goynters & foreplaine	6s	1 brose stock, 2 squares and gorges	1s 6d
3 smothing plaines &		1 brod ax and 1 fro	2s
a draing knife	3s 6d	holfost	1s 6d
2 Plains and 2 revolving plains	10s	hamer	1s 6d
4 round plains	5s	6 gouges	2s
3 rabet plains	4s	9 Chisels	5s
3 holou ploins	3s 6d	2 ogers & 1 draing knife	3s
9 Cresing plains	10s 6d	1 bench hook, 2 yoyet irons	1s
6 torning tools	9s	a gluepot	1s 6d

It looks very much as if this list, like many others dealing with craftsmen's tools, was compiled or copied by a clerk with only a vague idea of the objects he was recording or their uses. Some of the entries can only be interpreted by assuming that a joiner must have had certain tools to be able to do his work and the puzzles or "whatsits" are the gaps in the context. The bench planes include a roughing plane and two jointers, usually one about 30" long for shooting glued joints and a shorter trying plane about 22"-24" for truing up workpieces, with a selection of smoothing planes suitable for different types of finish. The "2 revolving plains" are almost certainly a misreading for "grooving planes" or plows; the "2 Plains" in the same line a couple of moving or side fillisters, in addition to the three fixed rabbet planes lower down. The hollows and rounds have now been given their proper names, even if not yet joined in holy matrimony. The "9 Cresing plains," like the hollows and rounds, with an average price of just over a shilling each, would be various moulding planes, mainly side beads and reeds, and possibly including one or two ogees or ovolos. The term "cresing" has not so far been encountered on this side of the Atlantic in connection with planes, but the tinman's "creasing irons" are used to make something like beads or flutes in metal.

The "brase stock" is probably connected with the "3 bits" in the previous line. It is either an early form of brass-plated bitstock or even a very early use of the present term "brace." Either alternative would be very surprising as early as 1675. Personally, I would prefer the second solution. As "6 gouges" are mentioned separately lower down in the list, the "gorges" must be something else. The most probable solution, taken with the "2 squares" preceding them, is "gauges," always a difficult word to spell in any case, although the clerk got "squares" right, for the first time in these lists, as far as I remember. The "yoyet irons" with the bench hook, also a newcomer, are very likely a misreading of "joint irons," either a pair of cramps or joiner's dogs.

In the same year, 1675, Joseph Carpenter, a turner and wheelwright of Rehoboth, Plymouth County, who, according to Savage, was born in Horwell, England, died in Plymouth, Massachusetts, leaving a wife Margaret and the following tools:

Spokes and Timber for Worke	5'	a square and spokeshave	6'
2 Broad Axes	15'	2 bursses 1 Great gouge	18'
a Crosscutt Saw & a Hand saw	10'	5 augers 3 wrybitts	1:01:00
2 Adds	10'	2 pair of plyers	3'
a frow and a holdfast	5' 6d	3 plaine Irons	2' 6
2 beatle rings & 4 wedges	10'	wimble bitts and a sett	4' 6
2 hatchets	5'	2 Smale saws and a scribe	4'
a viz	2'	a hamer and a m[ ]laing tool	2'
a spokeshave and Gripers	5'	a manderell	1.06
a [ ]arre and screw & bo[ ]		red occur & Glew	1'
& turning tooles	16'	smale nailes for wheeles	1'
plow plaines & a Joynter	1: 5:00	a burning iron	1'
a Greate Joynter	5'	2 pairs of Compasses & prickers	2'
a [ ]ave screw & turning tooles	1: 4:00	Working benches	5'
		6 Spinning wheels	1:10:00

In spite of his name, there can be no doubt that Carpenter was a wheelwright and turner. This was an unusual combination of trades although most wheelwrights did a bit of turning at times. He must have prepared a good deal of his own timber, much of it riven, judging from the frow and the wedges, but apparently he had broken his "beatle" but had not had time to make another on which to fit the rings.

The line beginning "a []arre and screw & bo[]" could be interpreted as "an augre and screw & box," as in the Fellows list below, for making his own wood screws, which would probably be useful on the spinning wheels. His planes were confined to plows and two jointers, but wheelwrights and wagon builders do most of their shaping with their axes and spokeshaves, mentioned on two separate lines. One of these may have been a drawing knife. As a rule, in the seventeenth century the spokeshave was associated more closely with coopers, in spite of its name. The "[ ]ave screw and turning tools" is a bit of a mix-up. It may be another version of the screw tools line, with one set of turning tools in the box or drawer with the taps and dies, and the others in a rack on the lathe itself. The entry "2 bursses 1 Great gouge" is most interesting. The bruzz or bur chisel is a large, stout chisel with a V-shaped blade, indispensable to a wheeler for chopping out the mortices for his spokes. R.A. Salaman notes on p.135 that "bruzz" is equivalent to "birse," which, according to Wright (1898) is a Northern (English) dialect name for a triangular chisel. Possibly Carpenter was himself a north countryman. The "Great gouge" in the same line is also a typical wheelwright's tool for boxing the hub and other jobs. The "wrybitts" in the next line are small twist bits, similar to those now known as "Swiss" or "German" gimlet bits. A modern use of the same term occurs in a recent "Arnold & Walker" catalogue, where the timber dogs with one end twisted round in a different plane to the other are called "wry dogs." There are also some "wimble bitts" a few lines further down, but no wimble anywhere to put them in. Another typical item of equipment belonging to a wheelwright is the "sett," usually called the "spoke set" or "set stick," fully described in the invaluable Salaman on page 511, for checking the "dish" on the spokes of the wheel. The timber scribe in the next line is also described in Salaman on pages 484-486. The "hamer and a m[ ]laing tool," which may well be "marking tool," is also used for marking timber. The "burning iron" is described in Mercer under "Wheelwright's Burning Rods." It is remarkable that Joseph Carpenter should have been using these very special wheelwright's tools over three hundred years ago.

In 1694 a set of tools belonging to John Fellows, a deceased joiner of Philadelphia, Pennsylvania, was appraised by another joiner, Abraham Hooper, who is said to have come from Bridgwater in Somerset and had emigrated before 1687. There is no record of Hooper in the Somerset County Archives Office, but he was undoubtedly a man of

2 grind stones	£ 1. 4. 0	One Shave, One Drawing knife,	
1 Frame Saw	15.0	one rasp & pr nippers	4.6
3 []king Chissells and one Leathe	2. 0.0	3 Augers, 2 Mallets,	
Six Saws	10.0	one pr Compasses	3.6
a pcell Turning Tooles	7.0	2 Glewing Jointers and	
40 Moulding Plains	2. 4.0	Large Compass	7.6
9 other plains	9.0	One Glew pott & one melting Ladle	3.6
9 Plain irons	5.0	One Screw Auger and Box	3.0
Five Chissells	6.0	2 small files and one Saw plate	2.6
12 Fourmers	3.0	2 Hold fasts & one bench Hook	6.0
5 Mortis Chissells & 3 Hamers	3.9	One plow ls 6d, 2 hatchets	
Seven gouges	1.9	one ass & one plain board	7.0
2 Wimble braces some Bitts	3.2	8 Bedd Screws	2.0
		25 lb of Glew	1. 5.0

the West country. Coopers were always known as "hoopers" in this part of England, especially in and around Bristol, and the surname is still very common here. The tools list is as shown on the previous page.

Evidently Fellows did a good deal of his own turning, like most joiners in this period, and had his own "Leathe" and plenty of turning tools to use on it. The undecipherable word ending in [ ]king in the third line is probably "hooking"; these are usually listed as "hook tools" and were in common use before the advent of the modern powered lathe.

The "40 Moulding Plains" are a good indication of how far the new joinery methods had developed. The "9 other plains" possibly represent the bench planes. There is also a "plow" further down, probably an oversight, and a set of plane irons in reserve.

The inclusion of "12 Fourmers" among joiner's tools is unexpected. The old-fashioned "former" was a large carpenter's roughing chisel and as a rule they had only two or three at most. This is very likely an error for "firmer" chisels, from ⅛" upwards, especially as the average price was only 3d, whereas the preceding "Five Chissells" averaged over a shilling each, possibly the missing paring chisels as opposed to the "Mortis Chissells." The "2 Wimble braces Some Bitts" gives us another variation of the old boring-tool problem—or what might be called the boring old problem of the wimble. This solution makes the best of both worlds and is the main reason why I prefer the "brace" version in the Cole inventory above. According to the *Oxford Dictionary*, the word "brace" as a carpenter's tool first occurred in 1567. It must have been a very shy word, as it makes only very rare appearances in the 17th century, as in these lists and in England it ousted the "wimble" only towards the end of the 18th century, and then mostly in the official toolmaker's lists.

The "Screw Auger and Box" takes us another stage further in cleaning up the problem of this particular "whatsit," which has puzzled both the clerks and ourselves. The "Saw plate" with the two small files, no doubt saw-files, would be the uncut blade of a spare handsaw. In the early days of the wide-bladed saws they were often described as "plate saws" to distinguish them from the narrow webs of the framed saws. The "two hatchetts one ass" are obviously two side axes and an adz, but I have no idea what a "plain board" could be, unless it is some kind of shooting board, possibly for mitres, or perhaps a sticking board for moulding sash bars.

The tools in the inventory of Charles Plumbley, a joiner of Philadelphia, Pennsylvania, which were appraised on the 15th December, 1708, by Joseph Phippen and John Jones, a fellow joiner, show how comprehensive a first-class joiner's kit had become in the early years of the 18th century. Of the timber listed, over 60% of it was walnut, which was still very much in fashion at this time, although good deal of oak and pine was used and a small amount of mahogany.

The tools are as follows:

5 Hand Saws	1. 8.0
4 Tennant Saws	16.6
3 beau Saws	7.6
3 Small Saws	3.0
33 formers & broad Chizells	1. 0.0
14 Gowges	6.0
5 Mortice Chissells	4.2
9 Wyreable bitts	4.6
4 Center bitts and 1 Dott bitt	5.0
2 Wimble stocks 1 Iron 1 Wood	4.6
1 Drawing knife 15d 2 Hatchetts 5s	6.3
2 Long Plaines 1 Jack Plaine 2 Swich blocks 3 Smoothing plaines new	12.0
1 Jointer Yellow Jaunders 7/7 2 ditto beach 7/6	15.0
3 Long plaines 7/6, 3 Jack plaines 4/6; 3 Strike blocks, 4s & 3 Smoothing plaines	19.0
5 Hammers 7s 3 Raysors 3s	10.0
1 Morris Fraime	6.0
1 Chalk Role 10d 1 Hand Vice 2/6	3.4
3 Gimblets ls 2 Punches 4d; 2 Saw Setts 10d 3 Compasses 18d [?]	4.5
4 Wooden Squares 2/8 2 Levels 1 Setting Square 2/6	5.2
The best bench 13/ 1 Ditto 10s	
1 Ditto 7s	1.10.0
One Wheel and Layth &c	1.10.0
22 Hollows and Rownds at 15d each	1. 7.6
9 0 Gs at 15d; 6 belexions 14d pr asticle 15d 1 Sash plain 15d 1 groove plain	1. 2.0
3 Rabbitt plains 2/6, 2 Philisters 2/6, 1 pr Inch 1 pr 1/2 inch 3/4	8.4
2 Ploughs 6/8 1 Revele Plain 2/ 1 Small Cornish 3/	11.8
2 Rownd Smoothing Plaines & 1 Hollow ditto	2.0
1 Upright Smoothing Plaine 3 Half Uprights ditto	2.8
4 Old Chizzells 20d 10 Augers 15s 3 Screws & nutts 9s	1. 5.8
8 Turning Gowges 8s 8 Turning Chissells 8s	16.0
10 Ivory Turning Tools 8s 8 Turning hooks 8s	16.0
5 Iron Turning Mandrells at 5d	2.1
1 looking Glass	16.0
2 Iron hold fasts 4/6; 6 Glew potts wt 41 lb	17.1
1 Small Vice 5/ 29 pr Snipe Bills 7/3	12.3
2 Rules 5/6 17708 Spriggs & Tacks at 4/2 per m	3.19.3
8 Turning Strings 16/ 4 pr Fish Skins 10/	1. 6.0
22 New Wimble bitts 11/ 7 large 1/2 round files 7/	18.0
9 Small half Round files 10d; 8 Hand saw files 5/4	12.1
4 small Hand saw files 2s 12 molding plaine irons 5/	7.0
3 Plow Irons 1s	1.0
2 Oyle stones 5s 1 lb Beeswax 18d	6.6
7 New Plaine Irons 5/10, 3 Tooth Plaine ditto 2/6, 1 Saw plate 10d	9.2
3 Old Paring Chissells 15d 2 New Varnis brushes 2s	3.3
7 plaine Irons 3/ 2 Tooth Plaine do. 1/8	4.8
76 lb of Glew 14/ 4:8:8 1 pr Calipers 20d 1 pr pinchers 10d	4.11.2
3 Rasps 3/, 2 Wood files 2/ 2 Quarts Varnish 25s	1.10.0
2 pine chests	1. 4.0
1 large Frame Saw	9.0
1 Grindstone	7.0
6 Veneaireing Screws	6.0
2 Ordinary benches in old house	11.0

There are reasons for thinking that either Plumbley or John Jones, who helped to check the list, or both, were English or had a recent English background. They had also very likely come across Moxon, whose *Mechanick Exercises* had been circulating in parts from 1683 and published as the *Doctrine of Handy-Works* in 1703, only five years before the date of the inventory. Quite a number of the tool names are those first found in Moxon and usually attributed to him in the dictionaries. It is rather amusing that in some cases they repeat Moxon's errors and obscurities, but in others help to clear them up.

Taking each group of tools separately, all the saws in the list are those mentioned in Moxon, including the "Large Frame Saw" near the end, but set out in a more methodical way. It was probably overlooked because it was hanging up in a dark corner of the shop. Moxon had a lot of trouble with his saws because the only illustrations he had available were those in Felibien's *Principes de L'Architecture* of 1676. These were, of course those used by French and other Continental woodworkers of that period and still are to a large extent. They are, apart from the crosscut at "N," various types of framed saws with narrow blades or "webs." Yet Moxon seems to have been aware of the existence of the new broad bladed hand and other saws which had been developed by English sawmakers from about the middle of the 17th century if not earlier. He got his artist to include one in the bottom right hand corner of his plate of joiner's tools, but it has no reference letter and there is no mention of it in the text. The old-fashioned narrow-bladed handsaw with the tanged and ferruled handle is clearly shown at "E," not "D," and the new illustration, although poorly drawn, does bring out the essential difference between the new and the old. The handles were shaped from a flat blank and fitted to the blade with screws or rivets. The clue is given by the inclusion, among the miscellaneous tools at the end of the list, of "1 Saw plate 10d," as in the Fellows list fourteen years earlier. The best and earliest illustration of the new pattern is in Ralamb's book on Swedish shipbuilding, published in 1691, where it is specifically described as "An English Handsaw." The nearest Moxon gets to this is his reference to the "Tennant-Saw," which, "being thin, hath a Back to keep it from bending," which clearly describes a special wide-bladed saw, of which Plumbley apparently had four, one on each bench and one spare. The "3 beau Saws" in the next line are obviously Moxon's "Bow Saws," the English version of Felibien's *"Scie a tourner"* or framed saw which can be turned to any angle for cutting curved work. As far as is known, this is a term first found in Moxon and still new enough in America for the clerk to think it might be French. What the "3 Small Saws" at 1s each are is anybody's guess, but they could be small backed saws or some kind of pad or compass saw.

The confusion in the Fellows list between "Chissells" and "Fourmers" is repeated here and again Moxon is of little help. He illustrates the "former" at both C3 and C4, but the text only refers to C4, which is in fact the old "former." However, he calls it the "Skew former" and comments that it is "seldom used by Joiners," but does not give the real reason—that it was a carpenter's tool. On page 76, he also calls the chisels C1 with parallel blades "Formers, because they are used before the paring Chissells, even as the fore Plane is used before the smoothing Plane." He was probably very proud of this, but somebody must have been pulling his leg, for the two types have entirely different functions. Unfortunately, both Fellows and Plumbley took his word for it. It will be seen later that the proper name for these chisels is "firmer chisels," for the good and sufficient reason that they are stouter and "firmer" than the thinner paring chisels and can be used with a mallet; cf. the German *Stemmeisen*. In Plumbley's shop there were 33 "formers & broad Chizells," or a set of 11 for each bench. Fellows had 12, but although I noted that this was far too many for them to be "formers," I missed the real significance of the number. The penny dropped when I looked at my own set. I have ten firmers, from $\frac{1}{16}$" to $1\frac{1}{2}$", but I think one or two have been pinched; I am sure I had the round dozen at one time. Plumbley's assessors found "3 Old Paring Chisells" later, probably one for each bench but borrowed from time to time as and when needed. I had only two or three myself, mainly for staircase work.

Incidentally, this is the first mention of "2 pine chests" in these joiner's lists. They were beginning to find this the best way of storing the ever-increasing number of moulding planes for the new style furniture-making and house joinery. There are no fewer than 79 planes of one sort or another in this list, and naturally take a bit of sorting out. There is first a special group of new planes in reserve, but they have the same names as the sets of five bench planes for each bench: a jointer, a long plane, a jack plane, a strike block, and a smoothing plane. The boss's jointer was made of "Yellow Jaunders," whatever that may be, possibly a special bit of hardwood he fancied for the job. The other two were the same price, only of "beach," as most of the others were. By this time the bench planes were beginning to be organized mainly on their length: the jointers for shooting glued joints, 28" to 30"; long planes 22" to 26", for trying or truing up workpieces; fore or jack planes from 16" to 18" for preliminary or rough work; strike blocks about 12" for truing mitres or end grain; and smoothing, from 7" to 9". The use of the term "Jack" instead of "Fore" and the inclusion of the "Strike Blocks" show the influence of the *Doctrine of Handy-Works*, where these terms are first used. Indeed there must have been some argument in Moxon's time about the exact meaning of the new term "Jack," as he goes out of his way to explain that it was the carpenter's name for the joiner's fore plane. Actually, both carpenters and joiners had always called them fore planes up to this time, but Moxon had no way of knowing that. Eventually the "fore plane" disappeared altogether in England, but in the New World the name was retained to distinguish the roughing planes at the upper end of the range 16" to 20" long. As late as 1958, the No. 6 Stanley was still called the "Fore Plane."

The moulding planes are also dealt with in a more systematic way than previously, beginning with a fairly large group of hollows and rounds. Each bench had three sizes of ogee and two "belexions," the latter very likely common or square ovolos. The special use of the term "bolection" for a large moulding, usually some kind of ovolo, round the panels of external doors and rebated over the styles and rails, was a later development. In Pl. 6 of his section of bricklaying, Moxon describes an "Ovolo or Boltel" with only one fillet. The only other moulding planes listed by Plumbley are the "asticle" (astragal?), a "Sash plain," probably a sash ovolo, and the "Small Cornish" two lines lower down. The ordinary moulding planes, including the hollows and rounds, are given a value of about 15d each and must have been more or less the same size, but the "small cornish," what we would now call a crown moulding, is priced at 3s, which suggests that it was larger and handled like a jack plane. There were also a dozen "Molding plane Irons" at 5d each in reserve, known to the trade as "soft blanks." They were ground or filed to the required profile and hardened and tempered by the friendly local blacksmith.

A clear distinction is made between the three "Rabbitt" planes at 10d and the two "Philisters" at 1/3d each, the latter being fitted with adjustable fences screwed to the sole, which was one up on Moxon, who had neither the tool nor the name. It is rather odd that in American usage this is usually "filletster," which was assumed to be the original form of the word, later shortened (in England) to "fillister," but surviving, like other old tool names, in the New World. Yet the first time the word appears (in Philadelphia, Pennsylvania) it is spelled "Philister"! Perhaps the "later" form was earlier than we thought, and possibly a bit of folk etymology. The cryptic "1 pr Inch 1 pr 1/2 inch 3/4" in the next line, assuming four planes at 10d each, may be interpreted as two pairs of matched tongue-and-groove planes, the sizes referring to the thickness of the timber.

The two "Ploughs" (as in English) are 3/4 each. They must have been fairly elaborate tools to be as expensive as this, but the irons lower down are "Plow Irons" at 4d each. The "Revele" plane sounds like a large rebating plane similar to the type now known as the "panel plane" or if skewed, the "badger," also handled as a jack like the "Cornish" plane which follows. A tool like this would be useful for finishing the large rebates in door- or window-frames (casements). The "2 Rownd Smoothing Plaines & 1 Hollow" were compassed smoothers for cleaning up circular work and the same price as the ordinary smoothers. The "Upright" and "Half Upright" planes in the next line may be connected with the two lots of "Tooth Plaine Irons" at 10d each among the miscellaneous tools lower down. Included with his timber, Plumbley had "a parcell of Olive Wood & other Veinarys" and also "6 Veneaireing Screws" at 6s, which accounts for the toothing planes.

Although he had ten augers and three "Gimbletts," a good deal of his boring was done with his "2 Wimble Stocks 1 Iron 1 Wood." Here he seems to have quarreled with Moxon, who called the joiner's wooden brace the "Piercer" and his blacksmith's brace the "Wimble." For use in these, Plumbley had "4 Center bitts and 1 Dott bitt" at a shilling each. This is early for centre bits and the "Dott" bit is so far unidentified, as are the "9 Wyreable bitts" in the previous line at only half the price. This may be another version of the "wrybitts" in the Carpenter list of 1675, probably general-purpose twist or gimlet bits. A reserve of "22 New Wimble bitts" also at 6d each, is listed further down. Whether the "Morris Fraime" at 6s was some kind of drill-stand for boring mortices is a matter of speculation; it is difficult to think of any other explanation, and in view of their other gear, I would not myself put it past them. The "3 Screws and nutts 9s" after the augers shows a further stage in the nomenclature of the screw tools in the previous lists.

Plumbley and his men must have done quite a lot of their own turning; his "Wheel and Layth" was "praised" at 30/, and he had full sets of turning gouges, chisels, hook tools, and mandrells, as well as another set for ivory turning, which in England at this time, was usually regarded as a special trade. The "Turning Strings" for driving the lathe at 2s each were expensive items. The most interesting entry connected with the turning section is the "Looking Glass" at 16s. In the inventory of household goods and effects there is a "Looking Glass 30s 2 Smaller ditto 3s. . . 1.13.0. The presence of a mirror in the workshop can be explained only as an attempt to improve the lighting, especially with the ivory turning.

Another very unusual entry is that giving the "4 pr fish Skins," costing no less than 10s, used for cleaning up before the general adoption of glasspaper. But to me, the most interesting problem presented by this list is "Who made all the planes?" My own view is that although by this time specialist plane makers may have been working in some of the larger towns in England and very likely in America, they spent most of their time on moulding planes and such special tools as plows and fillisters, hollows and rounds. The bench planes were still being made very largely by the men themselves, like Plumbley's jointer of "Yellow Jaunders." No bench planes by an English maker known to have been working before about 1700 have so far been identified in any collection known to me. This is quite understandable, as no special tools are required, whereas for moulding-plane making as a commercial proposition, the tradesman must have a large amount of gear, such as mother planes, patterns, jigs, and so on, which would be useless to a working joiner once the tools were made. In some family shops, as with the Dominys, one of the sons became the plane maker. In other cases, a man who had the skill and a flair for this kind of thing, broke away and concentrated on plane making in his own shop. It seems likely that the date of this inventory was more or less when this transition started to become effective.

The following tools are listed in the inventory of the estate of Jacob Leavitt, joiner, of Fairfield, Connecticut, dated June 23, 1759:

1 Steel Plate Saw	16.0	1 Glew Jointer	3.0
1 Do 6s, 1 Duftail Saw 4s	10.0	1 Long Plain	1.6
1 Compass Saw	6.0	1 Do. 1/, 1 Foreplaine	2.6
2 Broad Hatchets	4.6	1 Smooth Plain	9
1 Glewpot & Kettle	14.0	1 Do. Rownd Face	9
1 Holdfast 2/6 1 Goos 1/6	4.0	1 Looking Glass	9

1 Schew Plain	1.6	4 Turning Chizzels 4/		4.0
1 Round Plain	9	3 Turning Gouges		2.6
1 Hollow 9, 1 Round 1/6	2.3	4 Chizzls 3/		5.8
1 Small do. 1/, 1 smnll do. 8d	1.8	4 Plaine Irons 2/8		
1 large do. 2/ 1 Small Hollow 1/	3.0	1 Hammer 1/ 2 Rasps 1/4		2.4
1 Hollow plain	1.6	3 Wood files 1/6 3 files 6d		2.0
1 foreplain 8d 1 Ogee 8d	1.4	1 Former 6d Gimbletts 1/4		1.10
1 Filister 6d 1 do 6d	1.0	2 Squares 8d 2 Guages 6d		1.2
4 Old Plains	1.0	3 Prs Pinchers 2/3		2.3
1 Iron winch for a Grindstone	2.0	1 pr. Compasses		4
1 Adds 1/ Turning Screws 1/	2.0	Whetstones 6d		6
2 Bitstocks & 10 bitts	9.6	2 Frows		3.0
Table Plains 1/ Mortis Chizzel 6d	7.0	1 Lath		2.6
		1 Good bench		10.0

We have now arrived at the "Steel Plate Saws" hinted at in the earlier lists. Evidently Moxon's advice was now being taken seriously. The first one would be a handsaw and the second a backed tenon saw, followed by the smaller dovetail saw, also with a back. The "Compass Saw" which follows is strictly a carpenter's tool. From the price given, the appraiser was using the carpenter's name to describe a framed turning or bow saw.

The four standard bench planes are represented: the "Glew Jointer," two "Long" planes, a "Foreplaine" and a "Smooth" plane, as well as a compassed smoother and another odd "foreplaine" further down. One of these must have been a jack plane, but Leavitt reverts to the older usage.

The "Looking Glass" which follows is priced at 9d, but in Leavitt's household goods and chattels he had "1 looking Glass 15/-" and another "Small Looking Glass 8/-." These prices compare with the similar entries in the Plumbley inventory. At these rates, a ninepenny mirror would have been about the size of those in a girl's handbag and not much use in a joiner's shop. The price is suspiciously close to that of the moulding and other planes among which the "Looking Glass" occurs. One possible solution is that this was a plane for making looking glass or picture frames. One of the standing puzzles of the plane collections both in this country and in the United States is the occurrence of a number of moulding planes with a complicated profile, usually a combination of ogee and one or more astragals, which have no "spring"; i.e., they work straight down like a side bead or ordinary astragal moulding. One clue to these may be found in the list of moulding planes in Bennett's *Artificers Lexicon* (London, 1857), which includes a "Picture Frame" plane at 5s (about the same price as the Quirk Ogees & Astragals in the same list), and another "Italian Picture Frame" at 5s 6d. The profiles of these are not known, but they may well have had a shape similar to those found with no spring, the earliest known so far being a John Jennion ca.1740 and a Mutter about 1760-70.

The "Schew Plain" at 1s 6d, the same price as the "Long" plane above, may well be a panel-raising plane, which is usually skewed; several examples of these are known by Wooding, John Cogdell, and other early 18th century London makers and are the nearest they seemed to get to making bench planes, the stocks being about 8" long and 3" x 3" in section. Leavitt had very few moulding planes—five rounds and three hollows and one ogee—together with the "Table Plains" lower down, at 1s, which, with a "Mortis Chizzel" at 6d, amounted to 7s, which does not add up, which is very rare indeed in these inventories. Their spelling may have been rather fanciful but their arithmetic was usually pretty sound. What he probably had was three pairs of table planes for different thicknesses of table tops, priced at 1s 1d each, which, as they had to be carefully matched, would not be excessive. He also lists two "Filisters" (still no filletster), but no ploughs or grooving planes, unless they were the "4 Old Plains" in the next line. The boring tools are "2 Bitstocks & 10 bitts," which introduces the term which is still the standard name in the New World, although the English "brace" is now getting more common there.

The "Goos" with the holdfast was the tailor's iron of that name, very useful for veneering on small work. The only other tool name which is new is that of the "Scribing Gouges," now the usual name for incannel gouges, often used for scribing in sash work.

One puzzle in this list is the entry "Turning Screws 1/-" after the "Adds." Leavitt possessed a lathe and a small group of turning tools, but it is doubtful whether the turning screws had anything to do with this. It may be a garbled version of "turn screws," the original and proper name for what is now known as the "screwdriver." If so, it is the earliest known mention of this very useful tool. This conjecture may be confirmed by the entry among the sundries "2 Gross Skrews" priced at 10s. Wood screws, which were still handmade at this time, were naturally very expensive and were used only on special work.

After the great fire in Boston in 1760, Daniel Ballard III, a shop joiner, put in a claim for losses sustained, including his tools. The prices shown bear no comparison to those in other lists, but Benno Forman says that this was a period of severe inflation.

The Ballard tools of 1760 are as follows:

Three hammers new handills	15.0	4 Oge plains at 15/	3. 0.0
Large hand Saw	4.10.0	4 beed plains at 13/	2.12.0
Tennant Saw	4. 0.0	5 filisters at 10/	2.10.0
long Compass Saw	1. 2.6	4 quarter rounds at 10/	2. 0.0
2 Small ones	1.10.0	a Sett of Sash tules	2. 5.0
large Celection plain	3. 0.0	4 Mortis Chisels at 6/	1. 4.0
Smale do.	2.12.0	6 gaiges at 7/	2. 0.0
4 rabbit plains at 13/	2.12.0	4 Mallets at 6/	1. 4.0

One pair of groving plains	1.10.0	a plane 60/ to fluting tules 60	6. 0.0
one Nosing plain	1. 5.0	Six rounds at 13/	3.18.0
2 doz 1/2 of pairing		6 hollers at 13/	3.18.0
2 rasps at 5/	10.0	Crucked tules	5.10.0
3 Squairs	3. 0.0	2 fore plains	2. 0.0
1 Spoke Shave	2. 0.0	6 Smoothing plains at 18/	5. 8.0
a drawing Nife	15.0	one long plain	2. 0.0
3 pr of Cumpasses	15.0	3 rub stones	2. 5.0
1/ doz of gimblets	15.0	1 grin stone	2. 5.0
two new plain irons	9.0	a frame for the grinstone	2. 5.0
two bivils at 15/	1.10.0	3 astoul planes at 30/	4.10.0
4 rules at 11/	2. 4.0	a panil plain fraimd	3. 0.0

Although Ballard was a ship joiner, apart from the three "Compass" saws, a long one and two small ones, he does not appear to have had any tools not found in the usual house-joiner's kit. His bench planes include a "ginter" at 25s in a supplementary list, a long plane, two fore planes and six smoothing planes. The moulding planes are headed by a "large Celection" plane at £3 and a small one at £2.12s, which may be variations on Plumbley's "belexions," but the prices are higher than any bench planes listed. They may have been large and small cornice mouldings handled as jacks, similar to the Saxby plane (ca. 1750) in the Victoria and Albert Museum at South Kensington. There are in addition "4 Oge" and the same number of "quarter rounds," which were probably ovolos, a term which, although known to Moxon seems to have taken some time to establish itself. Another moulding mentioned by Moxon but making its first appearance here as the name of the tool occurs in the entry "4 beed plains at 13/."

The "Sett of Sash tules" must also have been moulding planes, very likely two pairs of sash ogees or ovolos. The "Nosing plain" would be mainly for staircase work and there are six each of "rounds" and "hollers" at 13/ each. If they were planes, as they probably were from their position in the list, it would be interesting to know what the "Crucked tules" were in the next entry, or the "fluting tules" mentioned earlier. Only further research could settle a problem of this kind, but the "3 astoul planes" were almost certainly a near miss for astragals.

Ballard also had "4 rabbet plains" and five filisters (still no 'filletster'), but the most interesting tool of this group is the last, the "panil plain fraimd," a panel-raising plane with an adjustable fence and depth stops. No ploughs are mentioned, but there is "one pair of groving plains," which may have been plows, but are more likely to have been a matched pair of tongue-and-groove planes. Apart from the half dozen "gimbletts," there are no boring tools, which must have been an oversight.

The following tools are listed in the inventory of the estate of Joseph Gavet of Salem, Massachusetts, taken by Nathaniel Gould, another shop joiner, on 19 October, 1765.

One Joiners Shop & Board House so called £30.0.0
Joiners' Tools viz:
One Whip Saw 8/, one Cross Do [saw] 24/
2 Finerings Do [saws] 26/8 1 Small Do [saw] 8/-
2 Steel Plate hand Saws 10/8, 1 Pannel Do 4/
2 Berminsham Do 2/ 9 Turning do 18/8
2 Duftail Do 4/8 8 Turning Chizels 6/
6 Turning Gloves 3/4 One Turning lock 26/8
one Large Turners Wheel 12/ 13 Maundals 4/
6 Augres 8/ 2 pr Canipers & 2 Center Punchers 1/4
5 Setts of Screw tools 10/ 3 Setts Larger do 6/8
One Glue frame 1/4 one Ridling Sive 8d
1 Ladder 9d one pair Steps 1/1
1 Grindstone and frame 2/8 3 Boxes 1/4
1 Glue Pott mould 6d one small grindstone 10/
9 Glue pans 6/ one Ladle 5d one knife & Gouge 3d
One Frame for drying of Glue 2/
One Sliting horse 3/ 3 Work benches 32/
8 Forms 8/ 8 Joynter Stocks and Irons 12/9
3 Fore Plain Do. 5d 6 Smoothing Do. 4/8
3 Jack Do. 2/8 9 Graving Plow ditto 6/0
8 Round Do. 4/- 8 Hollow Do. 4/8
5 Rabit Plains 3/4 8 Filister Do 3/2
21 Crees Do. 12/ 4 Jack Do. 2/8
2 Mitre Boxes 8d 12 Screw Presses 12/
1 Iron Vice 10/8 one Do 2/8
1 Broad Axe 2/8 1 Hatchet 4/
1 Adds 8d one Frow 2/8 1 Do 6d
2 Polishing Irons 1/4 one pair 2/
one Drawing Knife 8d 4 Mallets 2/
2 Holdfasts 4/8 5 Pairing Chizels 2/8
7 Mortis Do 3/9 16 Firmers 5/9
7 Gouges 2/8 one Small Drawing Knife 2d
6 Center Bitts 4/ 4 Pin Do 2/8
1 Bitt Stock 8d 2 Small Trammels 1/7
2 Saw Setts 10d 1 pair Nippers 3d
3 Hammers 2/ 3 pr Compasses 11D
12 Squares 3/4 four Scrapers 1/1 one Chalk Line 1d
14 Files 4/8 one Chimney Grate 12/
1 pair Iron Doggs 8/ 10 Tonges 1/4 3 Plum Bobbs 1/2
6 Rules 4/ One Screw Press 2/8 one sett of Stamps
14 Weights 4/8 one pair Scales 10d
Whet Stones 8/ one small Brass Kittle 2/
140 lb Finering Weights £2.6.8
3 Glue Potts 21/4 one pair Bellows 1/4
1 Colouring Cag 1/4 3 Oyle Potts 10D
1 Iron Skillet 3d one Oyle Pott 2d
One Oyle Bottle 2d two Earthern Porringers 1d
One hand Brush 5d 30 Patterns 4/
1 Candle box 1d 2 Stamps 1/4

1 Paint Board & Muller 1/ One Saw Peice 6/
1 Large Rub Stone 2/ 1 Large Iron Kittle 5/4
Sundry Pieces of Maple £1.8.10
Ditto of Oak 8/3 Ditto of Pine 11/
Ditto of Walnut 73/2 Do Mehogany 10/8
Hatchet 4/ a Parcle of Axhandles 4/9

Note: The original inventory is not on file; this is the Clerk of Court's Record Copy, E P. 432, 419-422 and contains a number of errors of transcription.

Joseph Gavet (1699-1765) was a joiner and furniture maker, the son of a Huguenot, Philip Gavet, possibly a mariner. Joseph may have been an apprentice of one of the Symonds brothers of Salem, John (d. 1728) or Thomas (d. 1758), both of whom were joiners. Joseph Gavet's son Jonathan (1731-1806) was a shop joiner and turner. Two of Jonathan Gavet's sons, Jonathan II (b. 1761) and William (b. 1766, d. 1857), were turners. Nathaniel Gould (d. 1761) who took the inventory, was a Salem shop joiner and among his effects there was a copy of "Chipendale's Designs." It is clear from this that the Gavets, father and sons, were well in the forefront of their trade and the tools a representative selection of the types used at this time, a few years before the Declaration of Independence.

Taking the saws first, there is a whipsaw at 8s and a cross saw at 24s, three times as expensive. The whipsaw is, of course, a two-man cross-cut, the second one being very likely a large frame saw for ripping. Gavet must have sawn his own veneers; the two "Finering" saws were small framed saws, probably used on the "Sliting horse" further down the list, as shown in Diderot and Roubo, who would have called it a "chevalet." The small veneer saw may have been a fret saw. The hand saws are now "Steel Plate" saws, as well as the smaller "Pannel" saw, one of the earliest mentions of this tool. There is also a "Saw Peice" at 6s in the list, similar to the uncut "saw plates" in the earlier inventories. The "2 Berminsham" saws may well have been made in Birmingham, England, possibly backed tenon saws, like the two "Duftail" saws below, the "9 Turning" saws being various sizes of framed bow saws for general use.

After the "Duftail" saws there is a self-contained group of tools relating to turning, which, after correcting the "errors of transcription" in Benno Forman's note, would read: "8 Turning Chisels; 6 Turning Gouges; One Turning Stock (Lathe); one large Turner's Wheel and 13 "Mandrells." The "Calipers" and center punches would also be used with the lathe. The eight sets of "Screw Tools" gives the final solution of the "augers and screw box" problem in the Carpenter list and Plumbley's "Screws and nutts," which we have been chasing steadily through the material. All of these men made their own wooden screws, from the bench screw down to the small thumbscrews on tools such as gauges and also for spinning wheels and other needlework equipment for their girlfriends. From the entries that follow it also looks as if Gavet's men made their own glue from start to finish, including the mould for casting the glue "potts," which at over 7s apiece must have been expensive items.

There were three benches, as in the Plumbley shop, and 83 planes, or four more than Plumbley's, to use on them. The bench planes themselves start, as usual, with "Joynters," of which there were eight, some of which must have been the "long" planes of the earlier lists. There are three each of both "Jack" and "Fore" planes, the price of the latter being probably an error for 5s. For the 21 moulding planes the older general term "Crees" planes is revived, and another four of them were the size of jacks, as was suggested with some of the other lists. For the larger mouldings and general work, the shop had a set of eight pairs of hollows and rounds, what was later known as a "half set," either the odd numbers or the evens. The "9 Graving Plows Do" were three sets of fixed or adjustable grooving planes and the five "Rabit" planes were supplemented by eight "Filisters." In view of this, it would be of interest to find out exactly when they started to call them "filletsters." No sign of it yet.

Although Moxon mentioned the mitre box, these two are the first to have been noted in an inventory. The "12 Screw Presses" at 1s each were probably the tools we now call "hand screws," made of wood and no doubt in the shop, and used with the veneering. There is another "Screw Press" further down the list at 2s 8d, very likely some form of sash or framing cramp. The two pairs of "Polishing Irons" recall the "Goos" in the Leavitt inventory, but the Gavets also had a large number of "Finering Weights."

The chisels are now in their proper quantities and order: "5 Pairing Chizels" at 2s 8d; "7 Mortis" at 3s 9d; and "16 Firmers" at 5s 9d, with no mention of "formers," which resolves this aberration satisfactorily.

The boring tools include "6 Augres"; one "Bitt Stock"; "6 Center Bitts" and "4 Pin Bitts," the latter being the term we used fifty years ago for the spoon bits for boring holes for pins of wood driven through the joints of doors, sash and other framed work.

Another special joiner's tool mentioned here for the first time is the "Scraper," after the squares, but although there were two rules to each bench, the gauges seem to have been overlooked. Perhaps they found them difficult to spell. The list of tools ordered from England in 1760 by Wm. Wilson, a dealer of Philadelphia, for one of his customers, Richard Johns, reproduced in Charles F. Hummel's *With Hammer in Hand*, (Winterthur, Delaware. 1968) is roughly contemporary with the three inventories dealt with above and the tools named are similar. The curious thing is that although I was under the impression that I had read this fascinating book several times and studied it very carefully in detail, I cannot remember noting the fourth item on the list "2 Picture Frame Planes." If I had, it would have given me a reference to these tools nearly a century earlier than that from Bennett's *Lexicon*, on which I had relied in my comments on Leavitt's "Looking Glass." There are, in fact, several references in the Dominy accounts to making and repairing looking glasses, which were more likely in those days to have figured in household effects than picture frames, being essential to the well-being of the better looking and more important half of the human race, both in England and America. Mr. Hummel was apparently unable to find any planes in the Dominy workshop corresponding to these imported tools, but I venture to suggest that Nos. 75 and 76 are possible candidates.

The plane maker John Ridgus and the saw maker White have not been so far identified in England, but Robert Moore, three dozen of whose "best plain Irons" were ordered, is now known to have been the son of Robert Moore, whitesmith (edge toolmaker) of Birmingham. In 1721 he was apprenticed to Jos. Green of Birmingham, ironmonger. His working career would have been about 1730 to 1770. A comparatively large number of planes fitted with his irons have been reported, mainly by London makers known or thought to have been working before 1770, and it seems from the Wilson list that his reputation was as high in Philadelphia in 1760 as it was in England. Perhaps no apology is needed for connecting the Declaration of Independence with things like these which helped to make life easier or even possible in the pioneering days in America. Not, in any case, to readers of *The Chronicle*, who are in many cases descendants of the early settlers and their successors and are members of an Association devoted to recording, preserving and studying these everyday things. From their forebears they have inherited an intuitive feeling that these things matter, if only to keep the grand events in proper perspective.

Another point, not wholly unrelated to the above, emerges from the study of these humble records, carefully compiled by barely literate men to whom the idea that they were writing history could never have crossed their minds. It is the almost total inadequacy or even irrelevance of much of what we have been taught to regard as the official version of how and when these things originated. For my part, I have long ceased to feel any surprise at all that certain tools or methods of working, which according to the "authorities" could not possibly be dated earlier than, say, 1780, were known and widely used a century or more earlier. Most of the accepted sources were hopelessly out of date even in their own time, depending, as most of them did, on books written by other "scholars" right back to Pliny the Elder and beyond. Not that I have anything against Pliny, except that he lived about 2,000 years ago.

The traditional method of compiling books of this kind absolved them from going into the existing workshops and finding out for themselves, but even when they did try this they often met another obstacle. The workmen themselves also had their "traditional" view that some of their practices were still "mysteries" and not to be divulged to strangers. There was also the natural inclination of a craftsman when confronted by an old gentleman with a notebook asking silly questions about the tools, to give him a few silly answers. Take Moxon's description, occupying three whole pages, of the method of preparing workpieces in his section on "Joinery." The most important part of the process is not mentioned at all, nor the tools used for it. The joiner who explained this at such inordinate length kept that part under his hat, having carefully locked the drawer in which he had hidden the tools to avoid awkward questions. We should not, perhaps be too harsh with Moxon—after all, he was only a printer. In this connection we might recall the notice that is said to have been common in the saloon bars of the Wild West in the pioneering days: DON'T SHOOT THE PIANO PLAYER, HE'S DOING HIS BEST. This would also, of course, apply to later historians of tools and craftsmen.

REFERENCES

Dictionary of Woodworking Tools, R. A. Salaman, Revised edition, The Taunton Press, Newtown, CT, 1990.

Mechanick Exercises or The Doctrine of Handy-Works, Joseph Moxon, Reprinted from the 1703 edition, The Astragal Press, Mendham, NJ, 1989.

With Hammer in Hand, Charles F. Hummel, University Press of Virginia, Charlottesville, VA, 1968.

Mr. Hewlett's Tool Chest—Part I

By JAY GAYNOR

December 1985

Jay Gaynor is Curator of Mechanical Arts at Colonial Williamsburg. He spent a good part of 1984 preparing exhibit material for the opening of the DeWitt Wallace Decorative Arts Gallery at Colonial Williamsburg. The tool chest about which he writes below is part of that material.

The survival of a group of precisely datable and well-documented English tools from the third quarter of the eighteenth century is important, first and foremost, to the history of English tools in England. But, since English tools were widely used in British America and served as the basic patterns for many American-made tools, such a survival is also extremely important to the study of early American technology and the history of American tool manufacture.[1]

This importance is further enhanced by the rarity of tools from this period. Apart from archaeologically recovered examples and certain marked types such as molding planes, very few English and American handtools verifiably made prior to 1800 survive. Precisely datable examples are even rarer. And, while a number of contemporary publications provide documentation of continental tools, detailed descriptions and illustrations of English and American tools are possibly more scarce than the tools themselves.—CHRONICLE ED.

On Saturday, February 13, 1773, William Hewlett, a London ironmonger and brazier, sold a small mahogany tool chest. It contained fifty-six tools and an assortment of nails, screws, and other hardware. A bill of sale and a copy of the billhead from Hewlett's shop accompanied the purchase.

The history of the chest from that Saturday until 180 years later is unknown. For many years, it probably was kept in a well-to-do London house, where it was carefully cared for and its contents seldom used. By the mid-1950s, it had passed into the hands of a London collector of clocks where it stayed until the Colonial Williamsburg Foundation purchased it in 1957.

When Colonial Williamsburg acquired the piece, there was a growing interest in technology, and the chest was thought to be a worthwhile addition to the collection of antique objects. Still, in the late 1950s tools were not high on many collectors' or museums' lists of priorities, and the chest resumed its quiet existence, reposing in the Governor's Palace for nearly a quarter of a century. It drew little attention beyond that of the museum's product development staff, who had the piece copied with modifications to be sold as a silver chest. Not until several years ago did we realize the rarity and value of the small treasure we owned.

Figure 1A

The chest and its contents are possibly the earliest known examples of a "Gentleman's Tool Chest." Christopher Gabriel and Sons, the London planemakers and tool dealers, stocked chests of this general description in the 1790s,[2] and chests of the same basic design were standard items illustrated and described in nineteenth century Birmingham and Sheffield pattern books.[3] The chests were offered in several sizes, and the larger the size, the greater the number and variety of tools the chest contained. The more complete chests even included tools for gardening. These toolkits, as their name suggests, were designed for use by the "Gentleman" homeowner or shopkeeper, and the tools and hardware they contained were intended to handle minor jobs around the house or business.

William Hewlett, as an ironmonger and brazier, was a logical purveyor of such a chest. Ironmongers' establishments were to some extent the eighteenth century equivalent of modern hardware stores. In the late seventeenth century, Joseph Moxon notes that saws could be purchased from the ironmongers in Foster Lane.[4] Robert Campbell, writing in *The London Tradesman* in 1747, says that "the Braziers and Ironmongers Shops are generally united," and in addition to iron, steel, and brass hardware and equipment for the kitchen and house, the homeowner could find there "Locks of all Sorts, Hinges of various Kinds and different Materials; Chases and Handles for Cabinet-Work, Nails, Wood-Screws, and generally all Sorts of Brass and Iron Work that are useful for Furniture, or any part of Furniture. The Brazier, or Ironmonger, neither makes nor is supposed capable of making all the different Articles in his Shop; . . . his Profit arises from the Difference between the buying and selling Prices."[5]

The billhead which Hewlett included with the chest gives us quite a bit of information about him and his business. It was printed originally for Hewlett's predecessor, James Buttall, and by adding a few handwritten lines to it, Hewlett adapted it to suit his situation. It reads (with the handwritten insertions in italics): "London *Feb. 13, 1773* Bought of *Wm. Hewlett Successor to* James Buttall, Ironmonger and Brazier, At the Gridiron, Near Hungerford Market, in the Strand."

This making do with Buttall's billhead suggests that Hewlett had taken over the shop only very recently. In 1762, one Thomas Buttall was the proprietor,[6] and James, who eventually took over the business, was doubtless a relative. In the available directories, Hewlett first appears in the 1774 one for Sheffield. There, under the heading of "A List of All the London Merchants &c.," he is listed as an ironmonger at 484 Strand. In 1775, he appears in the London directory, and he continues to be listed until at least 1812,[7] indicating that he probably operated a successful business.

The Gridiron, as the shop was called, was located in the western end of the Strand, one of the major commercial thoroughfares of eighteenth-century London. It was not far from Charing Cross and close to Hungerford Market which was noted during the period as a source for household provisions.[8] If the tool chest is an indication of the

Figure 1B

*Figure 2. The billboard is now glued to the inside of the tool chest lid.
Originally it probably was given to the purchaser together with the bill of sale.*

quality of stock that Hewlett and his predecessors carried, they must have catered to the carriage trade.

The bill of sale supports this assumption. On it Hewlett or his clerk carefully listed the chest, the hardware mounted on it, and the items it contained. Each entry is priced. But nowhere is the purchaser mentioned. Since the purchaser's name would have been important if the goods had been custom ordered or purchased on credit, the chest and its contents were likely an over-the-counter item sold for cash. Further, since the total price was £6.6.0, the equivalent of three to six weeks' pay for a London journeyman cabinetmaker at mid-century,[9] the purchaser was someone of means—indeed, probably a gentleman. From the bill of sale we know the original contents of the toolbox as well as the name and price of each tool included:

Bought of Wm. Hewlett/Successor to James Buttall

1 Very neat mohogany Chest	1.15.	8 Chizles & Gouges	5.
Brass Locks hinges & handles	10. 6	6 Files	2.
Nails Screws & brass hooks	10. 6	1 Pair Pinchers	1.
A mohogany Jackplane	3.	Hand Crow & Chizle	1. 8
A Smoothing ditto [plane]	2. 3	Bench Vice	7.
Mallett	1. 6	Hand Vice	3. 6
1 Turkey Stone	4.	Beak Iron	2.
Rule	1.	Pair Shears	.10
Bevil & Square	2.	Screw Plate	2.
Gauge	. 9	Cutting Nippers	2. 6
Copper Glue pot	3.	Rack Compasses	2. 6
Hatchet	2.	A Pair Common [Compasses]	. 6
Hand Saw	4.	2 Pair Pliers	1. 6
Dovetail Saw	5.	3 Turnscrews	1. 8
Frame Saw	2.	4 Awls	.10
Key hole Saw	. 7	6 Gimlets	1.
Claw Hammer	1. 9	1 Punch	. 2
Rivetting ditto [hammer]	1. 6		
			£6. 6. 0

The separate entries for the chest and its hardware are logical. While Hewlett would have supplied the handles, locks, and hinges from his own stock, he would have either contracted with or employed a cabinetmaker to produce the box itself. Thus, for his own bookkeeping, it would have made sense to charge separately for the items. Since the tools and hardware in the chest are itemized separately, it is likely that Hewlett assembled them himself rather than buying the chest and tools as a pre-packaged group for resale.

Yet, the types and quantities of hardware and tools included with the chest were very similar to those included with chests listed in Birmingham pattern books of the early nineteenth century.[11] This and the likelihood that the chest was an across-the-counter item indicate that probably even as early as the 1770s, these chests were a well-established type. This development of standardized kits of tools to satisfy a genteel market—a market that is not the natural one for trade tools—is an early indication of intentional marketing efforts on the part of tool sellers.

There can be little doubt that this chest and its contents were selected to attract the affluent customer. While its basic design is that of nineteenth-century boxes, the chest's detailing and hardware are earlier. Hewlett described it as "very neat," meaning that it was made in the elegant but simple style in vogue in London during the 1770s. Judging by the chest's construction and finish, a competent London cabinetmaker produced it. Unfortunately, the piece is unsigned, and his identity remains unknown.

The chest is 24" long, 11 11/16" deep, and 10 1/4" high. It is made principally of mahogany with deal and oak secondary woods. The outside top and front are veneered with figured mahogany, while the case back, sides, and moldings are solid mahogany.

The top of the chest lifts to reveal a 1 3/4" deep till divided by thin mahogany partitions into one large and sixteen smaller compartments. The seven compartments along the back of the chest are only 3/4" deep, and the six gimlets may have been stored by sliding their

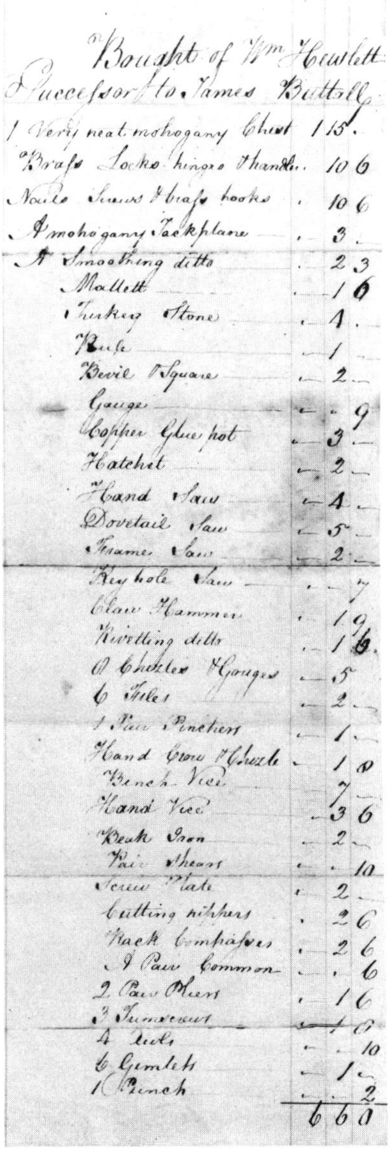

Bought of Wm Hewlett
Successor to James Buttall

1 Very neat mohogany Chest	1	15	.
Brass Locks hinges & handles	.	10	6
Nails Screws & brass hooks	.	10	6
A mohogany Jackplane	.	3	.
A Smoothing ditto	.	2	3
Mallett	.	1	6
Turkey Stone	.	4	.
Rule	.	1	.
Bevil & Square	.	2	.
Gauge	.	.	9
Copper Glue pot	.	3	.
Hatchet	.	2	.
Hand Saw	.	4	.
Dovetail Saw	.	5	.
Frame Saw	.	2	.
Keyhole Saw	.	.	7
Claw Hammer	.	1	9
Rivetting ditto	.	1	6
8 Chizzles & Gouges	.	5	
6 Files	.	2	.
1 Pair Pinchers	.	1	.
Hand Crow & Chizzle	.	1	8
Bench Vice	.	7	.
Hand Vice	.	3	6
Beak Iron	.	2	.
Pair Shears	.	.	10
Screw Plate	.	2	.
Cutting nippers	.	2	6
[illegible] Compasses	.	2	6
A Pair Common	.	.	6
2 Pair Pliers	.	1	6
3 Turnscrews	.	1	6
4 Awls	.	.	10
6 Gimlets	.	1	.
1 Punch	.	.	2
	6	6	0

Figure 3. The bill of sale. No purchaser is named.

blades through slots into the space beneath them. While the large compartment was likely intended for tools, the smaller ones were designed to hold the nails, screws, and other hardware that came with the chest. Both drawers are made entirely of mahogany. The 2⅛" deep upper drawer is divided into five compartments with a shallow tray covering the right-hand one. The unpartitioned bottom drawer is 3¹⁄₁₆" deep.

With the exception of a few shortcuts, the chest is constructed in the manner of good quality London case furniture. The sides of the case and drawers are dovetailed. There is a full dustboard between the two drawers, and "kickers" installed in the case above the drawer sides prevent the drawers from tilting downward when they are opened.

The brass hardware also is of good quality. The lifting handles on the chest sides are heavy and very business-like. The drawer pulls have relief decorated bails with their original tinted varnish finish. On the inside of the drawer fronts, the countersinks for the nuts securing the drawer brasses are filled with putty. This unusual and nice detail results in smooth interior drawer surfaces. The till and drawers all have locks, and the butt hinges of the lid have stops incorporated in them so that the lid is held open in a vertical position.

Except for a few minor repairs and replacements, the chest is in superb condition with well-patinated exterior surfaces and an interior that appears to have received very little use. Its owners must have only rarely carried it about or even opened it. Many of the tools have qualities that also would have appealed to the dilettante. Bright finishes, a liberal use of mahogany and brass, and an occasional "BEST" stamp characterize them. Since there are so few other documentable tools from this date with which to compare many of the Hewlett tools, it is difficult to determine what the inclusion of these particular pieces in a gentleman's chest implied about their quality. It is likely that most are basically typical tools made for general use; although, if several different

Figure 4. The chest is built like good quality London case furniture with a full dust board separating the drawers and "kickers" along the case sides which keep the drawers from tilting when they are opened.

qualities of tools were available, those with the highest finish were selected, and special attention was lavished on the materials of wooden components.

Twenty-two of the fifty-six tools survived with the chest. They are the mahogany jackplane, a 24-inch folding rule, a square, a gauge, a hatchet, the claw and riveting hammers, two gouges and one chisel, one file and an unmounted file handle, a small watchmaker's beak iron, a pair of jeweller's shears, a turnscrew, two bradawls, and five gimlets. About half of the tools have makers' or sellers' marks; none were marked by their owners.

It is a mystery why these particular tools survived while others that would appear to be less useful and, therefore, less likely to be lost or broken—the screwplate, for instance—did not. Numerous scenarios have been concocted to explain the situation, but it is likely we will never know the true story. Equally mysterious, especially in light of the only partial survival of the kit of tools, is the condition of the items that are present. There are signs of use here and there. The hatchet has a badly nicked edge. The claw of the hammer has been battered. The turnscrew blade is chipped. Chisel and gouge handles have bruises where they have been struck. Overall, however, the condition of the pieces is virtually new. Whatever cleaning or polishing was done over the years to maintain the tools has had very little effect upon their surfaces. Metal pieces are bright with their original forged, filed, ground, and polished surfaces intact and only slightly mellowed by oxidation. Wooden pieces are clean with what appears to be their original finishes.

This condition, along with the information we have about the history of the tools, provides us with an exceptional—possibly even unique—opportunity to study tools of this period. The group of tools is small, especially when compared to the several hundred pieces in the Benjamin Seaton chest dating from 1796.[12] The Seaton tools survive with comparable information and in similar condition. But, the Hewlett tools are twenty-four years older than the Seaton ones and include a number of items that are not among those in the Seaton chest.

Notes - Part I

1. Charles F. Hummel discusses the dependence of early American artisans on English tools in "English Tools in America: The Evidence of the Dominys," *Winterthur Portfolio 2* (1965):27-46.

2. W.L. Goodman, "Gabriel & Sons, Stock Inventories," *Chronicle 36* (Sept. 1983):53-61.

3. See, for example, Joseph Smith, *Explanation or Key To the Various Manufactories of Sheffield* (Sheffield: Joseph Smith, 1816; Reprint, South Burlington, VT.: E.A.I.A., 1975), number 557; and the variety of pattern book and trade card illustrations of chests reproduced in Kenneth D. Roberts, *Tools for the Trades and Crafts* (Fitzwilliam, N.H.: Ken Roberts Publishing Company, 1976). Roberts' book contains a wealth of documentation concerning 19th-century gentlemen's tool chests.

4. Joseph Moxon, *Mechanick Exercises* (London: Daniel Midwinter and Thomas Leigh, 1703; Reprint, Morristown, N.J., The Astragal Press 1989), p. 95.

5. Robert Campbell, *The London Tradesman* (London: T. Gardner, 1747; Reprint, New York: Augustus M. Kelley, 1969), pp. 177-178.

6. Ambrose Heal, *The Signboards of Old London Shops* (London: B.T.Batsford, Ltd., 1947), p.105.

7. *Sketchley's Sheffield Directory* (Bristol:1774), p.75. *The London Register of Merchants and Traders For the Year 1775* (London: W Harris, 1775), n.p. *Kent's Directory: 1812*, p. 148. From 1779 on, Hewlett is listed at 460, Strand. Colonial Williamsburg does not have a complete run of London directories. The earliest one after Kent's 1812 is Kent's 1825. and it does not list Hewlett.

8. Alison Adburgham, *Shopping in Style* (London: Thames and Hudson, 1979), p. 48.

9. Campbell, *London Tradesman,* p. 171-172. The spellings of both gimlet and pliers on the bill of sale are those given in Samuel Johnson, *A Dictionary of the English Language* (London: 1755).

10. The spellings of both *gimlet* and *pliers* on the bill of sale are those given in Samuel Johnson, *A Dictionary of the English Language* (London: 1755).

11. See Roberts, *Tools,* pp. 4652.

12. See W L. Goodman, *British Plane Makers from 1700,* 1st ed. (New York: David McKay Company, Inc., 1968) and *Arnold & Walker Catalogue 5* (Needham Market, Suffolk: 1977), pp. 2-9. The tools of the Benjamin Seaton chest were purchased in 1796, and they, too, have survived virtually unused. The chest is in the collections of the Guildhall Museum, Rochester, Kent, England.

Mr. Hewlett's Tool Chest—Part II

March 1986

We are continuing to study the Hewlett tools at Williamsburg and plan to publish a thorough discussion of them and their historical context. In the meantime, however, the following descriptions provide brief introductions to the individual pieces.

"A Mohogany Jackplane"

The design of the jackplane stock is typical of English planes of this period (Figure 5). It is 13¹⁵⁄₁₆" long and 2¹¹⁄₃₂" wide. The heel and toe have flat vertical chamfers terminating in clean, well-defined lambs' tongues, while the upper edges of the stock have rounded chamfers. The open tote is offset ⅛" to the right. The wedge has a rounded top with its front edges broken by a narrow, flat chamfer.

The single, round-top, 1¾" iron is stamped "P. LAW" with a crown and is bedded at a 49° angle. Philip Law was an edge tool maker working in Coalpitlane, one of Sheffield's six principal streets, in 1774.[13] Like a number of the edged tools in the chest, the plane iron has a laminated construction. The blade is made principally of iron with a steel "bit" welded onto it to form the cutting edge. As Joseph Moxon recommended,

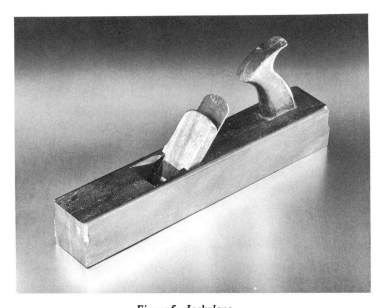

Figure 5. Jackplane.

the blade was sharpened so that the cutting edge is slightly convex.[14] Only the bevel, back, and lower edges of the iron were ground; the remaining surfaces were left black.

Although the dense mahogany stock is unmarked, the deftness of its manufacture as well as its "mainstream" design suggest that it was the product of one of the many specialist planemakers working in London at the time. If so, the absence of a mark might indicate that the plane, made of mahogany rather than beech, was a special-order item for the wholesale market. In that market, retailers commonly bought unmarked goods and subsequently marked them with their own names.[15]

"Rule"

The 2-foot, 2-fold rule is made of boxwood with brass joint pieces and end bindings. Both sides are engraved with the same 24-inch scale, reading from right to left, and calibrated in full inches, halves, quarters, and eighths.[16] One side is marked: "C. Stedman Best Box Rule LONDON," flanked by two rudimentary "roses."

The rule came from the shop of Christopher Stedman. He advertised on a trade card printed sometime before the late 1750s that he made and sold "At the Globe on London Bridge. . . all Sorts of Mathematical Instruments, for Sea or Land, Gauging, Surveying, Measuring, Geometry, Navigation, Arithmetick, &c. Wholesale or Retail."[17] The instruments illustrated on the card include a folding rule, fraught with artistic inaccuracies, but generally of the same form as this one. Stedman's shop and the other buildings on London Bridge were removed in 1758, and sometime before 1761 Stedman moved about a half-mile northeast to a location near the East India House on Leadenhall Street.[18] This placed him about two miles east of Hungerford Market and Mr Hewlett's establishment.

Although Stedman's rule has only a simple, linear scale and is much less complex than the rules with Gunter's Line of Numbers and Lines of Timber and Board-Measure, it is otherwise probably a typical English joint-rule of this period.[19] Each leg of the rule is ⅝" wide and slightly less than ¼" (.19-.20") thick. The brass bindings and hinge parts are made from ⅟32" sheet brass. The bindings at the end of each leg consist of two pieces. one inlet into each side of the leg, held in place with a single rivet.

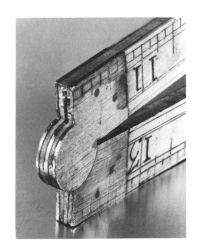

Figure 6. The brass center leaf of the rule hinge is attached to the upper leg and pivots in a saw kerf cut in the lower leg.

The joint is of a type commonly seen on early rules and dates back at least as far as the 1670s. It is made of five pieces of sheet brass, held to the rule legs with rivets (Figure 6). This results in a wood-to-brass joint rather than the brass-to-brass joint typical of more recent rules. The hinge pin is a brass rivet.

The calibration is typical of English rules of this period. It uses three lines scribed parallel to the edge of the rule with transverse lines of varying length indicating divisions. Oddly, on this and other early rules, the calibrations do not extend all the way to the rules' edges, decreasing the accuracy of measurements that can be made with them.

While a gauge probably was used to scribe the lines running parallel to the rule's edges, the inconsistencies and run-overs of the division lines indicate they were cut individually using a less-exact method. Many of them are slightly askew to the left, possibly indicating some idiosyncratic aspect of the cutter's working method, his sloppiness or ineptitude, or an unsquare marking guide.[20] The numbers are individually stamped with one stamp serving as both a six and a nine. Stedman's mark also appears to have been made with individual letter stamps.

The sheet brass was finished with a fine abrasive leaving a satin surface with parallel scratches running in the direction of the legs. Brass as well as boxwood appears to have been finished with a varnish or shellac.

"Square"

One characteristic of many small squares made in the latter part of the eighteenth century is the nearly equal lengths of their legs. The mahogany and brass square in this kit has this feature, but the proportions of the beam and blade widths to their lengths and the decoration at the end of the beam give it a very different appearance from that of commercially made squares dating from the end of the century. Its design is more closely related to that of all-wooden squares. It might be a "transitional" type—possibly even a "London pattern" which fell out of favor as sturdier designs took over the market. Or it may simply be an inexpensive tool, made of fancy materials. It is slightly out-of-square. (4⅞" wooden beam; 5½" brass blade).

"Gauge"

The marking gauge is a straightforward one, crisply made entirely of mahogany (Figure 7). The oval fence is locked in place by a captive wedge similar to that used on English planes with sliding arms. The beam is square in section with rounded corners and ends. The scribe is a small iron or steel blade-shaped point set into the beam. Like the square, the gauge bears no calibrations or marks. This leaves unanswered questions

Figure 7. Marking gauge.

concerning Hewlett's sources for the tools and whether they were "standard" commercial items. (beam 6⅞" LOA [length over all])

"Hatchet"

Even the simplest gentlemen's tool chests at the turn of the century contained a hatchet. This one is an example of a type that was in style from at least the second quarter of the eighteenth century until sometime into the nineteenth, when the Kent-style hatchet with its rounded blade shoulders became the "common" British hatchet.[21]

The head of the hatchet has a well-defined, although short, poll, pointed lugs above and below the eye which are nearer the poll than the blade, and a symmetrically flaring blade with straight shoulders (Figure 8). Its edge is beveled on both sides for chopping rather than paring work. Although the sides are ground, making this a "bright" rather than a "black" finished hatchet, the head is otherwise very cursorily finished. The handle is made of what appears to be a fruitwood. It widens at the upper end to fit the eye and has a pronounced caulk, a form which was in use from Moxon's time well into the nineteenth century.[22]

A hot-stamped, two-part mark is on the blade. A "4" indicates the hatchet's size, and "DRAKE•MAWE & Cᵒ" presumably designates either the hatchet's maker or its retailer. No information has been found about this firm or a likely Drake or Mawe. The hatchet is probably either a Birmingham or a Sheffield product. (handle, 12⅜" LOA)

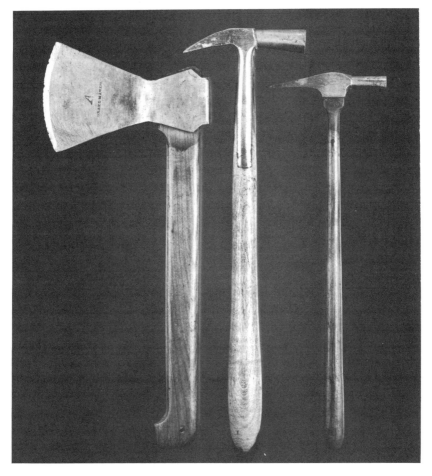

Figure 8. Hatchet, claw hammer, and riveting hammer.

"Claw Hammer"

The claw hammer is a light, general purpose hammer with a strapped head and lignum vitae handle. The early nineteenth-century pattern books describe hammers of this type as either "Gentlemen's" or "Tool Chest" hammers.[23]

The strapping and attachment of the head to the handle follow a design commonly used on several types of eighteenth- and nineteenth-century English hammers. The hammer head has no eye, and the straps are an integral part of the head forging. The handle is fitted between the straps and inletted to receive them. Two iron rivets pass through the straps and handle, securing the head. The handle, a standard shape for hammers of this type, was turned. Although the hammer head is finished bright, the clean up is coarse and imprecise with an overall effect similar to that of the hatchet. File or rasp marks remain on the handle where it was fitted to the head, and vise jaw marks just below the straps may have been impressed there during the fitting. (14$\frac{1}{16}$" LOA)

"Rivetting Hammer"

Riveting hammers of this form were made for use by clock and watch makers, but because of their design and small size they would have been well-suited for a number of light jobs and aesthetically pleasing as well as appropriate tools to include in a gentleman's kit. Both the head and handle of this hammer are very similar to ones illustrated in Plate 6 of the John Wyke catalog of watch and clock tools.[24] This pattern of riveting hammer was a standard Lancashire product and continued to be manufactured into the twentieth century with only minor modifications in form.[25]

The hammer is beautifully forged and finished. Both the head and the peen are steeled. The round head and its face are lathe turned, while the top of the peen and hammer sides are finely ground. All other visible surfaces are crisply cleaned and polished. The long, slender handle is made from the same wood as the hatchet handle, and it also is well shaped, although not quite so highly finished as the head. (12$\frac{3}{16}$" LOA)

"8 Chizles & Gouges"

The chisel and two gouges in the chest are the earliest known documented examples of types common in the late eighteenth century (Figure 9). English tanged chisels with bolsters which can be dated to the period around 1770-1800 often have a distinctive shape differing from the shape of later examples. The bolsters are relatively large. The shanks are long, continuous sweeps. Except on narrow tools, the transition from the shank to the blade proper is distinct but not sharp, and the blades flare, increasing in width toward the bezel. In side profile, the chisels taper from a thick section at the bolster to a delicately thin edge. (Beveled-edge, tanged chisels were not standard English tools of this period.)

This design may have been used much earlier than 1770. There is evidence in Moxon and other sources that chisels with well-defined shanks and "shoulders" (the transition from the shank to the blade) were being made long before this date.[26] Many English chisels that apparently date to the late seventeenth or earlier eighteenth

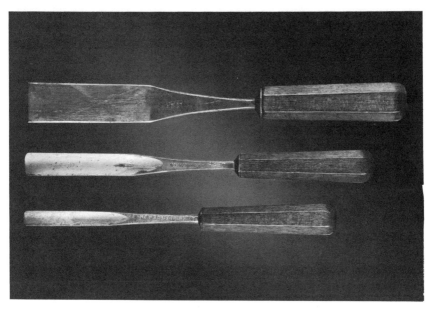

Figure 9. Newbould chisel and gouges.

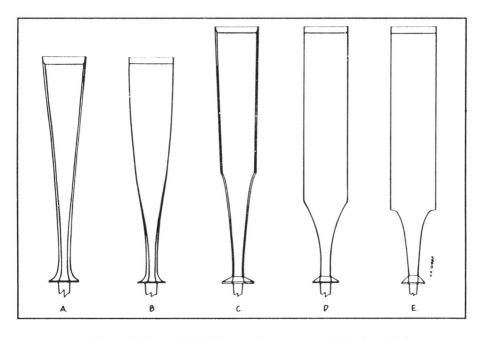

Figure 10. Tanged chisel shapes. Approximate periods of popularity:
A and B, late seventeenth to mid-eighteenth centures; C, late eighteenth century;
D, early nineteenth century; E, mid-nineteenth century to present.

centuries, however, have a "continuous sweep" profile (Figure 10A) or its re-curved variation (Figure 10B).

Sometime after 1800, chisels of the shape illustrated in Smith's *Key* became standard (Figure 10D). The shank is shorter relative to the overall blade length, and the shoulder becomes much sharper. Some blades, at least, continue to be flared. Finally, by the mid-nineteenth century, parallel-sided, heavier chisels with shorter shanks and sharp shoulders become the norm. Gouges follow the same basic styles (Figure 10E).

There are many exceptions to this progression, but it does illustrate a general trend. It also raises several questions. Why do the changes take place? Is the impetus for them basically stylistic, or do they relate to the technology of the tools' manufacture? Are they related to the way in which the chisels were used?[27]

The distinction between "firmer" and "paring" chisels and gouges of this period is rather muddy, but since the Hewlett tools are relatively light-weight, they were probably intended for paring. The chisel has a 1¼" blade, while the gouges are ½" and ¾". All are stamped "NEWBOULD" and "BEST." The mahogany handles are of a typical pattern for this period: octagon in section and tapering from their butt ends to the bolsters.

Although the tools were marked by or for Thomas Newbould, shearsmith and edge tool maker, located, like Philip Law, in Coalpitlane, Sheffield[28], the marks were struck on the blades after they were finished, and each of the three "NEWBOULD" marks was made with a different stamp. Other type C chisels were "cold" stamped, and it could have been a common practice. On the other hand, it could mean that Newbould bought the tools from someone else and marked them afterwards, that the makers from whom Newbould bought them marked them with their own "NEWBOULD" stamps but only after they were assured that Newbould would be the purchaser, or that Newbould himself made the tools, but waited to mark them until he was certain that he wanted his mark on them. The "BEST" designation also raises questions. Although the term was used a great deal to describe tools during the late eighteenth and early nineteenth centuries, its exact meaning remains unclear. Was it a matter of material, construction, or finish? Whatever it referred to, the "BEST" stamp on these Newbould tools appears to have been an afterthought. (chisel, 12" LOA; ½" gouge, 9⅞" LOA; ¾" gouge, 11¹⁄₁₆" LOA)

"6 Files"

The lone surviving file is a 5" pitsaw or frame saw file. It is a blunt or untapered file, and its section approximates a semi circle.[29] It is single cut on both its convex and flat surfaces, and, characteristic of these files, it is left untoothed for a short distance at its point. Files of this particular pattern were intended to sharpen larger frame saws and pit saws, but a small example such as this one was a good general purpose tool appropriate for a gentleman's kit.

The blades of half-round and similar files of this period usually do not taper evenly into the tang. Like this example, they are "broken down" at the heel of the blade so that one side of the tang is a continuation of the flat side of the blade while there is a step from the convex surface of the blade down to the other surface of the tang. The tang of this file is stamped "DUBLIN" (with a crown?). Since the mark does not appear in the 1774 Sheffield directory, the file is probably from either Lancashire or Birmingham.

The file is fitted to a mahogany handle with a brass ferrule made from sheet, wrapped and soldered into a tube. There is also a slightly larger, unmounted handle of the same design in the chest. (10⁵/₁₆" LOA)

"Beak Iron"

Like the riveting hammer, the small beak iron is unsigned but can be attributed to the Lancashire area. It is virtually identical to one illustrated in the Wyke and Green pattern book.[30] When the tool was mounted in a vise, beak-iron end up, the center surface served as a small anvil, while the round and square horns were used to bend and shape small parts. The square horn is pierced with a small hole, which permitted work on the ends of pins or parts mounted on arbors and facilitated making right-angle bends in wire stock.[31] By turning the tool over, the round end could be used to work small parts. All working surfaces are steeled.

The beak iron is finished with the same meticulous care as the riveting hammer. Comparing these two tools with most of the others in the chest reveals why the Lancashire makers had such a reputation for quality. (3⁷/₈" horn tip to horn tip)

"Pair Shears"

The shears are a small pair of metal-cutting hand shears. They are marked "BOSTON" and were made by Thomas Hall of Sheffield. The 1774 directory lists Hall as a "jewellers sheer and edge tool maker," located in Peacroft, but no mark is given for him. In the 1787 Directory, he appears as an edge tool maker who "also draws Steel," with "BOSTON" given as his mark.[32]

The shears are made of iron with steeled blades. The blades are ground, and the remaining surfaces of the shears are filed or ground bright. The scrolls forged on the ends of the handles serve as a stop to prevent the shears from closing too far and also may help to keep the tool from sliding out of the user's hands. (6³/₈" LOA)

"3 Turnscrews"

One turnscrew remains with the chest (Figure 11). It closely resembles an early nineteenth-century Birmingham design labeled as a "Screw driver" in a circa 1800 Birmingham pattern book and a "Turnscrew" in the R. Timmins & Sons book.[33] It has a brightly finished blade with a filed reel decoration at its base. The shank, also finished with a file, is round. The flat, 3⁄8" wide blade is ground. The rectangular section tang tapers from the filed decoration to a point. It was left as forged.

The turned lignum vitae handle has flattened sides and is decorated with turned incised lines. Like the other turned handles in the chest it has a hole in its end from either a lathe center or as a result of tearout when turned parts were separated. All of these handles may have been stock items that Hewlett purchased from a turner and fitted to the tools himself. The turnscrew handle has a heavy brass ferrule made as a sandcast tube.

Finding cast brass ferrules on the turnscrew and one of the bradawls was unexpected. They may be early versions of the "Strong Brass Ferrules" shown as stock items in the Timmins book.[34] (6⁹⁄16" LOA)

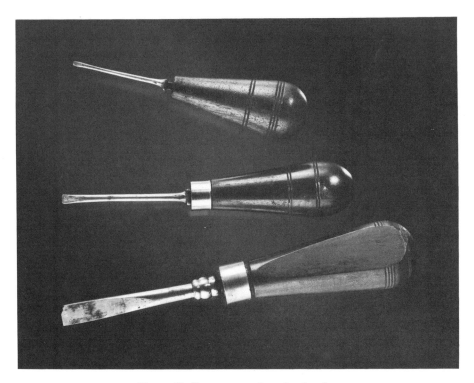

Figure 11. Turnscrew and two brad awls.

"4 Awls"

Two bradawls survive. Both have double-bevel chisel points, round shanks, and the abrupt, square bolsters characteristic of late eighteenth-century awls. The blade of the larger awl is approximately ³⁄₁₆" and the smaller is ³⁄₃₂" Both awls have turned, bulb-shaped, mahogany handles decorated with incised lines. The large handle is fitted with a cast brass ferrule like that of the turnscrew. The small handle has no ferrule. By the second decade of the nineteenth century, round-bolstered bradawls fitted with "London Pattern" or other thin-necked handles were standard.[35] (4³⁄₈" and 5³⁄₈" LOA)

"6 Gimlets"

The five gimlets in the chest are examples of the standard shell design illustrated in early nineteenth-century pattern books. They range in cutting diameter from approximately ⅛" to ⁹⁄₃₂", and are finished bright. Based upon the twist of their lead screws, they are probably "single." The gimlets have flat, tapering tangs, which were let through the turned, cigar-shaped boxwood handles and "riveted" or peened over. Copper washers, designed to prevent the peening from weakening the handles and available on gimlets by the late eighteenth century, were not used.

Four of the gimlets are stamped "T•SMITH" in a zigzag border on their blades and branded "TS" on the sides of their handles. Thomas Smith was a "gimblet" and drill bit maker working in Deritend in Birmingham in the 1770s.[36] The "TS" brand seems logically to be a mark of Smith rather than that of an owner. (2⅞" to 5⅛" LOA)

Other Tools

It appears that much could be learned by studying the tools that did not survive with the box. It might be possible to determine how the entire kit was intended to be used, and it would be interesting to see if illustrations could be found depicting the missing tools, so that, on paper at least, the kit could be reassembled.

Mr. Hewlett's tool chest, its contents, and the accompanying documentation fit well with a number of known facts and theories about tools and toolmaking in the 1770s. They suggest a number of new insights. But, above all, they seem to have the tantalizing and highly stimulating effect of posing several new questions for each one that can be answered. These questions reveal just how little we know about eighteenth-century tools and toolmaking. We have much to learn about tool designs and their sources, the technology involved in tool production, specialization, the development of toolmaking locales, concepts of quality, ordering and distribution systems, pattern books, and retailing. And one very basic question is yet unanswered: how well did the tools work? These tools and the paths down which they lead us may help answer some of these

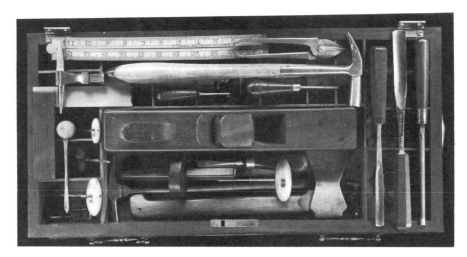

The till of the tool chest.

questions. Answers to others may never be forthcoming, at least not in detail. It is a shame we cannot ask Mr Hewlett.

Notes—Part II

13. *Sketchley's Sheffield Directory*, 1774, p. 21, and p. 9 of "A Description of the Town and Parish of Sheffield."

14. Moxon, *Exercises*. pp. 65-66.

15. See Campbell, *London Tradesman*, pp. 238-239, for a discussion of this practice by London cutlers bringing in "country" (Sheffield and Birmingham) goods.

16. Three other early (dated in the 17th century) English rules I have seen are calibrated right-to-left rather than the "standard" English left-to-right. Were both directions in use on early English rules with the left-to-right becoming standard in the 19th century, while Americans adopted and retained the archaic right to-left system in much the same way they preserved other outmoded bits of English custom?

17. H.R. Calvert. *Scientific Trade Cards in the Science Museum Collection* (London: Her Majesty's Stationery Office. 1971), Fig. 47.

18. *Encyclopaedia Britannica*, 13th ed., s.v. "London." E.G.R Taylor, *The Mathematical Practitioners of Hanoverian England 1714-1840* (Cambridge: Cambridge University Press. 1966), p. 247.

19. See the description of "The Carpenter's Joint Rule" in Edmund Stone's translation of Nicolas Bion, *The Construction and Principal Uses of Mathematical Instruments*, 2d ed. (London: 1758; Reprint, London: Holland Press. Ltd. 1972), pp. 14-15.

20. This is presuming that rules were calibrated in the late 18th century as described for the mid-19th century by J. Rabone, Jr. in "Measuring Rules," *Birmingham and the Midland Hardware District*, ed. Samuel Timmins, (London: 1866), reprinted in Kenneth Roberts' reprint of the 1892 John Rabone & Sons catalogue (Fitzwilliam, N.H.: Ken Roberts Publishing Company. 1982). pp 83-84.

21. See William Marples & Sons Ltd. 1909 catalogue (Reprint, M-WTCA., E.A.I.A., and Arnold & Walker, 1979), p. 45.

22. R. A. Salaman, *Dictionary of Tools* (New York: Charles Scribner's Sons, 1975), p. 48. Moxon, *Exercises*, plates 4 and 8 show variations on a hatchet and an axe.

23. A pattern book illustration closely resembling the hammer is number 512, "Tool Chest Hammer" in a Birmingham pattern book of tools and household utensils printed on paper watermarked 1798. The book is in the collections of the Essex Institute, Salem, Mass. "Gentlemen's" hammers, while of the same form, often have filed decoration on the straps.

24. Alan Smith, *A Catalogue of Tools for Watch and Clock Makers by John Wyke of Liverpool* (Charlottesville: University Press of Virginia. 1978), p. 41. This plate probably dates to before 1758-1759 (p. 15).

25. Theodore Crom, *Horological Shop Tools 1700-1900* (Melrose, Fla: 1980), p. 547.

26. Moxon, *Exercises*, plate 4. Two English chisels with long shanks and abrupt shoulders are in the collections of the Science Museum, South Kensington, London. One is dated to the 16th century; the other, signed, "MOOR[E?]," is thought to be 18th century.

27. The chisel-shape chronology illustrates the type of stylistic analyses we hope will be forthcoming for many early tool types. The best and most thorough 18th- and early 19th-century tool "typologies" done to date are those formulated by W L. Goodman and others for woodworking planes. Such typologies, while not exact, are nevertheless valuable aids to identifying and dating early tools. If functional changes accompany the stylistic ones, they also help determine changes in tool uses and working practices.

28. *Sketchley's Sheffield Directory*, 1774, p. 21.

29. W. T. Nicholson, *A Treatise on Files and Rasps*, Providence, R.I.: Nicholson File Company, 1878), p. 18. Similar files are illustrated in Smith's *Key*, numbers 790 and 791.

30. Smith, *A Catalogue of Tools*, plate 40, number 396A, p. 105.

31. *Ibid.*, p. 106.

32. *Sketchley's Sheffield Directory*, 1774, p. 21. *A Directory of Sheffield* (Sheffield: Gales & Martin, 1787; Reprint, New York: Da Capo Press, 1969), p. 4.

33. Number 543 in the Birmingham pattern book referred to in note 27, and number 243, plate 40, in Roberts, *Tools*, p. 93.

34. Roberts, *Tools*, plate 7, p. 55.

35. Smith, *Key*, number 546, and Roberts, *Tools*, plate 7, p. 55.

36. *Sketchley's Birmingham, Wolverhampton and Walsall Directory* (Birmingham: J. Sketchley, 1767), p. 28. Smith is listed in Birmingham directories until the early 19th century.

Acknowledgements

I wish to thank the many people at Colonial Williamsburg who have shared their expert insights regarding the tool chest, the documents accompanying it, and the tools with me. My special appreciation also goes to Hans Lorenz and Craig McDougal for the photographic work, Hannah Gibbs for the chisel drawing, and Mary Ann Williamson for her review of the manuscript.

Education of New England Apprentices during the Colonial Period 1600-1789

By WILLIAM J. LITTLE
September 1980

During the colonial period about one half million emigrants traveled to America. Although they were primarily English; Scottish, Irish, French, German, Dutch, Swedish and Finns were also represented within this rather large group. The primary motivation for emigration was to seek a second chance to make a livelihood, achieve social mobility and thereby provide security for themselves and their families.

The trip was not without dangers of the sea and of sickness. Blake, describing the tribulations of the journey to America of the ship *Unity* of London, bound for America, relates a measure of the suffering that took place[3].

> Here we have in essence some of the main causes why the immigrant suffered: storms at sea; poor food; superstition, anguish and fear; rotten and leaky ships; long delays in harbour and long voyages (sixteen to seventeen weeks in this instance); greedy crews and ruthless captains; overcrowding above and below decks; foul air, no ventilation; lice, disease and death. We cannot generalize. Conditions varied on every voyage. Some were short, undisturbed and peaceful; on others the conditions were appalling. Washing down the decks with vinegar could not check disease in overcrowded and undernourished conditions. Even the best "messes" between decks in the early eighteenth century only allowed the servant sleeping room of 6 ft. by 1½ ft.; and the widely advertised diets were defective in fruit and vegetables.

The source of emigrants during this period were:

1. In payment of goods; tobacco, lumber or sassafras roots (sale of redemption agreement).

2. Paid agents used as recruiters to entice dissatisfied laborers to travel to America for greater opportunity.

3. Kidnapping (agents seized stray children and drunkards, hid them from their parents or friends and "spirited" them aboard ship for America).

4. Convicts.

5. Those who had resources to pay passage or were willing to encumber themselves for passage costs. (11:47)

A study conducted by Menard (15:47) relates the demographic nature of emigrants during this period.

> As a group, the indentured servants were young when they emigrated. While they ranged in age from mere boys to the "old and decripit," the great majority were in their late teens and early twenties. Age on arrival was determined to be about nineteen. Probably most were from English families of the "middling sort," yeomen, husbandmen, and artisans, men whose expectations might well include the acquisition of a freehold or participation in local government.

To learn a trade and become a freeman offered the potential apprentice not only a method of learning a trade, but provided security in an unknown land. The status of freemen in America enabled journeymen to marry, hold a job, acquire title to property, vote, and to a large extent, achieve economic and social mobility. Although this concept lost popularity in later years, during the colonial period the process was strictly adhered to. Of those earning freeman status during 1700-1776 in New York, 49% were craftsmen (5:95). And further evidence of the importance of skilled craftsmen in the labor force is demonstrated by the large percentage of this category existing in 1774, 33% to 50% (5:96).

The difference between the indentured servant and the apprentice is not often apparent and the terms seemed to be used interchangeably within records and literature (14:63, 73) (24:70-71). Douglas offers five chief differences between the two classes of labor.

> 1. The apprentice was supposed to receive trade instruction, while the indentured servant was not. However, even here the popular confusion of terms was so great that we find many indentures specifying that the so-called servant is to be taught a trade. Obviously this is only apprenticeship in disguise.

> 2. The colonies prescribed the rudiments of a liberal education for the apprentice, while they required nothing of the sort for the indentured servant.

> 3. Since apprenticeship primarily involved "learning," the apprentice was generally a minor, while the indentured servant was usually an adult.

> 4. The apprentice was generally a child born in the colonies, while the indentured servant almost invariably came from abroad.

> 5. The unexpired serving-time of the indentured servant, like that of the slave, was transferable without consent of the servant (8:28-29).

The functions of colonial apprenticeship served a four fold purpose: as punishment for a debt, a penalty for idleness, a system of poor relief and the earliest form of colonial educational institution. This article is directed to the last of these purposes; the earliest educational institution of the colonial period.

For an examination of the education of apprentices, one must look to the relationship existing between the master and the apprentice. The association was more than simply employer and employee, or master and servant. This affiliation may be described as similar to that existing between father and son. Within this context of a family-like association, the apprenticeship as an educational institution will be examined. The industry and life of society are closely tied to its form of labor. It logically follows that the history of labor within early America is a fairly accurate index to the progress of society within the original thirteen colonies and in New England particularly.

Labor was basic to colonial economic and industrial development. Within the historical records of this period, notably apprenticeship agreements and related documents, are reflected the influence of European immigration, social conditions of the period, the need for education and the desire for economic and social mobility. Three documents in particular illustrate the development of labor through apprenticeship education as a process and as a social institution. The analysis of these records serves as the basis for this article.

Indenture agreement All forms of contract labor, whether voluntary or by judicial decree, were subject to an agreement, commonly called an indenture. The responsibilities of all parties were reduced to writing and often served as the only recorded history of early American labor. Areas covered in the agreement concerning the apprentice were: length of time of service, personal conduct (discipline), restrictions with regard to travel, marriage, informal and formal education, and personal care of the apprentice such as food, clothing, lodging and health.

Certificate of a Freeman At the successful conclusion of the apprenticeship period a certificate of a freeman was signed by the master and recorded in the town records. The privilege of freeman status of a town meant that its possessor was free to ply a trade or to engage in merchandising within the city limits (14:30). In addition, several other advantages accrued to freeman status, such as the right to marry, hold property and exercise responsibilities of citizenship, such as voting and holding public office.

Massachusetts Bay Act of 1642 This legislation specifically provided for education of children in formal educational institutions as well as informal arrangements such as the apprenticeship form of education. The Massachusetts Act served as the standard in other New England states and provided the foundation for future educational legislation.

These three documents form the historical perspective of apprenticeship education as it existed during the colonial period. Representative forms of these records follow in individual sections devoted to each topic.

APPRENTICESHIP INDENTURE

Apprenticeship is essentially a combination of education and industry. It is a process of learning by doing, under which a minor is taught the art of a trade by one who is at the moment engaged in it; the minor paying either in whole or in part for this instruction by the work done on objects destined for the master's consumption or sale (8:11).

The responsibilities of the master and of the apprentice were outlined within the indenture. This document was drawn in standard form, attested to (signed) by all parties and filed with the local government office. In later years, indenture forms were available in blank, similar to deed and will forms available today.

Researchers found that in the large metropolitan areas, the indentures were recorded and filed in a book reserved for this purpose, while in smaller towns and villages the recording of indentures could be found among an almost infinite variety of records. As such, the indentures reported have for the most part been taken from the records most readily obtained.

A representative form of indenture follows:

This indenture witnesseth that I William Potter	1
son of John Potter of Prouidence in the Colony	2
of Rhoad Island and Prouidence plantations,	3
(deceased) hath put himself and by these presents	4
with the free and full Consent of his mother	5
Jane Potter: Put himself an apprentis to Daniel Cook	6
of the same Town and Colony aforesaid Joyner	7
to Learn his Art After the manner of an apprentis	8
to serue him the said Daniel Cook his Executors	9
or administrators from the day of the date	10
of these Presents until he the said William Potter	11
shall attain and Com to the full age of	12
twenty one yeares: dureing all which term the	13
said Apprentis his master faithfully shall serue	14
his secrits Keepe his Lawful Commands Euery where	15
obey: he shall do no Damage to his master nor seene	16
to be don of others without Giueing notis there of	17
unto his said master he shall not wast his said	18
masters Goods nor lend them unlawfully to any att	19
Cards Dice or any unlawful Game he shall not Play	20
where by his said master may haue damage in his	21

own Goods or others he shall not Cummit fornication	22
nor Contract Matrimony with in the said term;	23
he shall not absent himself day nor night from his	24
masters seruis without his Leaue, Nor haunt aile	25
Houses or Tauerns; but in all things behaue himself	26
as a faithfull apprentis ought to do dureing	27
all the said term And I the said Daniel Cook	28
to promise and Ingage for myself my Executors	29
and administrators to Learn and Instruct my said	30
Apprentis William Potter in the	31
trade mistry or art of a Joyner in the best	32
manner that I Can within the said term; and	33
also Instruct him in the trade of a House Carpenter	34
as I haue oppertunity: and not put him	35
to any other seruis dureing the sd term without	36
his Concent, and also Learn or Cause him	37
to be Learned or taught to Reade English and	38
wright and Cypher so far as to keepe a Booke:	39
and to find and Prouide for him sufficient	40
meate Drink apparrill Lodging and washing	41
befitting an apprentis dureing all the said term:	42
Ane when the said term is Expired which	43
will be in the yeare of our Lord; one thousand	44
seauen hundred and nineteene or twenty; then to	45
sett him ffree: with as Good apparill in all	46
Respects fit for his body throughout as he now	47
is perticuliorly named on the back side of this	48
indenture, for the true performance here of Each	49
party binds themselues unto the other firmly by	50
these presents. In witness where of they haue	51
here unto Enter Changeable sett there hands and	52
seals this twenty ninth day of march anno Domoni	53
one thousand seauen hundred and sixteene (16:62-63).	54

From the indenture illustrated, examination reveals that apparently the letter v is missing from the entire document. The letter u served as a substitute. For example, serue, Prouidence, Euery, Giuing and haue are actually serve Providence, Every, Giving and have.

Apprentice provisions Lines 1 through 28 cover specific areas of discipline and moral conduct expected of William Potter, the apprentice. The contract was intended to educate the apprentice in the art or mystery of a given craft and to establish sound moral and spiritual values within the young boys. The relationship included room and board and the master was held accountable to the courts for mistreatment and physical abuse of apprentices. This agreement illustrates the use of the apprenticeship system as a social institution. William Potter's father is deceased, and his mother is providing for his support and education by arranging for his servitude to Daniel Cook.

Length of service This agreement calls for a period of service of three or four years (line 45). For the most part agreements required a time period of not less than seven years. Various attempts, including this agreement, were made to shorten this period, and thereby satisfy the labor shortage. In almost all cases this change did not produce capable and competent craftsmen. Extremes in the other direction existed. For example, one apprentice was seven years of age and would serve until twenty one years of age, a total of fourteen years! In general the most common period was seven years and a trade could be learned within this period.

Selection of trade Individual career selection was highly dependent upon the economic and industrial development of the specific area. The importance of making a good choice was heightened by the difficulties involved in changing one's mind later. After a man had gone through seven years of preparation for an occupation, he would not lightly undertake to learn another (17:68). This agreement is unique in that the master has two trades, joyner and house carpenter (lines 7 and 34). Often, farmers acquired out of necessity trades such as carpenter, smith and shoemaker. If a husbandman acquired proficiency in one of the trades, he was likely to ply it in addition to his farming (14:25) (5:27) (25:145).

In areas such as Boston, New York and Philadelphia, an aristocratic society developed and trades such as gold and silver smith, peruke maker (wig maker) flourished. In addition, a division of labor was possible due to the industrial development of the larger cities. For an example a wagon maker would have the following trades represented in his shop: blacksmith, ornamental iron worker, joiner, coachmaker, wood carver, upholsterer, wheelwright, leather worker, painter, japanner, and limner (portrait painter; coat of arms, scenes, etc.). Contrasted to this we find the wagon maker in Ipswich with the master craftsman performing all of the crafts illustrated.

The extent that the master was indeed a master of his craft and his ability to communicate his knowledge to his apprentice determined the level of knowledge and development possible for the apprentice. The degree that the apprentice was able to absorb this training served to perpetuate or retard the craft. The system seldom took aptitude into account, for colonial artisans were as yet innocent of that modern institution called vocational guidance. Prospective apprentices were never made to play with blocks or given I.Q. tests (5:132).

Discipline   The master was empowered by law and custom to discipline his apprentice and to compel the apprentice's compliance with the terms of the indenture. Disobedience, unruliness, laziness or any form of misconduct could be punished by scolding, deprivation, or whipping, except that the punishment must not be immoderately cruel or such as would permanently disfigure the apprentice. If the apprentice ran away and was apprehended, he was legally compelled to serve double the time of his absence, if more than twenty-four hours (14:75).

Samuel Magee, a cordwainer (shoemaker) of New York City, came before the General Quarter Sessions Court in 1731 to complain that his son Alexander, an apprentice to Thomas Hall, another cordwainer, had been "immoderately corrected" by Hall. The Court, after it had examined the marks of the immoderate correction upon the head, arms, and body of the boy, Alexander Magee, ordered his release from his apprenticeship (14:77).

Formal education   The education of William Potter is given on lines 37, 38 and 39. "Learn or Cause him to be Learned or taught to Reade English and wright and Cypher as far as to keepe a Booke (bookkeeping)" outline the responsibilities of the master, Daniel Cook. The goal of education in this respect was to develop the apprentice sufficiently to assume his role in society. Naturally all masters were not capable of teaching their apprentices to read. The records give abundant evidence of the illiteracy of masters in general. Most of them were obliged to make marks in lieu of signatures to legal documents and most of them must have resorted to the Town Clerk or to a schoolmaster in order to have the simplest kind of communication composed and written. Illiterate masters were obliged, therefore to send their apprentices to persons who could teach them, which, in most cases, meant that they sent the apprentices to schools (18:41).

The apprentices were not permitted to attend the free day schools operated by the towns since they were obliged to be working with the master during this period. Thrifty schoolmasters keen to take advantage of this situation opened evening schools (18:95).

The curriculum of the evening schools varied from the basic subjects of reading, writing, and arithmetic to more advanced subjects such as architecture, navigation, surveying, the projection of the sphere, the use of the globes, conic sections, gunnery, algebra, the theory of the pendulums, fluxions (mathematics; differential calculus) and numerous other vocational subjects (19:61, 63).

Most of the evening schools in the American colonies were kept only "during the winter." Although differences obtained in practice, the "usual" season was of six months duration from October to April, and comprised two terms. These three-month periods were known as quarters, and constituted the term-basis of all colonial schooling. With but few exceptions, the schools gave instruction every evening in the week, classes beginning, in most instances, at six and ending at eight or nine o'clock (19:20).

Provisions for master Areas covered in this indenture concerning the master were the protection of his business, such as keeping secrets, (confidentiality of processes, methods and records), conservation of tools, materials, finished products, through proper and prudent use and not selling or disposing of by gambling or otherwise, the master's tools, materials or finished goods by the apprentice.

Since manpower during this period was scarce, the apprentice type of labor served to supply manpower to individual craftsmen during the colonial period. The indenture was a document issued to protect the apprentice from abuses that occurred during the period of servitude. Abuses took the form of the apprentice performing farm and household chores or general labor, rather than tasks associated with the area of craft instruction. Often times the boy did not receive education in reading, writing and arithmetic. A further problem faced by apprentices was not acquiring a sufficiently high level of knowledge and techniques to engage in the craft at the termination of his indenture period.

Apprenticeship indentures were terminated by: successfully completing the time, by judicial degree resulting from mistreatment of the apprentice by the master, by not learning a trade or acquiring formal education, by the master moving without taking the apprentice along and by joint consent of the master and the apprentice.

CERTIFICATE OF A FREEMAN

Obtaining the status of a freeman enabled the craftsman to ply his trade, marry, hold property and to exercise the privileges of citizenship such as voting or holding a public office. The certificate was issued after the craftsman had taken the oath of a freeman. Although a nominal charge was made for recording freeman status, in cases of the poor or indigent the fee was waived.

Citty of   )
) S S
New York)

Isaac De Riemer Esq'r Mayor and the Alderman of the Citty of
New Yorke. To all to whom these presents shall Come Send Greetings
Whereas Thomas Evans Bricklayer hath made application to be made

a Freeman and Cittizen of the said Citty. These are therefore to Cer-
tifie and Declare that the said Thomas Evans is hereby Admitted
Received and allowed a Freeman and Cittizen of the same Citty to
Have, Hold Enjoy and Partake of all the Benefits Liberties Privileges
Freedoms & Immunities Whatsoever Granted or belonging to the
same. In Testimony whereof the said Mayor hath hereunto Subscribed
his Name and Caused the Seale of the said City to be affixed the first
day of October, Anno. Dom 1701. Annoq. Reg. Regs. Will, tertii,
Nunc. An. & c Decimo Tertio

                                    I.D. Reimer, Mayor. Will Sharpas Clk.
(14:37)

Although important in earlier colonial times, the procedure of freedom lost
popularity and eventually was no longer enforced after this period.

MASSACHUSETTS BAY ACT OF 1642

When viewed from the perspective of history, the passage of the Massachusetts
Bay Colony education acts can be seen as the product of a number of factors which
came together in the 1640's in such a way as to intensify the concern of the leaders of
the colony for the education of the populace. Together with the inherent practicality of
the Puritans, these factors resulted in the passage of these laws which embodied some
of the basic principles upon which the American system of education was to be built:
a. the state could require the education of its citizens, b. the state could require the
establishment of schools, c. the unit of administration was to be the local community,
d. the state could supervise and control the schools through public offcials, and e. the
financial support for the schools was to come from the people (6:103).

The educational legislation of these colonies shows that the various assemblies
sought two main ends, namely compulsory education and compulsory schools. The first
contemplated a minimum of education for all children, to be given by parents, masters,
or someone employed by them for this purpose. The Massachusetts act of June 14, 1642,
was the first general colonial act of this character. It reads as follows:

This court, taking into consideration the great neglect in many
parents and masters in training up their children in learning and labor,
and other imployments which may bee profitable to the common
wealth, do hearupon order and decree, that in every towne the chosen
men appointed for managing the prudencial affaires of the same shall
henceforth stand charged with the care of the redresse of this evill, so

as they shall bee liable to bee punished or fined for the neglect thereof, upon any presentment of the grand jurors, or other information or complaint in any plantations in this jurisdiction; and for this end they, or the greater part of them, shall have power to take accompt from time to time of their parents and masters, and of their children, concerning their calling and impliment of their children, especiallity of their ability to read and understand the principles of religion and the capital lawes of the country, and to impose fines upon all those who refuse to render such accompt to them when required; and they shall have power (with consent of any Court or Magistrates) to put fourth apprentice the children of such as shall not be able and fitt to employ and bring them up, nor shall take course to dispose of them, of such as they shall find not to bee able and fit to imply and bring them up, nor shall take course to dispose of them themselves; and they are to take care that such as are set to keep cattle bee set to some other impliment sithall, as spinning up on the rock, kniting, weveing pate, ect,; and that boyes and girles bee not suffered to converse together, so as may occasion any wanton, dishonest, or immodest behavior, and for their better performance of this trust committed to them, they may divide the towne amongst them, appointing to every of said townsmen a certeine number of families to have special oversight of; they are also to provide that a sufficient quantity of materialls, as hempe, flaxe, ect. may bee raised in their severall townes, and tooles and implements provided for working out the same; and for their assistance in this so needful and beneficiall impliment, if they meete with any difficulty or opposition which they cannot well master by their owne power, they may have recourse to some of the magistrates, who shall take course for their help and incuragement as the occasion shall require, according to justice; and the said townsmen, at the next Court in those limits, after the end of their yeare, shall give a breife account in writing of their proceedings herein; provided, that they have bene so required by some Court or magistrate a month at least before; and this order to continue for two yeares, and till the court shall take further order. (11:87).

Although apprenticeship was an English custom, the colonists saw in this institution new and broader possibilities of its uses. In New England and New York, the first laws concerning education and the the first compulsory education laws were contained in apprenticeship enactments. The apprenticeship system took care of the entire problem of public elementary education during the colonial period. By the enactment of these laws the scope of apprenticeship was broadened to such an extent that it became a new, and peculiarly American institution (18:107).

BIBLIOGRAPHY

1. James Arnold, *The Shell Book of Country Crafts,* (New York: Hastings House, 1968)

2. Bernard Bailyn, *Education in the Forming of American Society* (Chapel Hill: The University of North Carolina Press, 1960)

3. J.W. Blake, "Going to America", *History Today,* Volume 8 (June, 1958), pp.394-403.

4. Daniel J. Boorstin, *The Americans, The Colonial Experience* (New York: Random House, 1958)

5. Carl Bridenbaugh, *The Colonial Craftsman* (New York: New York University Press, 1950)

6. R. Freeman Butts and Lawrence A. Cremin, *A History of Education in American Culture* (New York: Holt, Rinehart and Winston, 1953)

7. Lawrence A. Cremin, *American Education, The Colonial Experience* (New York: Harper & Row, 1970)

8. Paul H. Douglas, *American Apprenticeship and Industrial Education* (New York: Columbia University, 1921)

9. Edward Eggleston, *The Transit of Civilization From England to America in the Seventeenth Century* (New York: D. Appleton and Company, 1901)

10. Cheesman A. Herrick, *White Servitude in Pennsylvania* (New York: Negro University Press, 1926)

11. Marcus Wilson Jernegan, *Laboring and Dependent Classes in Colonial America 1607-1783* (New York: Frederick Ungar Publishing Company, 1931)

12. Henry J. Kauffman, *The Colonial Silversmith, His Techniques & His Products* (New York: Galahad Books, 1969)

13. Kenneth A. Lockridge, *Literacy in Colonial New England* (New York: W.W. Norton & Company, 1974)

14. Samuel McKee, *Labor in Colonial New York 1664-1776* (Port Washington, Long Island; Ira J. Friedman, Inc., 1935)

15. Russel R. Menard, "From Servant to Freeholder: Status Mobility and Property Accumulation in Seventeenth Century Maryland", *William and Mary Quarterly,* Series 3 (January, 1973), pp. 37-67

16. Perry Miller, *The New England Mind in the Seventeenth Century* (Cambridge: Harvard University Press, 1954)

17. Edmund S. Morgan, *The Puritan Family, Religion and Domestic Relations in Seventeenth-Century New England* (New York: Harper & Row, 1944)

18. Robert Francis Seybolt, *Apprenticeship & Apprenticeship Education in Colonial New England & New York* (New York: Teachers College, Columbia University, 1917)

19. _____, *The Evening School in Colonial America* (Urbana: University of Illinois, 1925)

20. _____, *The Public Schools of Colonial Boston (1635-1775)* (Cambridge: Harvard University Press, 1935)

21. Darrett B. Rutman, *Husbandmen of Plymouth, Farms and Villages in the Old Colony, 1620-1692* (Boston: Beacon Press, 1967)

22. Abbot Emerson Smith, *Colonists in Bondage, White Servitude and Convict Labor in America* (Chapel Hill: University of North Carolina Press, 1947)

23. _____, "New Light on Some of America's 'First' Families", *Journal of Economic History*, May 1942, pp.40-53

24. Warren B. Smith, *White Servitude in Colonial South Carolina* (Columbia; University of South Carolina Press, 1961)

25. Rolla Milton Tryon, *Household Manufactures in the United States, 1640-1860* (Chicago: The University of Chicago Press, 1917)

26. Louis B. Wright, *The Cultural Life of the American Colonies, 1607-1763* (New York: Harper & Brothers, 1957)

Journal of a Cabinet Maker's Apprentice

By WINIFRED C. GATES

June 1962 and September 1962

In 1844, while working as an apprentice in Greenfield, Massachusetts, young Edward Jenner Carpenter on the advice of his father, began a "Journal." He wrote, "Two years ago I came here to learn the cabinet trade & I Suppose I must stay till the 4th of August 1846, my 21st birthday — a long time to look ahead, but short to look back." This young apprentice was the eldest of seven children of a country physician, Dr. Elijah W. Carpenter of Bernardston. He was educated at Franklin Academy in Shelburne Falls, Massachusetts one of the best schools in that region. Probably because of financial reasons, he did not go on to college but was apprenticed in Greenfield to the son of his father's medical friend, Dr. Joel Lyons of Gill. Joel Lyons and Isaac Miles had been partners since 1839 in a cabinet firm employing one journeyman, two apprentices and at times other workmen.

Carpenter remained with Miles and Lyons for seven years. In 1849, he gave up cabinetmaking and went to Brattleboro, Vermont, where he established the first newspaper agency in that town, carrying on a successful business in newspapers, magazines, books, toys and notions. He became Brattleboro's first librarian, continuing as such for twenty-five years. He retired in 1894, going to Amherst to reside with his son until his death in 1900.

In his journal he recorded such items as the weather, reading, lectures, church-going, national events, and social doings in the quiet county-seat (population two thousand). Work at the cabinet shop received his earnest attention as he reported furniture that he was "at work upon."

Besides Carpenter there was another apprentice at the same time, Dexter from Gill; both lived in the shop building and their meals were taken at the Miles' home nearby.

"Mechanics" of all trades in Greenfield in 1844, worked every day in the week except Sunday, and from September until March, they worked every evening. Young Carpenter noted on March 15th: "We quit working evenings tonight; we shall not have to work any more evenings till the 20th of next September." Although evening work was expected by Miles and Lyons, the boys often "quit working this evening to go skating" or to some social event or Lyceum lecture. ". . . I went to hear Dr. Darling lecture again last night. . . He lectured on Phrenology . . . it was first rate . . . he examined the head of Sol Severance & told his character exactly . . . He gave another lecture this evening which I must attend he is so interesting. I don't know but the bosses will think it is too much to go every night in the week. I have worked but one night this week, and probably shall not another for dancing school is Friday night."

Holidays enjoyed by the "mechanics" were Fast Day in April, Independence Day in July, Muster Day in October, and by proclamation of Governor Briggs, Thanksgiving Day on the last Thursday in November. Christmas Day was not a holiday, nor was New Year's day.

There were no signs of labor unrest among the cabinet workers, but Carpenter reported discontent in the group of men who were employed at the "Green River Works" where cutlery products that rivalled the famous English cutlery of Sheffield, were manufactured. Because chisels were the first product made by this concern in 1834 (*Picturesque Franklin,* 1891. Wade, Warner & Company, Northampton, Massachusetts, p. 95, 106.), it was known locally for many years as the "Chisel Factory." Quoting Carpenter's diary for "March 28th, Friday (1844) ... The Chisel Factory hands struck today and swore they would not work any more till Russell would agree to pay them every month, he paid them up once in six months before. They took a game of Wicket ball out in front of the shop this afternoon, there were thirty or forty of them playing at one time." Saturday's entry, without comment, stated: "The Chisel Factory hands all went back to work again today on the old terms"!

Carpenter does not record what financial agreement the apprentices had with Miles & Lyons. In several entries, he noted: "If I have made no mistake they owe me $8.41." "Lyons & I settled today & found that they owed me $10.87." "I looked over my account with Miles & Lyons tonight & they owed me $11.14."

From March first through July, Carpenter made seven "Secrataries" and five "Bureaus." He worked from ten to fourteen days on each, depending upon other tasks assigned to him, such as "fetching in lumber," helping "Uncle Jo" Frost, the journeyman make coffins, or in his own words: "Dexter was carrying in boards and I had to kick the lathe in his place." All of his secretaries were made of butternut with mahogany front; "... I finished a butternut board end Secratary tonight except putting three locks on the drawers, which is extra. Cheap secratareis are in good demand."

The "Bureaus" also were made of butternut with mahogany veneered front. They varied in detail: "board end with Ogee drawer" "board end and three cornered drawer" "panell end with double Ogee top drawer, and the front runs clear through. It is a little different from any that I ever made. It is bureaus and Secrataries all the time. I have worked on them for about a year & I begin to think it is about time to learn to make something else."

"... I quit my bureau & went to work on an organ case that they are in a hurry to finish. I have been at work on it all the week." Following that, Miles gave him "a stint to make a bureau with a double Ogee top drawer in eight days." Carpenter gave an account of his progress — on Monday, "I put the carcass together & some more and got through by nine o'clock." Tuesday: "I got along very well on my stint. I have the drawer stuff planed out & the Ogee drawer front veneered." Wednesday: "I got the drawers veneered this evening." Thursday: "...got along pretty well on my bureau. I put the top on & got ready to make the top case." Friday: ". . . I have got my bureau along so far that I guess I can finish it by tomorrow & gain two days and a half. I worked till after 10 o'clock tonight." Saturday: "I finished my stint this afternoon about 3 o'clock & gained till next Tuesday night."

"For my next job, I have got to make 6 butternut tables and a larger one, 3½ feet by 4." The following day he "got out the legs for seven tables." That job was completed in two weeks. Then Carpenter made five bureaus, one after another, and a "mahogony Secratary; they have sold the one in the wareroom."

Helping Lyons with furniture for the Odd Fellows, making a small cupboard for a young law student, a cradle for a Chisel Factory chap, "some press boards to press paper with for L. Merriam & Company" (printers and booksellers), several card tables, a dining table, and a "dress table," occupied his working hours.

Carpenter referred briefly to his tools: "I spent some time this afternoon fixing my jointer." ". . . I made me a pair of try sticks" ". . . my oil stone had got most horribly worn out in the middle so I took some emery rubbed it down level, it took me about two hours." "I finished my fore plane iron & flung it aside & Lyons got me a new one."

"Uncle Jo" Frost, the journeyman at the shop, was busy much of the time making coffins. Often he needed help from the apprentices, and the proprietors as well, when there were several deaths in town on one day. Carpenter was assigned to make the small coffins. "I have made the bigger part of two coffins today. One was for. . . a child 3 or 4 years old, the other was for a boy. . ."

One Saturday evening, "Uncle Jo" arrived at the shop with the news that "they had to go to work at once on a coffin for Mrs. Murray, who died that evening. The doctor said that she must go into the coffin as soon as possible. It was then ten o'clock." ". . . I worked last night till after 3 o'clock," Carpenter wrote on Sunday, "and did not get up this morning till after nine. Miles worked so hard he got sweaty & caught cold & is almost sick tonight." On a warm Sunday in June, "Uncle Jo" was haying on the meadows. It had rained for two days since he cut his hay, and he thought it was his duty to open it and not let it spoil. "Sol Newton's girl died today & (with Jo away) I thought I should have to go to work on the coffin, but they (the bosses) concluded to start early tomorrow morning; so I went to meeting at the Methodist church . . . This morning (Monday) I got up about half past four o'clock & went to work on a coffin, we got it done about half past eight." Again, on a Sunday the entry read, "Uncle Jo", Dexter and myself have been at work all day a making coffins, we made one out of Bass for David Willard's wife, it is to be covered with Broadcloth, & one of Cherry for an Irishwoman by the name of Lewis." For working on Sunday, the boys were allowed their choice of half a day of free time, or credit for it on their accounts.

On his own time, young Carpenter made "three light stands, . . . they will pay me 50 cts. a piece for making them . . . which makes $1.50 for spending money." He also earned small sums doing odd jobs for his family and friends: "Zuba Woods paid me a dollar for the work-box that I made for her. She said she thought it was cheap at that." ". . . Last night, I sawed out some stuff for a couple of picture frames for my grand-father's family. Tonight, I veneered them. I put mahogony on one and zebra wood on the other." Later, in payment for them he received: "12½ cts. for one glass & I took one (glass) that they sent down which was rather small, to pay for the other. They said they would pay the rest in Maple Sugar next Spring." A typical example of New England "dickering"! Another evening, he made "a trunk Wood"(is this a whatsit?) for a friend, for which he received twenty-five cents.

During the spring building season, Carpenter mentioned two "raisings" which he attended. "I went up to H. W. Clapp's this afternoon to the raising of a barn & worked considerable hard & didn't get as much as 'thank you sir' for it. I went to the raising of Mr. Sylvester Allen's house. It is a large house & a good deal of work to raise it. Mr. Avery is boss of the job."

Miles & Lyons obtained their lumber stock from nearby sources. "Lyons went to Colrain today to see about some lumber. This forenoon, Miles went over to Montague City & got about 300 feet of the best up river pine that I ever saw; it is so good that I am afraid that I shan't have the pleasure of working much of it." Either Miles or Lyons went several times a year to Hartford to purchase supplies. "Miles started this noon for Hartford to buy hardware and mahogany, he went down on the Telegraph." The Telegraph, (*History of Bernardston*, Kellog, 1902, Greenfield, Mass., p. 59) was the designation for a fast stage line which passed through Greenfield to Brattleboro, carrying the New York mail. A speed of seven miles an hour was expected, and extra fare was charged for such rapid service. There was an ordinary coach line in addition to the Telegraph. Thus Greenfield people traveled until 1846 when the "Rail Road" from Greenfield to Northampton was built. On his return after an absence of four or five days, Miles reported that "he had to send to New York for veneers for there were none in Hartford that were good for anything."

Craftsmen in neighboring villages were often employed by Miles & Lyons for special services, as the following entries show — "Charles H. Dennison of Guilford, Vt. had his shop burnt up last Monday night. Miles & Lyons had a lot of 4-inch square scantling for bed posts, some of it turned, burnt. They think it was the work of an incendiary. Mr. Wilson came today to paint chairs for Miles & Lyons." After eight days, the work required was completed and Mr. Wilson returned to his home in Colrain, a small town in the hills eight miles northwest of Greenfield. Whenever there were chairs to be painted, Wilson was called in to paint them. Was he merely a painter, or was he a specialist, an expert in the art of stencilling and the skill of striping? Carpenter gave no details of Wilson's work.

The demand for furniture in 1844 and 1845 in the Greenfield area of the Connecticut River Valley, was evident. Some of the sales mentioned in the journal were: "Miles & Lyons sent off tonight a load of furniture to North Adams for H. L. Dawes, consisting of a Sofa, 2 tables, a Pillar Work table, Dress table, wash Stand, 3 Bedsteads, a Set of Chairs & a Rocking chair"; "... Russel Warren the tinpeddler was here today & brought a Butternut table & a couple of rocking chairs" ; "... Isaac Burrows of Bernardston brought $43.00 worth of furniture for his daughter who is to be married ... He bought a Bureau of my make worth 13 or 14 dollars"; ... "Miles & Lyons have sold off pretty much all the work in the wareroom. I never saw so little work in the wareroom. They have but 3 or 4 Bureaus that are not sold. They have sold 80 or 90 dollars worth today; they need a jour.(neyman) now if they ever did."

As their sales increased, Miles & Lyons found it necessary to expand their facilities. In August 1844, "they concluded a bargain & made out the writing for a piece of land on a small brook down at the end of the street (Main) for a small water power." The next day "there has been a man here to day a trying to sell them a horse power, but I guess he found it rather dull business." To enlarge their shop and employ more workmen, Miles & Lyons on June 2, 1845 — "brought Birge (F. A. Birge also was a cabinetmaker in Greenfield) out of business, that is they have taken all of his stock & 3 benches. The apprentice David Chapin (in fact, he had just completed his apprenticeship) is coming here to work. I guess Birge thought he had more irons in the fire than he could tend upon." A week later ... "Lyons was tearing out to make room for more hands." There were three new journeymen: S. White Dickinson, a man of some experience, proprietor Joel's brother Charles Lyons, and David Chapin. . . "Dick & David have been fitting up their benches today ready to go into it tomorrow. It makes a pretty good shop full, 4 jours, 2 apprentices & 2 bosses, & they have advertised for another apprentice.

"A German cabinetmaker has worked here a day or two, he is brother to Frederick Boehmer, the man who makes surgical instruments & all kinds of cutlery, ... he has been in this country about four years & learned his trade in New York City ... He says when I am 21 if I cannot find work, to come to New York No. 46 Beekman Street and he will find me work." (*New York City Directory* for 1844, lists the name of Frederick Boehman at 46 Beekman Street.)

Apprentice Carpenter's "Journal" ends abruptly on June twenty ninth 1845, with a short observation on the weather.

Workshop Folklore in England

By R. A. SALAMAN

December 1984

Mr. Salaman is the author of Dictionary of Tools used in the Woodworking and Allied Trades *and of* Dictionary of Leather-Working Tools, *both standards in their fields. He designed and equipped the Wheelwright's Shop in the Science Museum, London, England, and has lectured and contributed to many learned journals and reference books. This article, which was published in England by The Tools and Trades Historical Society and is reprinted here with their permission, was dedicated by the author to Mr. W. L. Goodman on the occasion of his 80th birthday.*

Ever since my boyhood, spent in a Hertfordshire village, I have enjoyed talking with tradesmen in their workshops. Thirty years later, after the second world war, when visiting a village wheelwright, I saw him smoking hams in the saw-pit over a fire fuelled by his old oak felloe-patterns. It came upon me then — as it must have done to many others at that time — that here was a way of life and work that was soon to disappear altogether. From that time I determined to visit workshops as opportunity offered, and to acquire examples and learn the purpose of every tool used. There was no difficulty about finding the tools, for they were mostly lying unused or piled up for scrap.

During these same visits I also made a point of asking tradesmen if they could remember any sayings or legends about their work, for I found that what they could tell me not only explained former workshop practice, but helped me to understand the less idyllic side of a tradesman's life: for example, the frequent poverty, and the lack of adequate schooling. During the following thirty years I visited over three hundred tradesmen. Of these, 86 were smiths and farriers; 62 were wheelwrights; 40 were carpenters and joiners (often undertakers as well); 33 were specialist woodworkers such as shipwrights, millwrights, furniture makers, coachbuilders, hurdlemakers or coopers; 39 were leather-workers including tanners, curriers, shoemakers, machine belt makers, saddlers and glovers. The rest were from various trades including basket makers, sailmakers, farm workers, horsemen, plumbers and builders. Most of the tools I collected are now in the St. Albans City Museum, or in storage at the London Science Museum. The workshop 'day books' that I found are in the St. Albans Museum and in the County Record Office in Hertford — waiting for someone to study them. I gave some eighteenth-century sailmaker's Day Books to the National Maritime Museum at Greenwich.

The stories and sayings I collected during my all-too-short visits fall into three main types: tales of hero figures, instructional rhymes and proverbs designed to help an apprentice to learn the trade; and recollections of childhood — particularly of the food their mothers prepared, or the lack of it.

For the purpose of this article I will deal only with the first two subjects, and will begin with the heroes.

During a visit to Cornwall in 1946, I visited workshops in the countryside near St. Austel and Mount Charles. While in that district I was told that quite recently, a man who explored disused tin mines had been asked by a countryman if he was looking for the 'old king.' He was told that people in those parts still believed that the old king (King Arthur) was sleeping in one of the workings and that he would 'waken' in time of Britain's greatest need.

I was told later that these same Cornish folk used to speak of an even older legend connected with the tin mines: that 'Joseph the Tin Man' (Joseph of Arimathea) was a merchant who visited Cornwall to buy tin, and that waterfront workers in Falmouth could point to the stone on which he first set foot.[1]

I never found any trace of the ancient patron saints, such as St. Crispin, who took to shoemaking to support himself while teaching the Gospel. But I came upon more recent heroes of another sort. One of them was a travelling blacksmith who called at forges and offered to make farrier's pincers from a pair of old rasps. He was known as Pincher Jack. I have a pair of these pincers: there is no doubt about their exceptional quality, fine workmanship, and beauty of form. (Fig. 1)

Fig. 1. A pair of Pincher Jack's pincers from Mr J.M. Kaye, Blacksmith, Staincliffe, Yorkshire. Made c.1905. (Drawing by Jack Laird). The drawing of the pincers is taken from DICTIONARY OF (WOODWORKING) TOOLS, page 296, Fig. 443.

I first heard of this extraordinary man in 1946, from Mr. W. Ward, a blacksmith in Harpenden, Hertfordshire, but I soon found that he was known and spoken about in blacksmiths' forges everywhere I travelled in England and Wales, and in places as far apart as Westmoreland and South Wales. I may not be believed if I say that these farriers spoke of Pincher Jack as they would of the Prince of Wales if he had knocked at the front door to ask for a glass of water. Of course, everyone admires skill, but I suspect that to these blacksmiths and farriers Pincher Jack was a champion who showed the outside world that the products of the country smith were superior to that of the factories that were putting hundreds of smiths out of business. Eventually I found Pincher Jack's real name and who he was. I sent the story to John Geraint Jenkins (now Director of the Industrial & Maritime Museum at Cardiff) who had it published in *Gwerin*, the Journal of Folk Life 1960, Vol. III. A rather longer version was later published in *The Village*, 1969 Vol. 24 No. 4.

Another 'hero' story I came across was about an Irishman from Cork, named James Sullivan, who could be called in to deal with a difficult horse. I was first told of this man in 1964 by Mr. A.E. Sands, a smith and farrier of Elstree in Hertfordshire, who said that Sullivan was known as the 'Whisperer' because of his manner of speaking to horses. Mr. Sands told me that the Whisperer sat up all night with a vicious or unmanageable animal, and that in the morning, the horse followed him quietly out of the stable. (I have since read an interesting but incomplete account of this man in *Modern Practical Farriery* published by William Mackenzie c. 1890). To anyone who had bought such a horse, the Whisperer was a saviour indeed: for by employing him a terrible loss could be avoided. Without a horse, many tradesmen (let alone the village doctor) could not earn a living, and having spent their money on a useless animal, could not afford another.

Many of the proverbs and 'wrinkles' I have come across relate to the care of horses. Anyone who was brought up before 1914 will remember how dependent we all were on the labour of horses — in the fields, on the roads, and at every turn in our daily lives. This will perhaps explain the reverence with which specialists like Pincher Jack and the Whisperer were held, and also the number of proverbs current about the care of horses. Proverbs spoken by horsemen and farm workers have been recorded in a series of remarkable books by George Ewart Evans about the life and occupations of people in Norfolk and Suffolk. I will recount only two sayings that demonstrate how the daily life with horses entered into the language of ordinary people.

One was told to me by Mrs. Hilda Everett of Hatching Green in Hertfordshire, the daughter of a horse keeper who worked for the Tollemache family in Suffolk. Mrs Everett told me that her father used to say 'A man works best full collar,' which I take to be a horseman's version of the saying that a man works best when fully stretched. The other was told me by a 92-year-old horsecollar maker, Mr. George Stace, of Birchington in Kent. He said that as an apprentice, his old master, seeing him talking to a customer, told him that he must learn to 'whistle and ride'. I was puzzled by this until Mr. Stace explained that customers used to stand talking to the men in the shop, and that his master wanted his men to work while they talked.

Turning to the instructional rhymes and proverbs, I will begin with the sawyers. Mr. George Casbon, a wheelwright of Barley in Hertfordshire (1946), told me about travelling sawyers who used to call on his father twice a year to saw his logs of elm, oak and ash into planks. (As commonly happened at that time, Mr. Casbon bought and felled his own timber from surrounding woods). The sawyers carried their 7-ft. pit-saws on their shoulders, with the tiller and box (handle) in a separate bag. One of their sayings is wise advice to anyone who has to work hard:

Strip when you're cold, and live to grow old.

Another saying went something like this:

Deal knots
and empty pint-pots
are two bad things for sawyers.

Deal knots can break a saw tooth — which reminds me of a story told me by a shipwright, Mr. M.G. Worfolk of Kings Lynn (1966). He said that he went to sit by his old master who was lying ill in bed. He asked him: 'Did you ever have any enemies?' and the sick man answered: 'Deal knots and rusty nails are the only enemies I've got.'

Mr. Casbon remembered yet another sawyer's rhyme which was probably intended as a humorous excuse for bad workmanship:

A sawyer's no robber —
What he took off one side,
He left on the other.

Anyone who has ever helped a blacksmith by working the lever of the old type pear-shaped bellows will understand the usefulness of this rule-of-thumb told me in 1949 by Mr. Bysouth, a blacksmith working in Braughing in Hertfordshire. (The operation is not so simple as it looks, and unless these instructions are followed, the fire won't get hot enough to give the iron what smiths call a 'good soak').

Up high,
down low,
Up quick,
Down slow;
And that's the way to blow.

Mr. R.C. Pond, a retired Norwich shoemaker, told me that as a boy he was taught the following rhyme as a guide to the use of the edge iron (known in the U.S.A. as a collice). It is used for shaping and burnishing the edge of the sole:

Wet it,
Sweat it,
Set it.

(i.e., 'wet' the leather; 'sweat' the edge with a half warmed iron to shape it; 'set' and burnish the edge with a hotter iron on wax.)

Another retired shoemaker, Mr. William B. E. Glasow, who for many years worked for John Lobb the London bootmakers, told me that the old hands used to say, 'A good shoemaker can make a shoe with a knife and fork' — a saying directed to the toolmakers who persuaded us to buy a special tool for every job. Mr. Glasow also told me that, as an apprentice, he was told 'Good leather needs little hammering; bad leather won't stand it.' This relates to the practice of hammering dampened sole leather to consolidate it.

The following instructional rhyme is the well-known guide for children who used to follow the farmer when sowing corn. The farmer carried a metal dibbler in each hand and, while walking backwards, he made indentations in the land at close intervals to receive the corn. (Owners of small fields preferred to 'dibble' rather than broadcast, for it was considered more economical.) As the children counted the grain into each hole, they were taught to 'say a poem', as my informant put it, to prevent wasting the seed:

Five seeds in a hole:
One for the rook,
One for the crow,
Two to die, and
One to grow.

Mowing with scythes was a crucial operation at harvest time and one of the sayings of the mowers was 'Hang your scythe in a tree.' The explanation seems to be that by leaving the scythe outdoors, the rust caused a slight serration on the edge which, after sharpening, would leave a serration that would cause the scythe to bite into the stems of the corn or grass without sliding off — just as we like a saw-edge on our bread knives today.

I will end this account with a workshop legend which I can only describe as a carpenter's dream come true. I was told about it by the late Mr. Stanley Ward, a millwright who lived in Letchworth in Hertfordshire. In his retirement, Mr. Ward became a professional 'browser' and undertook research in libraries for various scholars. He sent me one of his discoveries from the London Library — a reproduction of a Sixteenth Century woodcut in a book entitled *Canonical Histories and apocryphal legends relating to the New Testament*, executed by Angelo Della Croce, Milan, 1873 (See Fig. 2).

The wood-cut depicts a man and a boy standing at opposite ends of a plank of wood and holding it up between them. Both have haloes about their heads. The boy is standing on a little pile of earth and shavings to bring him level with his father. On the bench there is an axe, snap-line, square and auger. The identity of the man and boy is obvious, but why hold each end of a plank? I consulted Mr. W.L. Goodman of Bristol, an authority on woodworking tools, who has made a collection of pictures of the Holy Family in the carpenter's shop, and later made a search in the New Testament Apocrypha. There, the chapter entitled 'The Infancy of Jesus Christ' gave the clue.

The story begins by saying that the King of Jerusalem sent for Joseph and ordered him to make a throne 'of the same dimensions with that place in which I commonly sit.' Joseph obeyed and continued to work for two years in the King's palace before he finished. But when he came to fix it in its place, he found it wanted two spans on each side of the appointed measure. The rest of the story is told as follows:

'Which when the King saw, he was very angry with Joseph; And Joseph afraid of the King's anger, went to bed without his supper, taking not anything to eat.

Then the Lord Jesus asked him, What was he afraid of? Joseph replied, Because I have lost my labour in the work which I have been about these two years.

Fig. 2. Photostat of a l5th Century woodcut reproduced in the Canonical Histories.

Jesus said to him, Fear not, neither be cast down; Do thou lay hold on one side of the throne, and I will the other, and we will bring it to the just dimensions.

And when Joseph had done as the Lord Jesus said, and each of them had with strength drawn his side, the throne obeyed, and was brought to the proper dimensions of the place:

Which miracle when they who stood by saw, they were astonished, and praised God.

The throne was made of the same wood, which was in being in Solomon's time, namely, wood adorned with various shapes, and figures.'

What of the future? My advice to students of the history of trades and tools is this: go on asking questions. You may not hear tales of heroes, nor any more instructional rhymes, but you will meet people who can tell stories about their working lives that are just as interesting as those their grandfathers told me.

Footnote
1. This story would appear to be the tin-miner's version of two widespread legends about Joseph of Arimathea: that after the Crucifixion he brought the Holy Grail to Glastonbury; and that earlier he had brought the boy Jesus to Britain, a visit celebrated in William Blake's poem of 1804, which we now sing as 'Jerusalem':

And did those feet in ancient time
Walk upon England s mountains green?
And was the holy Lamb of God
On England's pleasant pastures seen?

Eighteenth Century American Plane Makers

By JOHN S. KEBABIAN

September 1970

Editors Note: Difficult as it is to believe, little interest was shown before the 1960's in the names imprinted on the front (or toe) of wooden planes. Rather, whatever limited interest there was in planes centered on the various types and how they were used, not on who made them. In 1966 Kenneth and Jane Roberts began their project to identify and codify American planemakers. One of the first to research the early 18th century American planemakers was John Kebabian, and to him, quite properly, came the distinction of finding the first documented American planemaker to imprint his name on his product, Deacon Francis Nicholson. In the following articles he set forth his discoveries and laid the groundwork for all that has followed.

At the auction sale of the Blake estate in Fitzwilliam, New Hampshire, in 1967, I acquired the following planes with maker's marks of ancient aspect.

1. Smoothing Plane, marked "CE CHELOR" "LIVING IN" "WRENTHAM"—3 marks in all. Owner's stamp "L B"

2. Molding Plane, marked "CE CHELOR" "WRENTHAM."

3. Molding Plane, marked "I NICHOLSON" "WRENTHAM." This Wrentham mark and that in Numbers 1 and 2 are identical.

Later acquisitions were:

4. Skew-blade, Rabbet Plane, marked "I NICHOLSON." The mark is the same as in Number 3, but here fresh and sharp—it is rather worn in Number 3.

5. Molding Plane, marked "BRIGGS" "IN KEEN."

6. Molding Plane, marked "I WALTON" "IN READING."

7. Plow Plane, the movable fence set-screw-secured, marked "JO FULLER" "IN" "PROVIDENCE"; the anchor-emblem of Rhode Island is part of the mark "IN." Owner's mark "L B", though not acquired at the Blake sale, and somewhat different from the "L B" of Number 1.

Rockwell Gardiner, who is familiar with the touchmarks (makers marks) on Eighteenth Century silver and pewter, and who has seen the Chelor and Fuller marks, has considered them to be of Eighteenth Century, but had no proof of this.

It occurred to me that genealogical records might be a key to the dating of some of these planes. Accordingly, I went to the genealogy and local history room of the New York Public Library, and found the following:

In the Wrentham, Massachusetts *Vital Records* to the year 1850, compiled by T.W. Baldwin, the marriage of Ceasar (sic) Chelor and Judith Russel is recorded, April 20,

1758. Births are recorded of eight children to Cesor (sic) and Juda (sic) Chelor: Beulah, May 10, 1760; Juda, April 25, 1762; Elisabeth, December 12, 1763; Asor, August 4, 1765; Alpha, March 15, 1767; Hepzibah, March 16, 1771; Askins, May 8, 1773; David, September 20, 1775. The mention of Cesor or Ceasar Chelor at the birth of David is his last appearance in the Wrentham records. He does not appear in the 1790 United States Census at Wrentham.

It is evident from the above that the plane making Chelor was active in Wrentham between 1758 and 1775, and that he was no longer there in 1790. No other person of that name appears in the Wrentham records. The marriage record of the Chelors states that they were both Negroes. Would Chelor, therefore, be the first American Negro Tool Maker?

Concerning the "I Nicholson" of planes Numbers 3 and 4, the Wrentham records are less informative. Since capitals "I" and "J" were both written as "I" in the Eighteenth Century, we may confidently assume that the plane-maker was one of four John Nicholsons in the Wrentham Vital Records. No other "I" or "J" Nicholsons there appear. The marriage of John Nicholson and Mary Ware is recorded on April 22, 1742. The death of John, son of John and Mary Nicholson, November 16, 1737 (either this or the marriage date seems to be erroneous); birth of John, son of John and Sarah Nicholson, October 12, 1737; of John, son of Captain John and Sarah, October 24, 1787. Considering that "I NICHOLSON" was using the same "WRENTHAM" stamp as Chelor, we may assume that they were roughly contemporary. The most probable of the Nicholsons to be the plane maker would, therefore, be the John, son of John and Sarah, born October 12, 1737.

A plane is reported in the collection of Paul B. Kebabian, South Burlington, Vermont, marked "I NICHOLSON" "CUMBERLAND"; Cumberland, Rhode Island is adjacent to Wrentham, Massachusetts, just over the state line. There is no "I" or "J" Nicholson in either Wrentham or Cumberland in the 1790 Census.

We can assume from the above with some feeling of certainty that the "I NICHOLSON" planes were made in the Eighteenth Century, from circa 1760 to before 1790.

Investigation of the Briggs clan of Keene, New Hampshire, in regard to plane number 5, was without result. The family was large and prolific; there is no first name or initial on the plane, and there were many Briggses there in both the Eighteenth and Nineteenth Centuries. While we may be of the opinion from the apparently early form of the stamp and the apparently early location "IN KEEN" that the plane is of the Eighteenth Century, there is no proof of this as yet.

The "I WALTON" "IN READING" is almost certainly one of that large family in Reading, Massachusetts. The *Vital Records of Reading*, to 1850, list 18 male names of I. or J. Waltons born from 1709 to 1760, and a Joshua

—*332*—

Walton born 1811. Fourteen I. or J. male Waltons married from 1724 to 1795; Joshua, married 1832; and the only "J" death is Captain John Walton, died 1785, aged 75 years. Captain Walton was the only one of these numerous I. and J. Waltons to die in his native town, a reflection of the rapid depopulation of rural New England in the late Eighteenth and early Nineteenth Centuries.

There is only one Walton recorded at Reading the 1790 Census, if we so read the Joshua Watton there (very probably a mispelling or mistranscription of Walton).

As the Walton mark is clearly too old in style and form to be applicable to Joshua, born 1811, married 1832, we may be almost entirely certain that the "I WALTON" "IN READING" plane maker's mark dates from the Eighteenth Century.

Information was relatively certain and copious concerning the "JO FULLER" of Providence, Rhode Island. In the *Alphabetical Index of Births, Marriages and Deaths*, to 1850, compiled by Registrar E.M. Snow, we find the marriage of  Joseph Fuller to Lucy Potter, September 30, 1794; marriage to Betsy Gonsalve, November 16, 1806; and death of Joseph, August 10, 1845, aged 72 years, born, therefore, in 1773. Further, Joseph Fuller was occupant of a house and shop, corner of Westminster Street and Richmond Lane, in 1798 (H.R. Chace, Owners and Occupants...Providence, R.I.).

And lastly, Joseph Fuller is in the 1824 Providence Directory as a "blockmaker" by occupation, on Friendship Street. "Blockmaker" would evidently be a maker of ship rigging blocks (pulleys)—such an occupation would be perfectly in accordance with his manufacturing planes when the block business was slow. No other "J" or "Jo" Fuller occurs in the Providence Alphabetical Index. We may conclude, therefore, that Joseph Fuller the plane maker is the person of these various records, and that he was active from 1798, and probably before, to 1824, and probably later. So while his mark may be of the Eighteenth Century, his period of activity goes well beyond in the Nineteenth.

Since recording the above, I have come across three more Eighteenth Century planes:

8. A molding plane with marks "H.WETHEREL" "IN NORTON." These marks are similar to the Wrentham marks described above, and we may assume that the "Norton" is in Massachusetts, near Wrentham. We find in the *Vital Records of Norton*, to 1850 (1906), the following H. Wetherels listed, the name being given several variant spellings:

1. Henry, born 1703, married 1727.
2. Henry Jr., married 1760.
3. Henry, born 1764, married 1783.
4. Henry, born May 15, 1779.

The 1790 Census lists one Henry Witherall in Norton. Curiously, there is not a single death record of an H. Wetherel, or variant name, up to 1850.

The Records list also two brothers, Hiram H. and Horace B. The former was born in 1809 and married in 1842. His brother Horace was born in 1807 and married twice, in 1828 and 1844. We can rule them out as possible plane makers since their occupation is given as manufacturers of straw hats and baskets in the *History of Norton* by George F. Clark, 1859. Further, the marks are clearly earlier in style than their life-span. We may, therefore, assume that the plane making "H. Wetherel" was one of the Henrys of the Eighteenth Century, though, of course, the third and fourth name could easily have worked into the Nineteenth Century.

9. Two other planes, a Tongue and Groove (not matching) bear the marks "A. SMITH" "REHOBOTH", Massachusetts near Norton and Wrentham, and the marks similar in style to those of Chelor, Wetherel, etc. Unfortunately, the *Vital Records*, 1897,  have too many A. Smiths (eight entries for Abiall or Abiel; three for Aaron; two for Amos; etc.), some of them born in the Nineteenth Century.

As can be seen, a great deal of further research could be done on this group of tool makers. Probably the town records would be helpful in identification and datings.

We may call these men real pioneers in the industrial history of America. Though they all were no doubt the smallest of small manufacturers, they defied the British Imperial policy and actual laws by which America was supposed to supply raw materials to the fatherland, with the processing of the materials into finished goods to be done there. The work of Chelor certainly, and some of the others probably, preceded the American Revolution which freed America finally from these mercantile laws.

I would be interested in reviewing the information on any of the above named planes, particularly any which are dated or which have marks varying from the above.

More Eighteenth Century American Planemakers

By JOHN S. KEBABIAN
June 1971

Recently the writer acquired an early plane with the maker "F*NICHOLSON" "WRENTHAM" . Consulting the *Wrentham Vital Records* to 1850, it was found that there was only one person of that name and initial to whom this mark could be attributed. Deacon Francis Nicholson is recorded as deceased on Dec. 7, 1753, in his 70th year (i.e., his birth date would be about 1683). Also recorded

are his marriage to Sarah Ware, June 6, 1722; the birth of a son, Francis, May 27, 1729, and the death of son Francis on August 20 of the same year. There are no other F. Nicholsons in the records. In view of this, we may attributed the "F*NICHOL-SON" planes to Deacon Francis; his dates would indicate that  they can be assigned to the first half of the 18th century. Our fellow-member, Dr. Barnet Delson, is the owner of a plane with these marks, with the words "LIVING IN" before the "WRENTHAM."

"I*NICHOLSON", certainly a John Nicholson, was the next owner of these two

*I *NICHOLSON/ WRENTHAM*

*I*NICHOLSON/IN/CUMBERLAND*

latter marks. The specimen reproduced is from the collection of P. Kebabian. It is not clear what the relationship of John Nicholson was to Francis; the Wrentham records are of no help on this point. What is apparently the same "I*NICHOLSON" mark also appears with the marks "IN" "CUMBERLAND" (coll. P. Kebabian), a Rhode Island address.

Finally, the "LIVING IN" and "WRENTHAM" marks appear with the name "CE*CHELOR" (=Cesar Chelor). Here the Wrentham mark is so worn that Chelor must certainly have been the last of these three users. Note that the mark is the same size throughout (23mm.long). Chelor (see above) appears in the Wrentham records from 1758 to 1775 and is the only person of that name to so appear.

CE CHELOR/LIVING IN/ WRENTHAM

Another early Massachusetts planemaker is "I*PIKE" "DEDHAM" (coll. P. Kebabian). Pike's hand was not very steady, he having struck the place-mark no less than three times in an unsuccessful attempt to produce one which would read clearly. On referring to the *Town of Dedham Record of Births, Marriages and Deaths* (Vol.I) and *Church Records* (Vol.II), we find only two Pikes to whom the tool can be attributed: (1) Jarvis Pike (II, 41 and 81) who appears in the records in 1726 and 1733, and (2) Jarvis Pike, Junior, presumably the son of the first Jarvis (I, 60, 63, 66; II, 47, 48, 50, 53) who appears in the records from 1732 to 1742. Clearly, the

I * PIKE/DEDHAM

"I*PIKE" plane must be assigned to the first half of the 18th century. "J" was regularly abbreviated as "I" at that time.

"H • WETHEREL IN • NORTON". In the article above it was suggested that this maker might be one of four Henry Wetherels in the Norton, Mass. *Vital Records,* those being the only "H.Wetherels" therein. The first was recorded as born 1703, married 1727; the second, married in 1760; the third, born 1764, married 1783; the fourth, born 1779. Since then the following has been learned: Donald Wing, of Marion, Mass., reports a plane so marked and with a date written on in ink, "1764". Further, the writer has acquired a plane with the identical name-mark but with the place-mark "CHATHAM". This is Chatham, Conn., now East Hampton, near Middletown. This mark was also reported by Robert Carlson of Deep River, Conn. The 1790 Census (page 81, Conn. volume) reports "Henry Witherel" as a resident of Chatham. The East Hampton (Chatham) *Early Records...of the Congregational Church,* p.120, reports the death of Widow Witherill on Oct. 30, 1820, aged 84, i.e., born 1736). There are no other Chatham Wetherels (or variant of this name) in the Census or *Early Records.* We may conclude that the plane-maker was the second Henry Wetherel of the Norton *Vital Records,* married there in 1760, living in Chatham, Conn.,

1790, and making planes in both those places. A clearly later mark, "H • WETHERELL CHATHAM" probably relates to this maker also, despite the extra "L" in the name. Mr. Carlson also reports a Wetherell mark with a Middletown (Conn.) place-mark. *[H. Wetherell was later found to be the son of H. Wetherell. Ed.]*

The recent book by Mr. and Mrs. Roberts has shown how valuable a source of information are the city directories. Names of the following plane makers were found by a reading of the *Philadelphia Directory,* 1785, by Francis White, and the *Philadelphia Directory* of 1800 by C.W. Stafford, plus reference to directories for intervening years. (This involved reading of about 10,000 names, in enlarged microfilm at the New York Public Library).

Thomas Napier appears as a plane maker in 1785, and from 1791 to 1800. In 1797 he also was "manufacturer of Fisher's pills." He had previously worked in Edinburgh, Scotland, 1774-1775—see Goodman, *British Plane Makers.*

William Brooks first appears as a planemaker in 1791, and in subsequent directories up to 1800.

William Martin appears as a plane maker in the directories from 1791 to 1800. The 1800 directory also contains the baffling listing (a misprint?) of a "Martien Mabsom & Co.," plane manufacturers, at 99 N.Third Street, also the address of William Martin.

John Butler appears as a house carpenter, 1785 to 1794; as a plane maker in 1796; again as a house carpenter, 1797-1799; finally as a plane maker and a house carpenter in 1800.

Melchior Deeter is listed as a plane maker in 1797 and 1798; as a grocer in 1799; again as a plane maker in 1800. Hibsam Martin appears as a plane maker in 1799 and 1800.

Few other tool makers appear in these directories, though such implements as axes may have been made by the numerous blacksmiths in these volumes. I have noted the following: Charles Cecil, house-jack maker, 1785; he next appears as a lock maker, 1793; then a whitesmith in 1794 and 1796. Nail makers (all from the 1800 directory) were William Baker, John Brown, Ezra Crane, John Gillam, Benjamin Linton ("cut nail manufacturer"). Charles W. Janson, in his *The Stranger in America,* 1807, comments on the very large quantity of nails produced by the Philadelphia manufacturers in this line. Francis Mason appears as a saw maker in 1799 and 1800. Francis Mahony and John Wells were proprietors of steel mills. Their product was no doubt the "blister" or "shear" type, there being no record of cast steel being made at so early a date.

A few odd occupations, all from the 1800 directory, are: Charlotte Genter, rake maker; John Mansell, rule maker; Michael Miller, wheat fan maker. Spinning wheels were made by James Lynch, Davenport Marriott and Thomas Millard.

It should be noted that the Philadelphia Directories did not appear every year in this early period. Ones which gave occupations of the people came out in 1785, 1791, 1793, and 1796 to 1800.

More on the 18th Century Plane Makers of Wrentham, Mass.

By JOHN S. KEBABIAN

March 1972

Inquiries made at the Probate Court of Suffolk County, Mass. (Boston), have disclosed the following. The last will and testament of Deacon Francis Nicholson was signed on April 1, 1752, and was filed, after his death, on December 24, 1753. In this will, the Deacon refers to himself as "Toolmaker." The inventory of his estate was made at Wrentham on December 18, 1753—his manufacturing tools are, unfortunately, lumped under one entry, valued at £32.

By his will, Francis Nicholson liberated "my Negro-man Caesar Chelo" (the final "r" of the name omitted): "considering his faithfull Service, his tender Care, & kind and Christian Carriage, I do set him free." Nicholson bequeathed to Chelor "his Bedstead, Bed & Bedding, his Chest and Cloathing, his Bench & common bench-tools, a Sett of Chizells, one Gouge, one Vise, one Scythe & tackling, & ten Acres of land. . . & one third part of my timber." Chelor himself (including his bequest) was valued at £160. The will also discloses that John Nicholson was the son of Francis Nicholson; he inherited from his father "all my Tools & timber, except what is before excepted," along with land and other goods.

Caesar Chelor died intestate in 1784; his age then must have been about 65-70. His estate was inventoried on August 17, 1784, his tools being given a lump valuation of 424 shillings, 4 pence. The administration document is dated August 30, 1784. Chelor is also identified in the documents as "Toolmaker."

The span of activity of these three persons thus is clearly shown to extend from the early years of the 18th century to 1784. All three are so far known from their marks on planes (see articles above). Whether as "Toolmakers" they produced other implements, is not known at this point.

Thomas Granford of London ET AL

By PAUL B. KEBABIAN

December 1975

Author's Note: It would not have been possible to prepare the following material on Thomas Granford and some of his contemporaries without the documentation provided by William L. Goodman in British Plane Makers From 1700, *(London, 1968), the information he has kindly supplied through correspondence, and his generous permission to use these sources liberally.*

The evidence offered by hand tools with makers' marks, of which the makers have been chronologically and often geographically placed by supporting manuscript and published records or other relevant facts, clearly suggests that name identifications on tools began to appear about 1700. Goodman's research into the manuscripts of the Joiners' Company of London in the Guildhall Library, together with known examples of tools, had shown Robert Wooding of London as the earliest British plane maker with a standardized name stamp used on his product. Wooding, according to the Joiners' Company manuscripts, was indentured to Henry Shawe, joiner of London, on 11 July 1693.[1] He was admitted to the Joiners' Company in 1710, and died prior to 3 December 1728 when his widow Ann took Thomas Phillipson as an apprentice.[2]

The Richard Nock gouge about which Goodman has written represents another very early example of dating,

Figure 1. Thomas Granford name stamp.

and in this instance by supporting evidence. In a collection of tools taken back to Russia in 1698 by the workmen of Peter I, following their study of shipbuilding in Holland and England, there is a gouge marked with the maker's forename and surname. This tool, forged by an English edge tool maker who was working prior to the 1698 date, again provides a benchmark for a dating of *around 1700* for the earliest known, maker-marked tools, based on our knowledge today.

Figure 2. Granford moulding plane.

In July 1975 this writer visited the Old Mill Shop of Stephen and Emma Sanders in Skowhegan, Maine, and acquired a moulding plane marked on the front end Thomas/Granford. The tool possessed characteristics of length, chamfering, and style of wedge which suggested manufacture prior to the 19th century. In addition, the size and style of lettering of the name Thomas Granford was reminiscent of the marks on both a Robert Wooding plane and one made by Thomas Phillipson (see below) in his collection. On return to Vermont, a quick reference to *British Plane Makers* revealed

Figures 4 and 5

that Robert Wooding, following his initial apprenticeship to Henry Shawe, was indentured to "Thomas Granford Citizen also & Joiner" of London on 28 March 1699.[3]

The evidence suggests that the Granford plane may well be of late 1600's manufacture, and represents the earliest known specimen of a plane with the maker's name struck with a conventional die in the characteristic toe, or front end position. We know that there was no specific Company of plane makers; the early London makers were members of the Joiners' Company, some of whom specialized in the making of planes or added this occupation to their formal craft of joinery. Undoubtedly the joiners of this era either made their own planes, or acquired them from a fellow joiner who made a particular practice of their manufacture.

Wooding, in his career between admission to the Joiners' Company in 1710 and death not more than 18 years later, took as apprentices William Flight, William Cogdell, and John Jennion—the latter two well-documented plane makers. Thus one may readily infer, if only on the basis of his own identified plane production and that of his apprentices when admitted to freedom of the Company, that Wooding himself had well developed the plane making art. What could be more likely than that he acquired his training and expertise in the making of wooden planes during the period of his apprenticeship with Thomas Granford? Since Granford was an established joiner in 1699, using an embossing die to mark his own work (which implies that he made planes regularly, quite apart from how many have survived), the Granford plane may thus be given the likely date pre-1700; almost inevitably it is pre-Robert Wooding.

A letter from William Goodman notes that he has located in the Guildhall records further information on Thomas Granford which identifies two men, father and son. A 5 March 1705 entry in MS 8052, Joiners' Company Apprentice Bindings, records that "Thomas Granford, son of Thomas Granford Citizen & Joyner of London puts *(sic)* to the said Thomas his father for 7 years." Goodman suggests that if Thomas junior was the eldest son, and of age (15-17 years) to be apprenticed, then Thomas senior—who could not have married until he completed his own apprenticeship—must have been about 40 when in 1705 he took his own son as an apprentice. He would thus have been born about 1665, started his own apprenticeship around 1680, and been some thirty-four years of age when he took Robert Wooding as an apprentice.

There is some possibility that the Granford plane could be a product of the junior Thomas Granford. The working experience of Robert Wooding, however, is highly indicative of an apprenticeship which led to plane making, and this apprenticeship was unquestionably under the elder Thomas Granford.

Thomas Granford's plane (Figures 1-2), made of beech wood, is 10⅝₁₆" long and approximately 3¹⁄₁₆" tall, and is finished on the top edges and end, down to the wider portion of the stock, with a chamfer some ⁷⁄₁₆" in width. The wedge looks to be contemporary, and may or may not be so. The iron is imperfectly struck with a maker's mark, the first three letters only, HIL(?) being at all legible. This mark may well be

Figure 6. Wooding plow plane. Note fence moulded at both ends.

Figure 7. Phillipson plow plane.

Figure 3. Mr. Lewington's "Grandfurdeus" mark.

HILDICK, a name found occasionally on irons fitted to early 18th century English moulding planes. Initials "A: L:" are formed on either side of the plane stock, in letters about 1⅛" high, by a small rectangular punch, and doubtless identify an owner at some point in the history of the tool. The plane cuts a hollow of approximately 1¹/₁₆" dimension.

A second plane marked with a form of the name "Granford" is in the collection of Neil Lewington of Ickenham, Middlesex, England. Lewington has provided a sketch of this tool, of which the dimensions are approximately the same as those of the tool in figures 1-2. On the right-hand fore part of the stock appears in embossed lettering THIS/IS GRAN/FURDEUS/MAKE (Figure 3). One may speculate that this early mark was made before plane makers stamped on the end grain, at the toe of the tool—a location which results in a substantially sharper impression.

Both Robert Wooding and Thomas Phillipson possessed name stamps of which the Roman capital letters in a serrated border ⅛" high are markedly similar to those of the Granford stamp. All three may well have been made by the same man (Figures 4-5). The Wooding plane came from the Albany, N.Y., area where it was reported as being in one family since the 1700's. The Phillipson plow was acquired at an antique shop only some thirty miles from Skowhegan, Maine, the source of the Granford tool.

In a letter of July 16, 1975, Goodman states that he now has location records of eighteen Wooding planes (three in the United States), and twenty-five Phillipson planes (nine in the United States).

During the past twenty-five years there has been a major development of interest in the identification of both individuals and firms responsible for the production and manufacture of hand tools. This interest stems from the concern of the collector, whether amateur or professional museum curator, to know not only how a tool was used, but also its date of manufacture, the locale where it was produced, and who made it. Once these identifications have been made, the tool becomes not merely a utilitarian—though not infrequently an aesthetically pleasing—collector's artifact, but also an historical record and an element in our understanding of social and industrial history. Further, identification of makers has significance in that it assists in the formation of collections based on local or regional criteria, and is now often important in determining price and value. It is one factor in the degree of sophistication to which the collecting and study of tools has advanced.

Although Mercer's *Ancient Carpenters' Tools* provides frequent reference to provenance in his characteristically rambling prose, e.g., "It [a level] was bought in 1916, from Miss A.H. Swartz, of Point Pleasant, Bucks Co., Pa., when found in an old carpenter and wheelwright shop among the many obsolete tools of her deceased father," he offers but meager information on makers and their marks. He whets our curiosity by writing, for instance, that an "axe hatchet" is smith stamped, but gives no clue to the name or mark of the blacksmith.

Kauffman, writing in the July 1953 issue of *The Chronicle* about names on the fore ends of planes and marks on plane irons, describes a plow plane which he had presumed to be of European manufacture, and notes that a cleaning of the tool revealed its Sandusky Tool Company mark. Rockwell Gardiner, in the June 1962 *Chronicle,* addressed the identification matter in "Marked Tools," and provided further stimulus to the study and research in makers' marks. Additional results of scholarly research by Hummel on the Dominy tools, Kauffman on axes, J. Kebabian on the Nicholsons and Cesar Chelor of Wrentham, William Hilton, Sayward and Streeter on Massachusetts plane makers, Bopp on the firm of Booth & Mills, and others have materially advanced our knowledge of the makers of tools. The major works on plane makers' identification and dating, however, are those of Kenneth and Jane Roberts *(Planemakers and Other Edge Tool Enterprises in New York State),* and William Goodman *(British Plane Makers From 1700).*

NOTES

1. Guildhall Library, London, MS 8051, Freedom Admissions, Joiners' Company of London, Book 2, 1687-1710 (cited in: William 1. Goodman, *British Plane Makers from 1700* (London: G.Bell,1968), p.130.
2. Op.cit.,p.131.
3. (cf. Note 1).

LETTERS

Dear Editor:

I have just been informed by Philip Walker that it has been discovered that R.W. Symonds, in an appendix to Sir Ambrose Heal's *The London Furniture Makers 1660-1840,* wrote "Thomas Granford, tool maker, who lived at the sign of the '3 Plane-makers' in Queen Street, near Cheapside, London, advertised in the Post-man during 1703, that he 'maketh and selleth all sorts of Joyners and Carpenters Tools, where any Artificers or Merchants may be furnished with greater or lesser quantities of the best sort, and to prevent abuses, I mark with my Name at Length instead of T.G.'"

I think this might be of considerable interest. This must be the man I wrote about as T. Granford II. From this information, it seems that he would already have been making tools in Queen Street when Wooding was turned over to him in 1699 and presumably, when the apprentice had finished his time, he stayed with him and eventually took his business over. This is probably why we have not heard much about his son, Thomas Granford III, who is known to have been apprenticed to his father in 1705.

The "3-Plane-makers" is a mistake, obviously; perhaps Mr. Symonds, like most of us, didn't realize they were as early as this. In any case, it means that Granford, and not

Wooding, was the founding father of the craft. I don't think this demolishes my theory about the "Gran Furdeus" planes; that could still stand up. What we must do now is find some hitherto "anonymous" planes of the Granford type, simply marked "T.G.". That will be the find of the century.

William L. Goodman
June 1980

Chronicle Eds. Note: Some comment on this important news: perhaps judgment should be reserved about the reading "3 Plane-makers" until we have obtained a copy of the original advertisement. Efforts to do this have been unsuccessful so far. As to the hypothetical "T.G." mark, it should be noted that Granford specifically states that he DID NOT use that mark.

Dear Editor:

. . . The London Furniture Makers 1660-1840 by Sir Ambrose Heal is available now as a Dover reprint. This work which is basically a partial list of London craftsmen in the cabinetry trade of the 1700's will, I am sure, prove to be a source of names of planemakers turned cabinetmakers and vice-versa, especially if Heal's full lists can be consulted. The book contains only a portion of the names he collected.

> THomas Granford , living at the Sign of the 3 Plane-makers in Queen-ftreet near Chesp-fide, London, maketh and falleth all forts of Joyners and Carpenters Tooles, where any Artificers or Merchants may be furnifhed with greater or leffer quantities of the beft fort, and to prevent abufes, I mark with my Name at length inftead of T.G. You may alfo be furnifhed with feveral forts of materials relating to building.

The Granford advertisement, from The Post-Man, May 6-8, 1703. The reference to "3 Plane-makers" (if it is to be trusted) should set us all searching for the other two. William Goodman thinks, however, that it should have read "3 Planes," as on the trade card of John Jennion, who worked later at the same address. (See next chapter for further discussion of this ad.)

To come to the point, the Granford advertisement does exist in the May 6-8 issue of *The Post-Man*, a copy of which is in the British Library in the Burney collection of early newspapers. . . I have checked to see if there are any earlier references to any planemaker in this or any other paper of the period and have turned up nothing. This appears to be Granford's only advertisement in the six years I checked fore and aft of 1703 (what a headache!). You were quite right in your reservation as it is indeed "the Sign of the 3 Plane-makers."

As for the initials "T.G.", the article in Heal's book with the Granford reference is subtitled "The Problem of Identification" and contains the interesting information "the branding of chairs and stools (with initials) was too widespread to be anything else but a trade custom. Chairmakers, particularly provincial chairmakers, continued to brand their chairs into the 18th century."

I myself have seen many ancient moulding planes that are branded on the side with two initials, usually about ⅝" high, which can lead one to all kinds of wild speculation of chairmaker/planemaker, etc. I shouldn't rule out Granford's early use of branded initials as his planes are usually stamped on the side anyway and he had not made the transition to endgrain stamping. I think the existence of a Granfordesque plane stamped "T.G." is an interesting speculation, and would form a link from branding to stamping. All we can surmise is that in 1703 Granford did not brand his initials. So, bring on the 17th century planemakers!

Mike Falkner
Llandovery, Dyfed
Great Britain
December 1980

Chronicle Ed.'s note: The side stamping to which Mr. Falkner refers applies apparently only to the mysterious "Granfurdeus" marks (see Goodman, British Planemakers, 1978, pp. 20-22). The one "Thomas Granford" mark known to your editor is struck on the endgrain of the plane nose. (See article by P. Kebabian above). Speculation aside, we do have Thomas Granford's explicit statement that "I mark with my Name at length instead of T.G.," which would appear to rule out both the initials and the "Granfurdeus" marks.

British Planemakers Before 1700

By MATTHEW CARTER

June 1983

In the first paragraph of the *Introduction to British Planemakers from 1700*, 2nd edition, Bill Goodman wrote "we have not found evidence of a professional planemaker working before 1700." At the time of writing (1978) the places in London we knew to search for such evidence were the invaluable records of apprentice bindings and admissions to the Joiners' Company, supplemented by the parish registers of City churches.

Since then another archive has been searched: the ledgers kept by the City's tax collectors. These records are in two groups: assessments for intermittent levies, mostly Poll Tax, kept in the Records Office of the Corporation of the City of London; and the regular annual valuations for Land Tax, in the Manuscript Department of the Corporation's Guildhall Library.

The ledgers seldom give trades (in the case of the Land Tax books, never) so they cannot support an exhaustive survey of a particular occupation at the time they cover; one of them, however, did yield vital evidence that broke the 1700-barrier.

In the ledger for the first quarter's Poll Tax assessed for Cripplegate Ward Within in April 1692—one of the few to give trades—"Plainmakr" is written against two householders: one with the familiar name of Thomas Granford, in St. Alphage Precinct; the other, hitherto unknown, Robt Hemings, in the Precinct of St. Albans, Wood Street.

Robert Hemings (also spelled Hemmings) remains a shadowy figure. He was made free of the Haberdashers' Company in 1676, but took no apprentice until 1688 when Francis Purdue, bound originally to a Joiner, was turned over to him. It looks as though Hemings had changed trade, without great success, since "poore" and a reduced assessment are written against his name in the 1692 ledger. He was still in the same precinct in 1695 and in less straitened circumstances according to the record of another tax raised to pay for the French war which lists him with wife Mary, four daughters, Purdue and a second apprentice, John Scoffield.

I have found no further record of Hemings, nor heard of a plane with his name or initials on it.

Of Thomas Granford, the second of that name, more is known. He was identified in the first edition of *British Planemakers* as the master of Robert Wooding, and in the second edition as responsible for the four known planes marked as Granfurdeus make. More Granfurdeus planes have been collected since 1978; I make the current tally eight confidently ascribed, with half a dozen putative examples besides.

The best preserved of these are a delight to handle. It is not fanciful to say they feel ancient and of a fashion anterior to Wooding's. This is partly due to their length, partly to their unique and consistent characteristic of a chamfered shoulder, and partly to a rather casual skill in the workmanship: wedge-heads roughly facetted with a chisel, chatter marks in the scooped notch of a wedge, the ridged pattern of a nicked plane-iron along the sides and chamfers of a stock.

It is not known how Thomas Granford II acquired his skill. There is no record of his having served a formal apprenticeship; he was admitted to the Joiners' Company

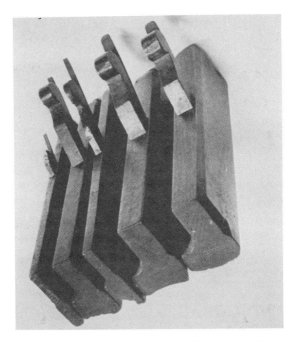

Figure 1. Four Granfurdeus planes and one by Nat Gamble (with broken wedge).
The ovolo was still in use in a country workshop until "collected" recently.

by patrimony, his father, Thomas I, having been himself a freeman of the Company. His mother, Ann, was present at his admission in the capacity of administrator of her late husband's estate. There is an inference that Thomas I's widow had carried on the business until her older son came of age to inherit both the business and the freedom of a City Company necessary to conduct it in his own right.

Other than the conjecture that his son and heir was trained at home, there is nothing to connect Thomas I with toolmaking; in fact little is known of him at all. He was at work by 1654 and dead by 1685. Part of his life was spent away from London; Thomas Harrison, apprenticed to him in 1659, was put to another master in 1662 to complete

*Figure 2. Three of the Granfurdeus planes shown in Figure 1.
The characteristic shoulder-chamfer can be seen.*

his term, and in the following year a more senior apprentice, George Hathaway, was admitted to the Joiners' Company in the absence of Granford, who was in Ireland.

Thomas Granford II was made free of the Joiners' Company on 1st February 1687. In September he took his first apprentice, Jonathan Massey from Dublin, who must have fallen by the wayside since he was not included in the Granford household assessed for Poll tax in June 1690. The Granfords—Thomas II, wife Elizabeth, son Thomas III and a servant—lived near the north end of Wood Street by London Wall, just beyond St. Albans Precinct where Hemings lived. They were taxed for ten window lights which suggests their house was of the most modest type allowed by the post-Fire 1667 Building Act, "The first, or least sort of house fronting by-streets, and lanes," not in Wood Street itself therefore, but in a turning off it.

By April 1962, the household had grown by a daughter, Elizabeth, and an apprentice, Joseph Collins, bound a year before. Like his predecessor Massey, Collins seems not to have stayed the course; Granford took a third apprentice, Nathaniel Gamball, in 1692. Gamble (as he spelled himself), evidently a more likely lad, was still present in 1695 and not replaced, or reinforced, until Granford took his fourth apprentice, Richard Nicholls, in 1698. In 1699 Robert Wooding was turned over to him, having served six years of his term to Henry Shaw. It would be nice to think that Wooding had shown an unusual aptitude for toolmaking and that his first master, unqualified to train him, passed him on to the recognized expert.

Between 1696 and 1703 (a gap in the tax records makes it impossible to be more precise) Granford moved to St. Antholin's Upper Precinct, Cordwainers Ward, which includes the Cheapside end of Queen Street. It was from here that he advertised in the Post Man of 1703, no doubt marking his move to new and better premises.

The advertisement has already been reprinted in the *Chronicle*, but it bears repeating for the light it sheds on Granford's business.

Thomas Granford, living at the Sign of the 3 Plane-makers in Queen-street near Cheap-side, London, maketh and selleth all sorts of Joyners and Carpenters Tooles, where any Artificers or Merchants may be furnished with greater or lesser quantities of the best sort, and to prevent abuses, I mark my Name at length instead of T.G. You may also be furnished with several sorts of materials relating to building.

Wood Street was the joiners' street; Samuel Pepys visited several shops there looking for a new table (24 August 1664). Queen Street, however, was much grander. It had not existed before the Great Fire and was a proud innovation of the rebuilt City, planned partly to avoid the former congestion of jumbled alleys between Cheapside and the Thames, partly to provide together with its continuation, King Street, a route as imposing as the names suggest from the river stairs to the City's Guildhall.

"In Queen-street near Cheap-side" was a good address. It is likely that by moving there a prospering Granford wished to attract wholesale business from merchants— Cheapside was one of the City's principal commercial streets—without losing touch with the local woodworking community. (I timed the walk between the sites of the old premises and the new: 5½ minutes.) The position of the new shop was evidently propitious; it remained the home of planemakers for 66 years.

Interestingly, in a later generation William Cogdell and Thomas Phillipson reversed the trend, serving their apprenticeships in Queen Street, but setting up independently in Wood Street. Perhaps by then Wooding's business was so dominant that it was prudent to put a little distance between the Three Planes and competing enterprises.

The change in emphasis in City plane-making from retail to wholesale custom gained ground with time. Granford still gave craftsmen precedence; he addressed himself to "Artificers or Merchants," whereas some 30 years later Jennion's trade card reversed the order, adding for good measure "& Others." (What others, Bill? Gentlemen amateurs? Tool collectors?)

A puzzle: according to the Post Man advertisement, Granford's sign was three planemakers. The sign of later occupants of the same house, Wooding and Jennion, was three planes. It may be that the name was changed, but there is something improbable about three planemakers as a device (three jolly workmen, their trade identifiable from the various obscure jibs, floats and cheeking chisels they brandish!). I suspect that three planes hung outside the house in Granford's day, as later, and that the copy provided to the Post Man's compositor read in fact "Thomas Granford, living at the Sign of the 3 Planes, Plane-maker, in Queen-street . . ."—slight variations between copy and print, particularly newsprint, being not unknown, then as now.

Wooding was admitted to the freedom of the Joiners' Company in 1704, qualified then to set up on his own account. Whether he did that immediately or continued to work as a journeyman for Granford for a couple of years I cannot establish; at all events I have found no separate entry for him in the Land Tax ledgers until 1706, when he is listed in the Precinct of St. Thomas Apostle and Trinity which comprised Queen Street and streets off it a few blocks to the south of the Three Planes in the direction of Thames Street and the river.

He was there for two years, in which he took his first two apprentices. In 1708 he married and moved towards Cheapside to within a few doors of the Three Planes. Granford and Wooding continued thus, close neighbours, until 1713 when Granford's name disappears from the tax books, Wooding's takes its place, and the house previously occupied by Wooding is listed as empty.

From this point, when he took over his master's business, Wooding stayed put. As was often the case, an improvement in a planemaker's circumstances was marked by an enlargement of his workforce: Wooding took his third apprentice in 1713, his fourth in 1716.

The only subsequent trace of Granford is the apprenticeship to him and his wife of his daughter Mehetaball in 1716. She seems to have been the only survivor past infancy of six children, all with old Testament names, born much later than Thomas III and Elizabeth.

Apart from Wooding, who needs no further comment, two of Thomas Granford II's eight other apprentices can claim to be considered as planemakers in a small way. The only known plane with THOMAS GRANFORD struck in the toe-end has been very reasonably attributed to Thomas III (*British Planemakers*, 2nd ed., p. 25).

Two planes are known marked NAT GAMBLE. One of these, a cornice plane, perhaps converted from a jack plane, I have not seen; the other is a moulding plane of the same length (a good 10½ inches) and style as Granfurdeus'. The date 1730 and other numbers, apparently random, are struck in the stock with a set of tiny figure punches of the kind used to calibrate mathematical instruments. At first I assumed the plane was new in 1730, but now I think the figure punches were new in 1730 and were tried on a plane already rather old-fashioned compared to, say, a contemporary late Wooding or early Cogdell. Nat Gamble's planes have in common with all the Granfurdei I have seen, and with one Wooding, irons marked with a crude embossed stamp HILDIK. Moulding planes of a length and workmanship similar to early Woodings turn up from time to time. They are often unmarked or defaced, but two have a legible mark IOHN DAVENPORT, and at least one of these has an iron marked HILDIK. This puts Davenport in the very exclusive company of the three other known Hildik-users and suggests a man of the generation of Gamble and Wooding.

I have not been able to pin him down in the records, although the name occurs in the City at the period and a John Davenport was made free of the Joiners' Company in 1680.

The Hildicks deserve study; the firm lasted until the 1820's.

Records of men in a calling as humble as planemaking who lived nearly three hundred years ago are necessarily meagre. What evidence has survived by luck must be interpreted largely by guesswork. At the present state of knowledge, Hemings and Granford look like pioneers; not, obviously, in making planes, but in making a business of it. And perhaps they were.

Seen from this distance, Hemings seems to represent the last of the old order, Granford the first of a new, in the transition from 17th-century anonymity to the 18th-century cult of personality in planemaking.

If planemaking indeed emerged as a distinct profession in London in the years after the Revolution of 1688, its development would be entirely consonant in its modest way with larger economic and social changes. The national prosperity grew rapidly, due mostly to foreign trade and in spite of war. London became less industrial and more commercial and its population increased greatly. It was a good time for private enterprise.

The following lists give the names of all known apprentices to the masters mentioned in the preceding article. The dates are of bindings.

To Thomas Granford I:
      1654 Ralph Rathborne
      1655 John Osland
      1656 George Hathaway
      1658 Thomas Harrison

To Thomas Granford II:
      1687 Jonathan Massey
      1691 Joseph Collins
      1692 Nathaniel Gamble(Gamball)
      1698 Richard Nicholls
      1699 Robert Wooding
      1703 Joseph Ames
      1705 Thomas Granford III (son)
      1706 William Hussey
      1716 Mehetaball Granford (daughter)

To Robert Hemings (Hemmings):
    1688 Francis Rurdue (Purdew)
    1692-5 John Scoffield
To Nathaniel Gamble:
    1705 John Gamble (Brother)
    1724 Edward Davis
    1730 Jonathan Gamble (son)

LETTERS

A Plane by Any Other Name. . .

Matthew Carter has made a valuable contribution to the history of early British planemaking, and to the Granford genealogy in "British Planemakers Before 1700," (*Chronicle,* June 1983). There appears to be a strange inconsistency, however, in one conclusion he draws concerning Thomas Granford planes.

I pass over the matter of attribution of the eight (or fourteen) "Granfurdeus" planes Carter counts. And although I am not one to seek a piece of wood from the True Cross, I am at a loss to understand the casual attribution of the one known Thomas Granford plane, so marked, to Thomas III—an individual for whom no firm evidence of planemaking seems yet to exist.

To the contrary—and as Carter and others have pointed out— Thomas Granford II in 1703 publicly advertised that he identified his product with his "Name at length," i.e., "Thomas Granford" (not "Granfurdeus," "T.G," or any other mark). This advertisement was printed two years before Thomas III, we are told, was apprenticed. Thomas Granford II was the planemaker to whom Robert Wooding was apprenticed in 1699; it is revealing to compare the makers' marks of Granford and Wooding from their tools. Bill Goodman's first attribution of the "Thomas Granford" plane to Thomas III in the second edition of his *British Planemakers from 1700* was made prior to Tony Barwick's bringing to light the 1703 Post Man advertisement.

The following quote from a letter from Bill Goodman, October 23, 1980, is pertinent: "I had a card yesterday from. . .They had got together at the Christie's auction on Tuesday last, and informed me. . .that one of the 'Gran Furdeus' planes on offer made £120 and the other £600. This puts the value of your genuine Thomas Granford up a bit."

Paul B. Kebabian
June 13, 1983

Persistent homonymity makes the Granford family tricky to understand. This tree summarizes what is known of it.

The Granfords

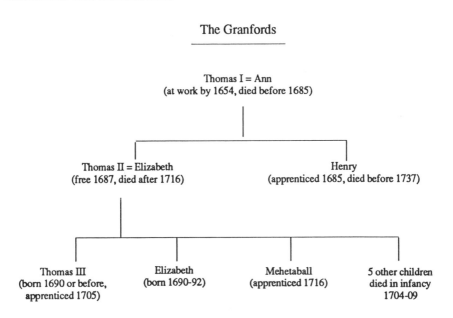

Thomas I = Ann
(at work by 1654, died before 1685)

Thomas II = Elizabeth
(free 1687, died after 1716)

Henry
(apprenticed 1685, died before 1737)

Thomas III
(born 1690 or before,
apprenticed 1705)

Elizabeth
(born 1690-92)

Mehetaball
(apprenticed 1716)

5 other children
died in infancy
1704-09

All About Those "Gunters"

By DAVID A. SPANG
March 1980

It is interesting to reflect on the thoughts that provoke one to look further into a subject. The seed for this short article originated from a casual comment noted in a recent sale catalog of tools. It referred to a Gunter's chain made by an unknown blacksmith of unknown past. The implication was that the blacksmith originated the chain as a land measuring device. The fact that the chain was 66 feet long and fit very comfortably into the English linear measurement system invited doubts as to its origin. The measurement table appeared too organized and orderly to be a chance happening. And why 100 links? It's a very odd multiple of 66 feet. In fact, each link computed to 7.92 inches long, a very strange number to be useful in any measurement scheme.

Early research quickly discovered that the Gunter chain originated with Edmund Gunter who turned out to be a renowned mathematician, not a blacksmith. This might explain the orderliness of fit within the measurement system, but what about the 100 links? Further research revealed that a number of devices, scales and mathematical instruments were associated with Gunter, some, like the chain, still bearing his name. I had to reach back to Gunter's original work before the rationale for 100 links of the chain became obvious. The selection of 100 units of measure was found to be part of an overall measurement technique proposed by Gunter.

I soon found these measurement techniques were interrelated on many of these tools. The discovery, acceptance, and use of his innovative methods of measurements, implemented through a variety of tools, will be the topic of this article. This relationship will be shown for those tools familiar to tradesmen and will limit extensive discussions on the astronomical and navigational tools which were also invented or expanded upon by Gunter.

Edmund Gunter, a renowned English mathematician, astronomer, and inventor, has left a legacy of tools and implements bearing his name. These "Gunters" were in constant use for over three hundred years by a variety of tradesmen/navigators and land surveyors. Twentieth century technology has supplanted the need for these tools, although I would suspect that the sector with a number of lines attributed to Gunter is still being used as an aid to navigation.

Edmund Gunter[1] was born in Hertfordshire in 1581. He was educated at Westminster School and from there was elected to Christ Church College at Oxford in 1599. After receiving the degree of Master of Arts, he entered into Holy Orders and in 1615 received the Degree of Divinity. He was particularly noted for his mathematical talents and soon received the professorship of Astronomy in Gresham College, London, where he greatly distinguished himself by his lectures and writings. He died in 1626 at the age of forty-five.

Edmund Gunter was fully aware of the difficulty encountered taking measurements or making computations in the existing English units of inches, foot, pace, perch, acre, etc. He noted it was particularly cumbersome to work problems when fractional measurements were involved, especially if the measurements were taken in different dimensional units (conversion to the same dimensional units is required before computations could be made; i.e., everything in feet or inches). He was also aware of the powerful new tool of logarithms recently improved on by Briggs in 1617.

He applied the logarithms of numbers and of sines and tangents to straight lines drawn on a scale or ruler, with which proportions in common numbers were resolved by the mere application of a pair of compasses. This method is founded upon the mathematical principle that the logarithms of the term of equal ratios are equidifferent. This was called "Gunter's Proportion."

Considering the computational problems of the English dimensional units, Gunter devised a measurement technique utilizing the base 10 whereby the inch, foot, etc. were divided into 10, 100, or 1000 equal parts. Gunter states that:[2]

"The ordinary measure for breadth and length are feet and inches, each foot divided into 12 inches, and every inch into halves and quarters, which being parts of several denominations doth breed much trouble both in arithmetick and the use of instruments.

"For avoiding whereof, where I may prevail, I give this counsel, that such as are delighted in measure will use several lines, first a line of measure, where every inch may be divided into 10 or 100 parts; secondly a line of foot measure, where in every foot may be divided in 100 or 1000 parts, both which lines may be set on the same side of a two foot ruler, . . ."

Having dispensed with the cumbersome inches and feet denominations, he carries this logic further and introduces the measurement by chain[3] and states:

"The like reason holdeth for yards and elles (45 inches) and all other measure divided into 10, 100 and 1000 parts.

"This being presupposed, the work will be more easie[4] both by arithmetick and the line of numbers[5], as may appear by these propositions."

Gunter then introduces the table on the following page.

This is the first place where Gunter proposes the use of a chain for land measure which is made up of 100 links each 7.92 inches long. He called each link a "centism" (100th part). Each chain is 792 inches long or 66 feet which is a cumbersome unit of measure. But Gunter designed the chain to be used without having to convert back to the foot. Each chain is 4 perches long[6], each perch is 25 links (centism) and an acre is made up of 10 square chains or 160 perches.[7] Therein lies the key to the use of the chain. As long as one took the measurements in chains and calculated the area in square chains, there was no reason to convert to inches and feet.

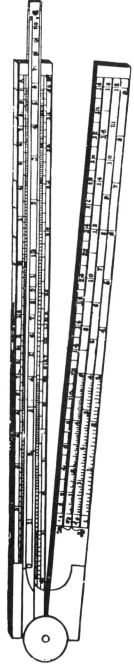

Fig. 1. "Gunter's Slide and Engineer's Rule" (Stanley Catalogue, 1892)

A Table for the Use of the Chain

	INCH	CENT.	FOOT	PACE	PERCH	CHAIN	ACRE	MILE
				LONG				
INCH	—	7.92	12	60	198	792	7920	63,360
CENT.	62.7264	—	1.515	7.575	25	100	1000	8000
FOOT	144	2.295	—	5	16.5	66	660	5280
PACE	3600	57.392	25	—	3.3	13.2	132	1056
PERCH	39,204	625	272.25	10.890	—	4	40	320
CHAIN	627,224	10,000	4356	174.24	16	—	10	80
ACRE	6272x10³	100,000	43,560	1742.4	160	10	—	8
MILE	4014x10⁶	6400x10⁴	2788x10⁴	1,115,136	102,400	6400	640	—

To review for a moment what has been proposed by Gunter. He first suggested that a decimal scale be developed wherein each inch will be divided into 10 or 100 parts. This line of measure is usually found on sectors and carpenter's rulers. It is referred to as a "line of lines" and identified on sectors by the abbreviation (lin.). In addition to making linear measurements, this line can be used in conjunction with dividers or compass when measurements or calculations are transferred, plotted or drawn. As suggested by Gunter, it is used for convenience and its total length has no significance or relevance. The only important factor is that it be divided into decimal parts.

Gunter next suggested that the foot could be better handled if divided into 100 or 1000 equal parts. The foot line and inch line could be placed on a two-foot rule which was the traditional length of the carpenter's rule at that time. The application of the inch divided into 10ths became quite common.

Measurements taken with either of these scales lend themselves to "easier arithmetick" and were in a form to be used directly with the newly defined logarithm scale or "line of numbers."

This logarithmic scale was referred to as "Gunter's Line." Multiplication or division of any two numbers only required the addition or subtraction of these numbers on the log scale by swinging a compass and reading the answer directly. This allowed for rapid calculations by a simple mechanical manipulation without having to memorize multiplication tables, etc. The "Line of Numbers" was readily accepted by the trades and became known as "Gunter's Line." This line is also found on the sector and identified by the abbreviation (Num.).

And finally, he suggested the construction of a 100 link chain for land measure which to this day is called the "Gunter Chain." Measurements taken with the chain were also in a form to utilize the "Gunter Line" or logarithm scale.

Gunter had now described a number of tools and techniques which had wide application in a number of trades. The line of numbers and his logarithmic scales of sines and tangents was transmitted to France by Edmond Wingate[8] who in 1624 published a French edition of Gunter's work. It can, therefore, be assumed that the information and use of these new lines was being widely disseminated. However, the 1645 and 1658 English editions of Wingate's work state that the reasons these lines are not being used more are: the difficulty of drawing the lines with exactness; the trouble of working thereupon by reason (sometimes) of too large an extent of the compass[9]; and the fact that a longer instrument is not readily portable.

John Brown in *The Description and Use of the Carpenter's Rule*[10] describes a carpenter's rule 24 inches long according to the standard at Guild Hall which has the inches divided into 8 parts or 10 parts as you please. He also describes the use of the line of numbers which he refers to as "Gunter's Line." He indicates that this line was first 2 feet long, but later was made 3 feet long stating that "the longer it is the better it is." Although both Wingate and Brown and others[11] recognized that the longer the length, the more accurate the computation, Wingate pointed out the difficulty of working a compass over a large expanse. Therein lies the reason for the last "Gunter" to be discussed, namely the "Sliding Gunters" or common slide rule.

Soon after Gunter announced the line of numbers, other mathematicians recognized the shortcomings of its use. The claim for inventing the "Sliding Gunters" or slide rule as it is known today created quite a controversy. The invention of the first slide rule was simultaneously claimed by a number of mathematicians, each claiming the invention was stolen from himself.[12] Both rectilinear and circular rules were invented about the same time. The rectilinear rule found more common usage. This rule is made up of two separate "Gunter Lines" which were manipulated by sliding one along the other. Therein lies its popular name, the "Sliding Gunter." Today we call this rule the slide rule. The "Sliding Gunter" or engineer's rule was available from most tool dealers (Figure 1)[13] and found wide acceptance within the trades. These rules are now collectors items.

During the literature search for information on "The Gunters," I noted that the use and the construction of the Gunter's Line was not very well understood. I would suspect that prior to the introduction of automatic scale engraving, a factory-made "Line of Numbers Rule" was too expensive for the ordinary worker. He therefore had to construct his own rule and scribe or stamp each individual graduation and numeric. This was not a particularly difficult task for the linear rule as the inch and foot standards were maintained at the various guild or trade organizations. But to construct the line of numbers required a conceptual understanding of its use as well as access to logarithmic tables.

One must remember also that even as late as the early 1800's mathematical concepts were still regarded as privy to the mathematicians and philosophers. This was pointed out in a rather detailed series of correspondence in the *American Mechanics* magazine in 1825.[14] In response to a query the writer presented the method of constructing a line of numbers. He first started by describing in great detail the constructing of a linear rule divided into 10th's. Using a compass he then showed how to add and subtract two numbers. He next indicated how to construct a logarithmic scale and by using the same principle as before with the compass, pointed out that you could now obtain multiplication, division and roots easily. At the completion of the discussion, the correspondent apologized for being "prolix" and stated that his "aim has been to explain the use of the scale in a familiar manner to the workman; and though it may lay me under the censure of many versed in mathematical knowledge, it will, I am sure, be excused by those who look to the motives which induced me to be minute, which is that practical mathematics may be extended as much as possible among the working classes. If this plan is persevered in, it will, I am confident, be of incalculable benefit to society in general . . ."

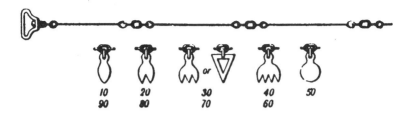

Fig. 2. Tally patterns.

Another "Gunter" which had significant usage with very well defined rules was the Gunter chain.

The usage followed very set and standard rules and procedures. Unlike the Line of Numbers, I have not found any reference which would indicate that chain and its usage was not understood. Most early texts on surveying presented the chain in a very straightforward manner and usually stated a series of rules to be learned, followed by simple examples.

The measurements taken were generally stated in chains and decimal fractions, the counted links.

Fig. 3. Chain bound for storage.

In reviewing a number of example calculations, a new term showed up which is the Rood[15] (quarter acre or 40 square perches). In describing the size of a piece of land, it was typically in acres, roods and perches.

When the land was uneven or hilly, a four-pole chain was too long to be convenient for the measurements. Surveyors therefore generally made use of a chain that was two poles in length and divided into 50 links. The measuring rules[16] were stated for both two poles and four pole chains. Surveyors were comfortable working with either length, but had to note which length was used, as both lengths were generically referred to as chains.

The Gunter chain is constructed of 100 pieces of straight wire, the ends are bent to connect to each other through small oval rings, which for flexibility should number three at each joint. The wire is of wrought iron or steel of from 8 to 12 gauge. The use of steel enabled the weight to be reduced without sacrificing strength. The ends of the chain consist of handles (sometimes brass) each with a swivel joint to eliminate twist. The total length is defined from outside to

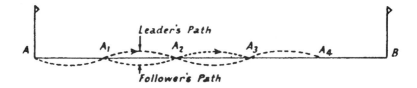

Fig. 4. Chaining a line.

outside of the handles. At every tenth link is attached a distinctive tag or tally usually of brass of the patterns shown in Figure 2. As each tag represents its distance from both ends of the chain, either handle can be regarded as zero, so a little care was necessary to avoid misreading. In taking readings between tags one counts the number of links from the previous tag, estimating fractions of a unit if necessary.

When stored, the chain is usually bundled as shown in Figure 3. To extend the chain both handles are grasped and the chain thrown out. The handles are then separated and the chain extended full length and inspected for kinks or twists which must be removed before the chain is used. To store the chain the process is just reversed. The handles are first brought together and the chain is gathered up starting at the 50th link and folding alternately two links upon each other until the handles are reached at which point the gathered chain is bound for storage.

Ranging rods, arrows and wooden pegs are used in chaining a line. Ranging rods are round wooden poles 6-12 feet long painted with characteristic red, white and black bands usually a foot long. Arrows are steel skewers about 15 inches long and ⅛" diameter. Ten of these arrows are usually found on a clip or ring which would attach to the chainman's waist to relieve his hands. Pegs are used at intersecting points of chains which require permanent marking and are driven into the ground by a mallet.

To chain a line the operation is carried out by a surveyor and two men called chainmen, one acting as the leading chainman and the other as follower. The chain is thrown out as before. The leader equips himself with 10 arrows and a ranging rod. The follower also takes a ranging rod. Figure 4 shows the process of measuring line AB having previously positioned rods at A and B. The leader drags his end of the chain forward to A and holds his ranging rod about one link short of the end. The follower holds his end of the chain firmly against rod A and sights down poles A and B signalling the leader to bring his pole at A into line with the poles at A and B. Once the alignment has been made, the leader straightens the chain past the rod by sending gentle "snakes" down the chain. The follower indicates the chain is straight and leader places an arrow at the end, outside and touching the handle at A1. The leader then drags his chain to A2 taking 9 arrows and his rod. The follower proceeds to A1 and puts his rod behind the arrow and the aforementioned straightening technique is repeated. The process continues with the follower picking up the arrow at A1 before he moves to A2. If the line measured is more than 10 chains long, the leader will exhaust his supply of arrows at the 10-chain point so that when the 11th chain is stretched out, the follower will have to hand over the 10 arrows back to the leader. This fact is pointed out to the surveyor who notes it in his field book. As you see, the process of chaining a line is quite simple and the factor of 10 is again used in the notations.

Chains can also be used for dropping a perpendicular from a point to a line or setting out a line at right angles to a chain from a given point on the chain.[17] The process is the same as you would follow if using a compass (the chain being used to swing the arc) and will not be discussed here. Standard error tables or allowable discrepancies in chaining[18] were developed which compensated for method error, wear, temperature, etc. Allowable errors were further tempered by common sense based on the terrain being measured.

Chains were also used by timber surveyors in estimating the wood contents of specified areas.[19] The chain was used as a straight line with timber tallies taken visually on both sides of it. When one wanted a more accurate measure of the available lumber each marketable tree was measured with calipers. This was done with some difficulty and required a number of men vs. the traditional "cruiser" who walked an area alone to make the estimate.

As has been shown, Edmund Gunter had his name associated with a number of tools which found considerable usage within the trades. His ideas were not necessarily to foster new tools, but to develop a method of "easie arithmetick" based on a decade or decimal system. He was the first to divide a surveying chain into 100 parts (centism) but cannot be given credit for originating the wire surveying chain or making it 66 feet long. The size of the acre was fairly well established by Gunter's time. There is no question though that his chain of 100 links became the most used.

The division of the inch and foot into 10 or 100 parts as indicated in his writing was never really directly attributed to Gunter and is still found on most rulers having multiple scales. Gunter never proposed a "Sliding Gunter" which employs two logarithmetic scales. He did suggest that the logarithmetic scale could be placed on a rule and used with the compass to make calculations. The origination of the "Sliding Gunter" was found to be others'.

Gunter contributed much to the trades in attempting to show that arithmetic could be easy with the use of his methods. He attempted to remove some of the mystery of numbers by employing simple techniques including the use of logarithms.

Although the shadows of the "Gunters" still linger, modern technology has all but eliminated their use. The hand calculator supplants the "Sliding Gunters" with greater accuracy, speed and flexibility. Steel tape and aerial surveying are now common methods of land measurement. The principles projected by Gunter are still valid today, but his instruments and tools no longer find a common need and are now only sought out by collectors.

BIBLIOGRAPHY AND NOTES

1. Hutton, Charles, *Philosophical and Mathematical Dictionary* Vol. 1, London, 1815, pp.621-623. Numerous references to Gunter were found. This reference was the most complete discussion on his inventions.

2. Gunter, Edmund, *The Works of Edmund Gunter Containing the Description and Use of His sector, Cross-Staffe, Bow Quadrant with a Cannon of Artifical Sines, Tangents and Logarithms* printed for Francis Eglesfield and to be sold at the Marigold in St. Paul's church yard, London, 1673, 5th Edition. Gunter's original work was written in Latin; English and French translations were soon available. The popularity of this work is demonstrated by the number of editions printed in a relatively short period. The particular reference quoted is from the second book on the cross-staff containing: "The Use of the Lines Thereon in Measuring of All Manner of Superficies and Solids as Board, Glass, Land, Timber, Stone and Gauging of Vessels as also in the Famous Art of Navigation."

3. Richeson, A.W., *English Land Measurements to 1800*, MIT Press, Cambridge, Mass., 1966, pp.77-78. The first recorded use of a chain for land measure was described by Cyprian Lucar in "A Treatise named Lucar Solace," London in 1590. Lucar's "Wyer Line" was 4 perches long, each link being 1 ft. long and further divided into ¼ and ½ perches. He further suggested that these divisions be distinguished from one another by being painted different colors. After Gunter, various length chains were proposed by others. *Oxford English Dictionary* (OED) notes under chain: "In 1669 the chains now used and in most esteem among surveyors are three. The First I will name is Mr. Rathburn's and that of Mr. Gunter . . .This year Mr. Wing hath described a chain of 20 links in a perch."

4. Richeson, op.cit., pp.44-45. Although Gunter wrote his original text in Latin, a number of his works were addressed specifically to the artisan or tradesman and to be used in practical applications. The trend of educating tradesmen in mathematics began in the later part of the 16th century where previous to his time mathematics was taught only to the clergy, nobility, and upper classes. It is interesting to reflect that we are still struggling with conversions to a decade system, viz . . .the meter vs. the yard.

5. Gunter, op.cit., the "Line of Numbers" or logarithm scale was first described by Gunter in Chapter 1 of the "First Book of the Cross-Staff" where he states under description of the staff "The line of numbers may be inscribed out of the first Chiliad of Mr. Briggs' logarithms . . ."

6. Richeson, op.cit., pp. 70-74. A surveying line of 4 perches long was first suggested by Valentine Leigh, London, 1577. This or Lucar's Solace (Ref.4) may have been the source of Gunter's chain of 4 perches or 66 feet long.

7. Richeson, op.cit., pp. 36-37. By the time of Gunter the acre had been fixed as containing 160 perches. Two lengths of perches were being used: that of 16½ feet for field measurements and 18 feet for woodland. In any respect the acre still contained 160 perches even though the woodland acre was larger.

8. Cajori, Florian, *On the History of Gunter's Scale and the Slide Rule During the Seventeenth Century*, University of California Publications in Mathematics, Vol.1, No. 9, pp. 187-209. Feb 17, 1920, p. 189 notes that English editions of Wingate's work were published in 1626, 1638, 1645, 1658.

9. Cajori, op.cit., p.189. Wingate had proposed these scales be placed on a ruler 66 cm long (26.5 in.) which would require a rather large compass and the chance for inaccuracies.

10. Brown, John. *The Description and Use of the Carpenters Rule together with the use of the line of numbers commonly called Gunter's Line*, London 1688, Chapters 1 and 2.

11. *Journal of Natural Philosophy, Chemistry and the Arts*, London, 1797. "A Method of Disposing Gunter's Line of Numbers, by Which the Divisions are Enlarged, and Other Advantages Obtained," pp. 372-375. This article by the editor William Nicholson is interesting in that it first states "Of the many ingenious instruments for computation which were in use during the last century, among mathematicians scarcely any are to be found at present except the sector and logarithmic scale of Gunter." Commenting on the 29.5 inch logarithmic scale recently published by Robertson, the article then develops a method of extending the scale of the Gunter's Line of Numbers to the equivalent of a scale of 60 feet on which four place accuracy could

be obtained. This was a 12 inch diameter circular instrument with a ten turn spiral which combined Gunter's Line and sector scales.

12. Cajori, op.cit. A lengthy discussion of this controversy is found in this work where the author points out that William Oughtred not Edmund Wingate invented the slide rule. The circular rule was described in print in 1632 and the rectilinear rule in 1633 by Oughtred. Wingate never claimed authorship of the slide rule, but since he was one of the first to write of its merits and his works were widely read he has improperly been given credit for its invention over the years.

13. 1892 Stanley Catalog reprint. As noted in the catalog, for an additional sum of $1.50 instructions could be obtained for using the "Sliding Gunter."

14. *American Mechanics Magazine,* Vol. 1, 1825, New York, Gunter Line, pp.14-15; Gunter's Scale pp. 231-232; 327-328. This magazine was the precursor to the *Franklin Journal* which started in 1826.

15. Rood: OED traces the Rood back to Old English. It was cited as still being used in the 1800's.

16. Gummere, John, *A Treatise on Surveying,* John Richardson, Philadelphia, 1817, pp. 81-83.

17. Bannister A. and Raymond S., *Surveying,* Sir Isaac Putnam & Sons, Ltd., London 1959, pp. 10-11.

18. Tracy, John Clayton, *Surveying,* John Wiley and Sons, New York, 1947, pp.60-61.

19. Graves, Henry and Ziegler, E.A., *The Woodsman Handbook,* U.S. Department of Agriculture Forest Service Bulletin 36, Government Printing Office, Washington, 1910, pp. 72-74.

Carpenters' and Engineers' Slide Rules Routledge's Rule

By PHILIP E. STANLEY

June 1984

Philip Stanley is the author of the book Boxwood and Ivory: Stanley Traditional Rules 1855-75.

Background

Probably the most elaborately marked and graduated boxwood rules ever made were the so-called "Engineers" rules manufactured between the early 1800's and the first decades of the 20th century. These rules were made in a number of different types, each type being known by its inventor's name (Armstrong's, Hawthorne's, Routledge's, etc.) and intended for a particular branch of engineering (see Roberts, Ref.1).

Of all the engineer's rules, the one most popular, and most frequently encountered today, is Routledge's engineers' rule, invented ca. 1811 by Joshua Routledge of Bolton, England. Routledge's rule was an attempt to provide the user with not only the means for performing rapid calculations (the Gunter's slide), but also much of the physical data required to work out the common engineering problems of that day. With Routledge's rule, a user could perform volume conversions, weight calculations, geometric analyses, and steam engine/pump computations (this last was a common type of problem 150 years ago; from its inception the steam engine had been used to drive mine drainage pumps, and was still more frequently used for that than any other purpose).

Physically, Routledge's rule was a 2-foot, 2-fold rule with slide, similar in construction to the more familiar carpenter's sliding rule, but differing from its counterpart in two significant respects: the drafting scales of the carpenter's rule were omitted, to be replaced by tables of physical and geometrical data and "Gauge Points," and the D scale of the Gunter's slide on the engineer's rule was arranged differently.

Using the data on the rule, and the Gunter's slide, the skilled user of the engineer's rule could solve whole classes of problems in mensuration, engineering, and mining without recourse to any other aids or tables.

The   Tables

The tables on the engineer's rule were 5 in number, and contained reference data organized as follows:

		SQUARE			CIRCULAR		GLOBE	
		FFF	FII	III	FI	II	F	I
	Cubic Inches	578	83	1	106	1273	1105	191
	Cubic Feet	1	144	1728	1833	22	191	33
TABLE	Wine Gals	134	1925	231	245	294	255	441
(1)	Ale Gals	163	235	282	299	359	312	538
	Imp. Gals	16	231	2773	294	353	3064	5295
	Water	16	231	2773	294	353	3064	5295
	Gold	814	1175	141	149	179	155	269
	Silver	15	216	261	276	334	286	5
	Mercury	118	169	203	216	258	225	389
	Brass	193	278	333	354	424	369	637
TABLE	Copper	18	26	312	331	394	344	596
(2)	Lead	141	203	243	258	31	27	465
	Wt. Iron	207	297	357	378	453	394	682
	Cp. Ir & Zinc	222	32	384	407	489	424	733
	Tin	219	315	378	401	481	419	723
	Steel	202	292	352	372	448	385	671
	Coal	127	183	22	233	280	242	42
	Fre Stone	632	915	11	1162	14	121	21

TABLE 1: Conversion factors for relating the volumes of various geometric solids (SQUARE [rectangular parallelopipeds], CIRCULAR [cylinders], and GLOBE [spheres]), to the various units of volume (CUBIC INCHES, WINE GALLONS, etc.). The relationships are expressed in reciprocal form with the decimal point omitted (e.g., a one foot cube (FFF) contains 1728 cubic inches; 1/1728=0.0005787. . .; hence the gauge point in the CUBIC INCHES row for the FFF column is 578).

TABLE 2: Conversion factors for relating the volumes of the same geometric solids as in Table 1 to their weight in pounds for various materials. Again, the relationship is expressed in reciprocal form with the decimal point omitted (e.g., a one-foot long, one inch diameter cylinder (F1) of water weighs 0.3403 pounds . . .; 1/0.3403 = 2.938 . . .; hence the gauge point in the WATER row for the F1 column is 294).

TABLE (3)	Polygons From 5 To 12 Sides	5	6	7	8
		1.72	2.598	3.634	4.828
		9	10	11	12
		6.182	7.694	9.366	11.196

C.P. IR: Cupola (Cast) Iron

Fre Stone: Freestone; an easily cut variety of sandstone

Fig. 2A Table 3

G. PTS. OF A CIRCLE AREA 7854 C & A 0795 C & D 3.141 SQR.I 141 S.E.A. 886 S.F.T. 115	PUMPING ENGINES	DIA	G.P.	DIA	G.P.	DIA	G.P.	DIA	G.P.
		3	165	10	183	17	528	24	106
		4	292	11	222	18	591	25	114
		5	457	12	264	19	661	26	124
		6	66	13	308	20	731	27	134
		7	89	14	358	21	81	28	143
		8	117	15	412	22	885	29	154
		9	148	16	468	23	97	30	165

Figure 3: Tables 4 & 5

TABLE 3: A table of the areas (in square units) or regular polygons (pentagon, hexagon, etc.) of unit side, for polygons from 5 to 12 sides (e.g., an octagon with one-foot sides has an area of 4.828 square feet).

TABLE 4: Gauge Points of a Circle. A table of numbers relating the diameter, area, and circumference of circles, and further relating them to the dimensions of their inscribed squares and triangles.

The significance of the various terms is as follows:

AREA: The area of a circle of unit diameter (Actual value: 0.078539).

C & A: The area of a circle of unit circumference (Actual value: 0.7958).

C & D: The circumference of a circle of unit diameter (This is Pi; actual value: 3.1416...).

SQR. 1: The diameter of a circle within which is inscribed a square of unit side (Actual value: 1.414...).

S.E.A.: The side of a square equal in area to a circle of unit diameter (Actual value: 0.8862...).

S.E.T.: The diameter of a circle within which is inscribed an equilateral triangle of unit side (Actual value: 1.155...).

TABLE 5: Gauge Points (G.P.) For Pumping Engines. A table of values which permit the computation of the required diameter of steam engine cylinder necessary to drive a pump of known diameter which is raising water a known height. This table presupposes steam pressure to operate the engine of 7 pounds per square inch, and the the cranks on the engine and the pump have the same swing.

The gauge points are calculated for a series of pump diameters (DIA), such that the square root of the product of the gauge point and the height is the diameter required for the pumping engine cylinder (e.g., In order to drive a 6-inch pump which raises water 4½ yards, a pumping engine run by 7 psi steam would be required to have a cylinder diameter of D = 6.60 x 4.5 = 29.7 = 5.45 inches).

In order to fit these tables onto one leg of the rule (an area of about 10½ inch by ¾ inch) some rearrangement of tables 1, 2 and 3 was required. These tables were run together vertically, as shown in Figure 2, and then arbitrarily divided (at the points indicated by the asterisk (*) into three pieces of uniform height. These pieces were then formatted onto the rule beginning at the left, and tables 4 & 5 placed to their right, as shown in Figure 4.

The Gunter's Slide

The D scale on the Gunter's Slide of the Routledge's rule was slightly different from the D scale used on the 2-fold sliding carpenter's rule.

On the carpenter's rule, the D scale was "folded" at the value 4; that is, instead of beginning at 1 on the left and progressing through 2, 3, etc. to 1 again on the right, the D scale began at 4 on the left, and progressed through 5, 6, etc. to 1 near the middle, and thence through 2, 3, etc. to 4 again on the right.

The purpose of this "folding" to the D scale on the slide of the carpenter's rule was to reduce the number of manipulations required in common operations. Folding accomplished this in two ways. First, a folded scale introduces the folding point as a multiplying or dividing factor when transferring points between the folded and unfolded scales during computations; a scale with the proper folding point can thus in many cases reduce the number of slide settings by one or more. The factor 4 appears frequently in many engineering problems; hence its choice as a folding point for the D scale on the carpenter's rule. Second, folding a scale has the effect of moving numbers near 1 & 10 towards the physical center of the scale. The physical location of the numbers involved in a slide rule calculation will often influence the number of steps required to perform that calculation; by moving the number 12 to the near-center of the D scale certain calculations, particularly those related to timber measurement, were thus made easier.

The D scale on Routledge's rule could not be folded, however. When Routledge computed the values for the various tables while designing the rule, he included in the

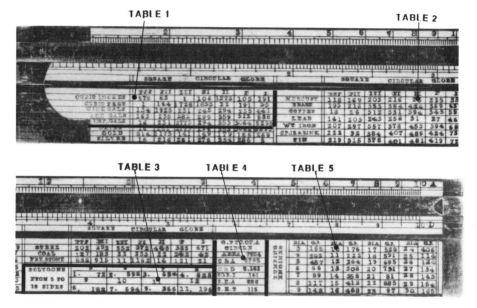

Figure 4: Arrangement of Tables on Routledge's Engineer's Rule.
(Rule shown in two sections)

gauge points all factors (such as 4 or Pi) other than the specific geometric dimensions of the shapes; the use of these values with a folded D scale would have introduced an unwanted factor and rendered the results useless. The table values could have been computed to allow for the use of a folded scale, but this would have effectively obscured their physical significance and recognizability.

Instructions for Use

Most of the large makers of Routledge's engineer's rule published instruction books on its theory and use. The earliest of these is the set of instructions written by Routledge himself, published about 1811, shortly after he invented the rule. These instructions were reissued several times over the next 25 years, the 4th edition appearing in 1813, and the 6th in 1823 (Ref.2). Other manufacturers who published such instructions were Hermon Chapin (1858, Ref.3), John Rabone & Sons (1867, Ref.4 through 1872, Ref.7), The Stanley Rule & Level Co. (1867, Ref. 5, through 1892, through 1902, Ref.6), Edward Preston & Sons (Ref.8), and the Chapin-Stephens Co.

(1914, Ref.9). It is interesting to note that the Chapin and Rabone instructions were exact copies of the Routledge 6th edition, even copying any typographical errors in the original.

These instructions begin with a description of the rule and an explanation of the theory and use of the Gunter's slide, beginning with multiplication and division, and then discussing "the rule of three" (proportions) and square roots. After this they illustrate the uses of the rule by means of a series of sample problems and solutions. There are sections on mensuration (calculating the areas and volume of solids), land measuring, liquid measure and cask gauging, the weighing of metals, gear and machinery problems, and pump/engine calculations.

The sample problems/solutions have a slightly archaic ring to those used to the jargon of modern engineering. A couple of examples from the Chapin instructions (Ref.3) clearly illustrate this:

"If a cask of the third variety is 20 inches at the head, 26 at the bung, and 29 inches long, what will be its contents in old ale and imperial gallons?

Set 29 upon B to the separate circular gauge-points for old ale and imperial measure (say 359 for old ale and 353 for imperial) upon A, and against 23.3, the mean diameter upon D, are 43.8 old, and 44.6 imperial gallons, the answer, upon C."

"What will be the diameter of a cylinder to work a pump, 12 inches in diameter, at 70 yards deep, and loaded with 7 lbs. on the inch?

In the . . .table (of gauge points for pumping engines) for a 12-inch pump is 264. Set 1 upon B to 264 upon A, and against 70 yards upon C, is 43 inches upon D, the answer."

Routledge's, Chapin's, and Rabone's instructions are all approximately 30-40 pages in length. The Stanley instructions are described in its catalogue as "200 pages long"; this would indicate that they had been completely rewritten, but until a copy of this edition is found and studied, this can only be a surmise.

Acknowledgements

The author wishes to acknowledge the assistance of Paul Kebabian and Ken Roberts, for allowing him to examine their copies of the Routledge, Chapin, and Rabone instruction books, and of Ken Roberts and Jim Hill for helpful criticism.

References

1. CARPENTER'S AND ENGINEER'S SLIDE RULES (PART I, HISTORY), Kenneth D. Roberts, P.E.; The Chronicle of the Early American Industries Association, Vol.36 No.1 (March 1983), pp.1-5

2. INSTRUCTIONS FOR THE ENGINEER'S IMPROVED SLIDING RULE, WITH A DESCRIPTION OF THE SEVERAL LINES UPON IT, AND DIRECTIONS HOW TO FIND ANY NUMBER THEREON (6th edition); J. Routledge, Engineer, Bolton; 1823

3. INSTRUCTIONS FOR THE ENGINEER'S IMPROVED SLIDING RULE, WITH EXAMPLES OF ITS APPLICATION; Published by Hermon Chapin, Pine Meadow, Conn., 1858

4. INSTRUCTIONS FOR THE USE OF THE PRACTICAL ENGINEER'S & MECHANIC'S IM-PROVED SLIDE RULE, AS ARRANGED BY J. ROUTLEDGE, ENGINEER; Published by John Rabone & Sons, Birmingham, England, ca. 1867 (Reprinted by the Ken Roberts Publishing Co.)

5. PRICE LIST OF U.S. STANDARD BOXWOOD AND IVORY RULES, LEVELS, TRY SQUARES, GAUGES, HANDLES, MALLETS, HAND SCREWS, &c, Stanley Rule & Level Co., January 1, 1867 (Reprinted by the Ken Roberts Publishing Co.)

6. STANLEY RULE & LEVEL CO., CATALOGUE NO. 28, Stanley Rule & Level Co., January 1902

7. CATALOGUE OF MEASURING RULES, TAPES, STRAIGHT EDGES, AND STEEL BAND CHAINS, SPIRIT LEVELS, &c., John Rabone & Sons, July 1892 (Reprinted by the Ken Roberts Publishing Co.)

8. ILLUSTRATED CATALOGUE OF RULES, LEVELS, PLUMBS & LEVELS, THERMOMETERS, MEASURING TAPES, PLANES, IMPROVED WOODWORKER'S AND MECHANIC'S TOOLS, &c. (6th EDITION), Edward Preston & Sons Ltd., July 1901 (Reprinted by the Astragal Press)

9. RULES, PLANES, GAUGES, PLUMBS AND LEVELS, HAND SCREWS, HANDLES, SPOKE SHAVES, BOX SCRAPERS, ETC. CATALOGUE NO. 114. The Chapin-Stephens Co., 1914 (Reprinted by the Astragal Press)

The English Carpenter's Rule
Notes on Its Origin

By PAUL B. KEBABIAN
June 1988

Paul Kebabian, a past president of the EAIA, is the author of American Woodworking
Tools *as well as many articles on early tools, trades, and industries. These two articles
are directed to the early history of the carpenter's rule; but also give insight into 16th,
17th, and early 18th century efforts to deal with the practical aspects of measurement.*

This limited study of the carpenter's rule was prompted by the acquisition of two
English boxwood rules with tables and graduations that were unfamiliar to the writer.
In an effort to determine an approximate date for the rules and to understand the
graduations, the investigation started with a 16th century source.

William Goodman, in his *History of Woodworking Tools*, and Raphael Salaman,
in the *Dictionary of Tools Used in the Woodworking and Allied Trades*, place the start
of industrial manufacturing of modern boxwood rules in the early decades of the 19th
century. This was the period during which the Rabone firm was established in Birmingham, England, and Belcher Bros. (1821), S. Morton Clark (ca.1833), Hermon Chapin
(1834), and E.A. Stearns (1838) were introducing factory production of accurately
graduated boxwood and ivory rules in the United States. Prior to this time, rule making
both in England and America had been largely in the hands of mathematical instrument
makers or individuals who made their own—often imprecise—rules for linear and other
measurement. Measuring devices for performing mathematical computation, for cask
gauging, &c were also the product of the mathematical instrument maker.

Some two hundred and seventy-five years before the period of factory production,
Leonard Digges, "Gentleman," published a significant contribution to the subject of
mensuration entitled *A Boke Named Tectonicon*.[1] Digges, who died about 1571, studied
at Oxford and subsequently became an accomplished architect, a student of optics,
mathematician, surveyor, and author of several books on meteorology, military science,
and on linear measurement and the measurement of solids. The *Tectonicon* had an
unusual publishing history: sixteen editions or printings were made between 1556 and
1637. Digges explained that his purpose was to provide the Landemeater (surveyor),
Carpenter, or Mason with a way to obtain "the true measurynge and readye accompte
of all maner of Lande, Timber, Stoone, Borde, Glasse, Pavement, &c." Portions of the
Tectonicon particularly relevant to the measurement of timber and boards, and the rule,
are "A table to finde the just Radix or Square of any Tymber," and chapters on the table
of timber measure, on superficial or board measure, and on the description and

construction of the carpenter's rule.

The emphasis placed by Digges— and later by Richard More[2]—on the need to determine cubic measurement of timber by one lineal foot units, and also to make superficial measurement of boards by the square foot, was based on the fact that wood sold to the carpenter was then, as today, priced by per foot measurement. Thus the carpenter needed to know, for example, what length of a timber of certain dimensions of width constituted a cubic foot. His rule, by means of specific tables and graduations, could provide the answer if the timber were in cross-section an equal-sided square, if his rule was accurate, and if he knew how to use it. But if the sides were of unequal dimension, Digges explains, the untutored were accustomed simply to add the width and depth dimensions and halve the result, thereby obtaining an incorrect figure for applying to the rule to find what length would give one cubic foot.[3] With a square timber of 8 inches on each side, Digges' table for timber measure reads 2 feet 3 inches as the length of one cubic foot. But if the timber were of unequal widths—as 4 inches and 12 inches—and

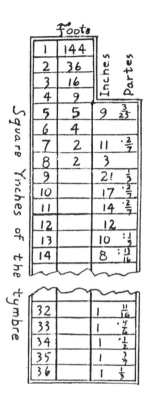

Fig. 1 Copy of part of the timber measure from the Tectonicon. For "Square Ynches" read "Inches Square." A slightly altered reduced copy.

if one added these and divided by two the resulting 8 inch figure would be grossly incorrect for determining cubic measurement by means of the carpenter's rule. For one cubic foot of the latter dimensions, the carpenter should get a piece some 3 feet long, rather than 2 feet 3 inches.

Digges provided a "Tabula Radicum" to give ready response to the question: What figure can be used with the table of timber measure to determine the length of a cubic foot of timber of specified, *unequal dimensions of width*. The coordinate of 4 and 12 and his Tabula, for instance, gives the figure "7." Opposite 7 on the table of timber measure is the answer 2 feet, 11⅔ inches. The single "pointe adjoyned" to the figure 78 of the Tabula Radicum was not a typographical aberration, but the introduction by Digges of one of the earliest examples of a qualifying mathematical symbol.[4] He

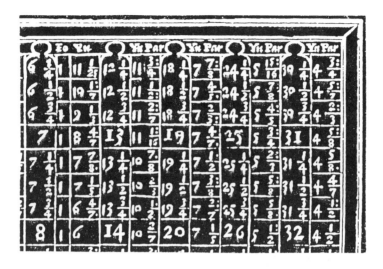

Fig. 2   A section of Digges' table of board measure. Reduction from negative microfilm.

explains that by one "prycke" beside a figure he means *a little quantity less than*, and by two points (:) *a fractional amount more than*. The square root of 4 times 12 is 6.928 or not quite 7, the figure to be used with the table of timber measure. It would not be possible to accommodate a series of whole and fractional decimal figures on a table, and of necessity, the figures of Digges' tables are in certain instances but close approximations.

Following the tables and discussion of timber measurement, Digges provides a square foot board measure table and instructions for its use. The columns are headed "Fo" and "Yn" for foot and inches, for widths of boards, glass, &c from ¼ to 12 inches by quarters of an inch, and "Yn" and "Par" for inches and parts of inches for material from 12¼ to 36 inches in width. The number of feet, inches, and fractions of inches in length that make a superficial square foot is given for each width dimension. Thus, opposite 6 inch width is the figure 2 feet, and opposite 25 inch width, the answer 5¾: inches (or slightly more than that length) per board foot.

Starting the chapter on the Carpenter's Ruler, Digges writes: "Because the effect of this Ruler is above declared by Tables, an instrument also wel knowen and commune amonge good Artifycers: I will not spende many woordes, in opening it. Behold the fyguers, and learne by them howe ye ought to make, and commonly to decke youre Ruler, bothe with Tymber and bourde measure."[5] We note that Digges uses the term *board measure*. Strictly speaking, a board foot is a measurement including thickness. That is, a piece of wood 12 inches by 36 inches represents 3 superficial square feet; if one inch thick, it is 3 board feet (or if two inches thick, 6 board feet &c). Popular usage through the centuries has given to the superficial measurement of lumber the term "board" measure. That Digges uses "bourde measure" in 1562 would indicate that 16th century measurement of sawn boards was then, as now, based on boards of 1 inch thickness. Edmund Gunter, however, writing in 1624, preferred the term "broad measure."[6]

Digges describes the ruler as "wel playned, twelve Inches longe, a quarter of an Inche thicke, and two Inches yn breadth. Truly yt were more commodious if it hath two foote in length." The 12 inch ruler he illustrates is scaled in inches, and half, quarter, and eighths of the inch. The timber measure is then added to the rule by writing or graving the proper figures, taken from his table of timber measure: near 1 inch on the rule is marked 144 (feet); 2 inches, 36(feet); 3 inches, 16(feet); &c to 12 inches, 12 (inches). The second figure represents the running length to make one cubic foot from 1 inch, 2 inch, 3 inch &c of square timber. Instructions are given for continuing the square timber scale on the rule for material up to 36 inches on a side, by inserting the numbers 13 to 36 on the ¼ inch thick side of the rule at their appropriate places according to the figures of the timber table. In a similar manner the figures for board measure are to be marked on the face of the rule, and continued along the second or opposite ¼ inch wide edge.

Digges does not stop at this point, but makes of his carpenter's rule a rather sophisticated instrument. He gives directions for marking on the back a "Quadrant Geometricall" with a 90° scale, for mounting a tiny plummet on a fine thread to hang from the quadrant, and attaching

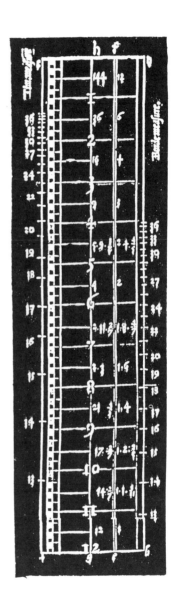

Fig. 3 Digges' carpenter's rule. Timber measure graduations are to the right of the line g h, and continue on the left edge of the rule. Board measure graduations are on the line f e, and on the right edge b d. Reduction from negative microfilm.

level sights "wel bored. . . made of wode, or rather metall," to be fastened on one edge of the rule when it is to be used for sighting.

Forty-six years following the first edition of the *Tectonicon,* Richard More of London published *The Carpenter's Rule.* This 1602 work is addressed to "The Worshipful, The Master, Wardens, And Assistants of the Companie of Carpenters of the Citie of London. . . " The author, himself a member of the Company, affirms that gross errors are constantly being made in the measurement of timber. The most common resulting in loss to the buyer are purchasing "waynie" or twisted-growth timber and measuring it as square, and secondly "taking halfe the breadth and thickness of a peece being added together, for the square thereof." More gives full credit at more than one point to Digges: "The ordinarie Rule which Carpenters, Shipwrights, and others doe use to measure Timber withall," he notes, "was invented and published by Paster Leonard Digges." However, he finds that most rules are "very false" and that "those divisions or strikes which are set on them for measuring of Boord and Timber, are not in their right places."[7]

More describes a variety of ways to measure boards and planks, and makes a case for using an arithmetical method. He provides instructions for measuring square, rectangular, round, oval, waynie, and taper growth timber. In chapter 12 of part 2, he includes tables for timber and board measure, with illustrations of the tables which are based essentially on those of Digges. More does not describe or illustrate a graduated carpenter's rule, but his references to Digges and his timber and board measure tables suggest that the construction of the instrument and its graduations and other markings had undergone no significant change from 1556.

The expansion of knowledge in the field of mathematics that took place in the 17th century, however, was to result in new scales being added to the carpenter's rule. John Napier published his work on logarithms in a Latin edition of 1614, followed by a translation in English in 1616.[8] Based on the work of Napier and other contemporary mathematicians, logarithmic lines of numbers were developed by Edmund Gunter, and in his 1624 work on the sector he described "The use of the line of Numbers in broade measure, such as boord, glasse, and the like," in "solid" or cubic measure, and in "Gaugeing of vessels." The invention of the rectilinear slide rule, using two juxtaposed rules with Gunter's lines, is generally credited to William Oughtred.[9] It is this invention that subsequently became an essential part of the carpenter's slide rule.

Coggeshall's *Timber-Measure by a Line of More ease* . . .(i.e., a modified Gunter's line) was made public in a 1677 pamphlet,[10] and Thomas Evarard's *Stereometry,* on cask gauging by use of his slide rule, was first published in 1684. Coggeshall's sliding rule is carefully described in the 1728 first edition of Chambers *Cyclopaedia*[11] and in subsequent editions at least through the 7th of 1751-52. It is this rule, in a two foot, two fold form with Gunter's line as modified by Coggeshall that persisted into the 20th century in both England and America.

The 1728 Chambers description of Coggeshall's sliding rule reads substantially as follows:[12] It is principally used in measuring the superfices and solidity of timber, &c. It consists of two rulers, each a foot long, which are framed or put together in various ways: sometimes they are made to slide by one another, like glaziers rules: sometimes a groove is made in the side of a common two foot joint rule, and a thin sliding piece put in, and Coggeshall's lines added on that side [the method used in 19th century American manufacture]: the most usual way is to have one of the rulers slide in a groove made along the middle of the other. Coggeshall's carpenter's slide rule utilized Gunter's lines for the A, B, and C scales but the fourth scale, "called the girtline, and noted D, whose radius is equal to two radius's of any of the other lines, is broken for easier measurement of timber..." The D line is graduated from 4 to 40, and is so illustrated on Coggeshall's rule in the plates of Chambers' *Cyclopaedia.*

The two English boxwood rules here illustrated in Figs. 4 and 5 bear graduations for board and timber measurement consistent with the tables developed by Digges. They also carry Gunter's line (two lines figured 1 to 10 as described by him in 1624 rather

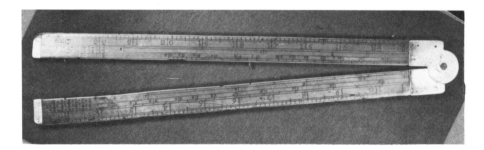

Fig. 4 The 2 foot, 2 fold boxwood carpenter's rule.

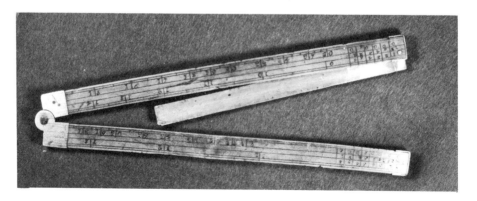

Fig. 5 The 2 foot, 3 fold rule of two hinged 9 inch legs and a 6 inch brass hinged extension.

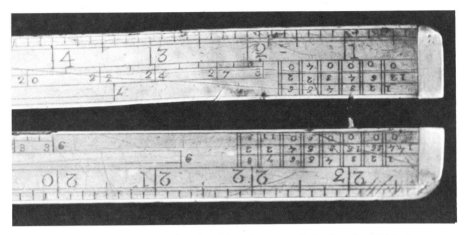

Fig. 6   Tables of timber (above) and board measure (below) on the 2 fold rule.

than Coggeshall's A, B, C, & D scales). To perform arithmetical calculations it is necessary to use dividers.[13]

The tables below, which represent those of the two rules, provide for superficial or board measure and for timber or cubic measure of wood up to 6 inches wide, and 8 inches wide, respectively. Both of the rules also include *lines of board and timber measure* which extend the tabular data for dimensions beyond 6 and 8 inches in width. As Digges and More explain in their books, the figures 1 to 6 of the board measure table represent inches of breadth, the second and third lines the feet and inches of length that make a square foot for the corresponding width figure on the top line. Thus, a board 1 inch wide requires 12 running feet to make 1 board foot; a board 5 inches wide requires 2 feet and 4 inches to make 1 board foot.

The line of board measure (graduated for 7 to 36 inch widths on one rule and 9 to 36 on the second) is used as follows: a figure for width, such as 11 inches, is located on the line. The rule is turned over, and on the 24 inch scale of the back side, and opposite the location of 11 on the first side, is found 13 1/16 inches. Thus a board 11 inches wide measures approximately 13 1/16 inches in length per board foot. That rule makers considered it necessary to provide for measurement of boards 30 inches and more in width speaks to the dimensions of trees being cut for timber (or perhaps being imported for sale) in England in the 16th and 17th centuries.

The top line of the table of timber measure reads 1 to 8, representing figures for square timber of 1 to 8 inches width on each side; the second and third lines are the number of running feet and inches required for one cubic foot. Thus a 4 inch square of timber requires a piece 9 feet long to yield 1 cubic foot; or if a 6 inch square, a length of 4 feet for 1 cubic foot. For square timber more than 8 inches on the sides, one uses the line of timber measure (graduated for 9 to 30 inch widths on the first rule pictured, and 11 to 30 inch widths on the second). In using this line one measures—using a second rule to perform the measurement—from the figure for width per side, to the end of the rule : for example, from 11 on the line of timber measure to the end of the rule measures approximately 14¼ inches. That is to say, an 11 inch square sided timber (or a timber of unequal width sides, the square root of whose product equals 11) contains 1 cubic foot for every 1 foot 2¼ inches of length.

These applications of the tables and lines are also explained in the editions of Chambers *Cyclopaedia* from the 1st edition of 1728 through the 7th in 1751-52 (and possibly later editions not available to the writer). Curiously, there is a substantial error in the example given for use of the board measure table which persists through all the editions noted, viz, "If a surface be one inch broad, how many inches long will make a superficial foot? Look in the upper row of figures for one inch, and under it in the second row, is 12 inches, the answer to the question." Unfortunately, the example errs not only by giving the wrong answer, but by asking the wrong question. The second row of the board measure table gives measurements in feet and not inches, and the question should ask how many feet make a superficial foot for a surface 1 inch wide, and the answer from the table is 12 feet. It is apparent that Chambers' work was considered a most authoritative source (as it was). *The Builders Dictionary* of 1734, borrowing virtually word for word from Chambers, contains the same error.[14]

That the 2 foot, 2 fold rule was in general use by 1728 is apparent from reading the descriptions of Coggeshall's sliding rule in Chambers, where this rule with Gunter's line is explained. Making use of the "common" 2 foot joint rule by inserting the slide with logarithmic scale is stated as one method of constructing it. One could conclude that between Digges and More (1556/1602) and the first edition of Chambers (1728) the 2 foot, 2 fold carpenter's rule had become a standard form of measuring instrument. Its illustration with tables and lines of board and timber measure in the 7th edition of Chambers is very similar to the rule in Fig. 4 above. In 1758 Edmund Stone's translation of Bion's book on mathematical instruments was published in a 2nd edition with Stone's "Additions of English Instruments." Stone's first example (Book 1, p. 14) describes the construction and use of the "Carpenter's Joint-Rule." Both in textual description and illustration in the plates, the rule is again similar to that of Fig. 4. Both rules are professionally made, and neither bears a maker's name. It would seem reasonable to suggest that these two English rules are of late 17th to mid-18th century manufacture, and that the 3-fold rule with 6 inch brass leg is probably somewhat later than the 2 foot example.

NOTES

1. Digges, Leonard. *A Boke Named Tectonicon*. London: Thomas Gemini, 1562. It is of interest that in his imprint Gemini notes that his location is within Black Friars and that he is "there ready exactly to make all the instrumentes apperteynynge to this Booke." The first edition published by Digges was dated 1556.

2. More, Richard. *The Carpenters Rule, Or, A Booke Shewing Many plaine waies, truly to measure ordinarie Timber*...London, 1602. Reprint. New York: DeCapo Press, 1970.

3. Digges. leaf 11 recto.

4. Cajori, Florian. *A History of Mathematical Notations*. 2 v.Chicago: Open Court Pub.Co. (1928-29) v.2, p.197.

5. Digges. leaf 17 verso.

6. Gunter, Edmund. *The Description and use of the Sector. The Crosse-staffe and: other instruments*. London, 1624. Reprint, New York: DaCapo Press, 1971. The first Booke of The Crosse-Staffe, p. 31-50.

7. More. p.2.

8. Napier, John. *A Description of the Admirable Table oe(sic) Logarithmes*. London, 1616. Reprint. New York: DaCapo Press, 1969.

9. Cajori, Florian. *William Oughtred*. Chicago: Open Court Pub. Co., 1916. p.47. See also; Nicolas Bion. *The Construction and Principal Uses of Mathematical Instruments*. Translated from the French...by Edmund Stone. 2nd ed., London, 1758. Reprint. London: Holland Press, 1972. Book 1, p.16.

10. Coggeshall, Henry. *Timber-Measure by a Line of more ease, dispatch and exactness, than any other way8 now in use, by a double scale*...London: Printed for the Author, 1677.

11. Chambers, Ephraim. *Cyclopaedia: or, An Universal Dictionary of the Arts and Sciences*...2 v. London: Printed for J. and J. Knapton (et al), 1728.

12. Chambers. Entry under: Sliding rule—Coggeshall's Sliding rule.

13. For an explanation of using Gunter's line see: *The Carpenter's Slide Rule, its History and Use*. Published by John Rabone & Sons. 3rd ed. Birmingham, 1880. Reprint. Fitzwilliam, NH: Ken Roberts, 1982. p.8-10.

14. *The Builder's Dictionary: or, Gentleman and Architect's Companion*. 2 v. London, 1734. Reprint. Washington: Association for Preservation Technology, 1981. Entry under: Rule.

The Early English Three Fold Ship Carpenter's Rule

Further Notes

By PAUL B. KEBABIAN

March 1989

The two foot rule illustrated in Fig. 1 was described in some detail in the previous article. Its two boxwood nine inch legs, a six inch brass hinged extension, 24 inch graduations, tables and lines of timber and board measure, and its Gunter's lines for mathematical calculations were identified. Two additional sets of graduations (on the sides reverse to those bearing the board and timber measures) were not, however, discussed. The reasons for these omissions were that the article was primarily concerned

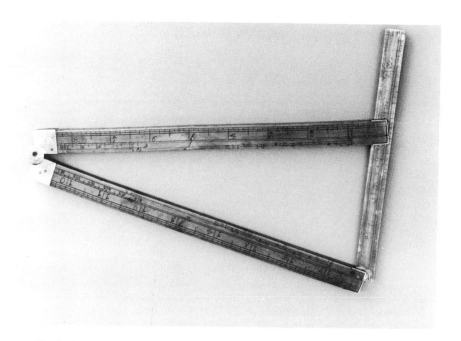

Fig. 1   The two foot, three fold shipwright's rule with sectors and octagon scale.

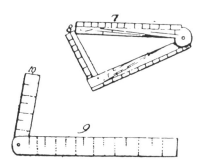

Fig. 2   Rules illustrated on plate K by Rålamb in 1691. 7 - 8 is the three fold "English" rule with sector, and 9 - 10 a "Dutch" bevel.

with the board and cubic measures of the two fold and three fold rules; that the writer was ignorant of the functions of the two sector lines on the rule; and that he overlooked the line of figures which constitute an early introduction of an octagon scale. It is these latter series of graduations—the sectors with nautical designations and the octagon scale—that clearly bring the tool from the general description of carpenter's rule to the more specific, i.e., ship carpenter's rule.

One of the earliest illustrations of a three fold rule of this type (Fig. 2) is in the 1691 Swedish work on shipbuilding by Rålamb.[1] Although his representation of the rule, no. 7 - 8 of plate K is not sufficiently detailed to show accurate graduations, it does clearly reveal two sector lines. The rule is described as 7 "Een Tumståk," *a (folding) rule,* and 8 "Tungan nti en Engelst Tumståk af 1½ fot lang," *Tongue of an English rule of 1½ foot length.* The description of length seems ambiguous (1½ rather than 2 feet), but it is likely that the author identifies at no. 8 the six inch tongue leg which is hinged to no. 7, a two fold part of 18 inches length,

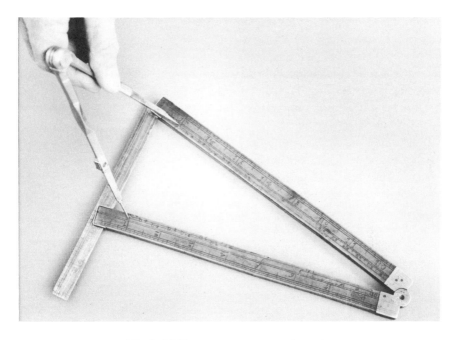

Fig. 3 Making a measurement on the sector.

rather than suggesting that the rule is 1½ feet long overall. The "English" attribution of this rule is of interest in the light of rule 9 - 10 (Fig.2) which Rålamb refers to as a "Dutch" rule (Hållenst Tumståk) and which appears to be a ship carpenter's bevel with graduated body and tongue. Nicolaes Witsen's Dutch shipbuilding text, *Architectura Navalis et Regimen Nauticum,* 2nd ed., 1690, contemporary with Rålamb, shows no three fold rule in the illustrations of shipwright's tools.

One end of one of the boxwood legs of the three fold rule is laterally slotted to a depth of about five sixteenths of an inch, into which the six inch brass extension can be snugly fitted. (see Fig.1) This gives the instrument rigidity so that dividers can be used to take off accurate measurements from the sector lines. The inner sector lines on each of the boxwood legs are graduated P, 3Q, 2Q, 1Q, and MH, representing Partners, third, second, and first Quarters, and Masthead. The function of this sector is to provide a series of diameter measurements for shaping the taper of a mast. Below decks mast dimensions are not taken into account by the sector measurements. The partners are the heavy timbers forming a supporting framework around the opening for a mast at the ship's main deck level. The mast head is here understood to be the top section of the lower mast (from which additional masts—topmast, topgallant, etc.—could be successively raised as required by the architecture and rigging of a particular vessel). The sector may be opened to an angle so that the distance between point P on each leg, using a pair of dividers, forms a convenient, arbitrary, transverse measurement, such as four inches. This transverse dimension, and the distances between the succeeding points Q3, Q2, Q1, and MH may then be measured with the dividers and laid off on a plan (see Fig.3).

Anderson states that the manuscripts of Keltridge and Battine on shipbuilding, written in 1675 and 1684 respectively, agree that the diameter at the head of a mainmast or foremast should be two thirds of the diameter of the mast at the partners.[2] Duhamel du Monceau, writing some eighty years later, gives the same ⅔:1 proportion for diameters of the mast.[3] It is of note that the figures obtained by use of the sector on the rule give a masthead dimension that is precisely two thirds of the diameter at the partners. (see Fig.4).

The second, or outer set of the two sector lines are designated S, 3Q, 2Q, 1Q and YA for Slings, third, second, and first Quarters, and Yardarm. The slings are the middle part of a yard, where it is suspended by rope or chain from the mast; the yard's function is to support a square or other shaped sail. Yardarm is the term given the end sections of a yard; yardarms are of somewhat arbitrary length, extending from the quarters to the two extremities of the yard, and in lateral dimension they may include the outer quarter. The second set of sector lines is used to determine diameter measurements at various points on the length of either half of a yard, from slings to yardarm end. The diameter of a yard at its

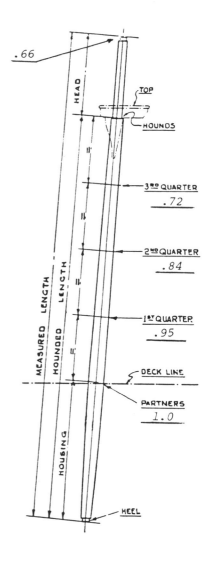

Fig. 4 Mast diagram from "Masting and Rigging" by Underhill. The figures for diameter proportions obtained from the mast sector of the rule, on an arithmetical base of 1 at the partners, have been added at partners, quarters, and masthead.

ends should be one third of the diameter at its slings, according to Keltridge.[4] Again, the results of measurement by the sector are quite consistent with Keltridge's 1675 dimensions. For example, a three inch setting on the rule at the sector points for S gives a transverse measurement of only slightly more than one inch at the Y graduations. A work dealing with more modern sailing ships by Underhill includes a table of diameter dimensions for timber yards that are proportionately of somewhat greater size at the yardarms than the Keltridge figures.[5] For example, a yard of twenty-one inch diameter at the slings measures nine inches at the yardarm, or about 3 to 1.28, rather than 3 to 1.

It is probable that in shaping a spar, 20th century shipwrights carry on long-standing traditional techniques. A log is squared by the broadaxe, further reduced to an octagonal spar by hewing, and rounded by the mast drawing knife and spar plane.[6] Anderson, in

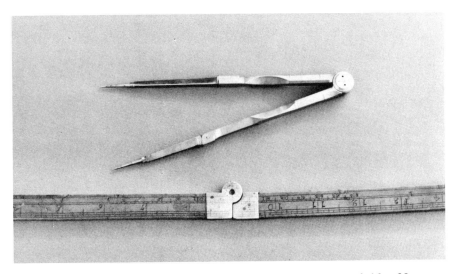

Fig. 5 The octagon scale, to left and right of the rule joint, scaled 0 to 28.

describing Lord Pembroke's ship model of 1692, the 1701 model of the *St. George*, and others of the early years of the 18th century, reports that the lower yards had an octagonal center section at the slings.[7] An *octagon scale* is included on the three fold rule (Fig.5). It would have been an obvious convenience for the shipwright to measure and lay out construction lines to reduce the square of a yard to an octagon, while portions beyond the center were rounded and tapered to dimension. The eight square line of this three fold rule is graduated with figures 0 to 28, and is of the type referred to as an "M" or middle scale on 19th century rules.[8] A dimension taken from the scale would be measured to right and left from the mid-points of the square sides of a yard, rather than from the corners, to mark the layout lines.

To review the data of the three fold rule of Fig.1, it bears tables and lines of board and timber measure developed during the 17th century. Two Gunter's logarithmic lines on the outer narrow edges of the rule are consistent with that mathematician's descriptions published as early as 1624. On the other hand, the rule does not incorporate the more convenient sliding lines of numbers which were developed in the latter part of the 17th century by Oughtred, Coggeshall (1677), and Everard (1684). It incorporates sectors for the taper of masts and yards, the octagon scale, and two foot scale graduated in inches for linear measurement.

Further utility of the rule as a shipwright's tool has been suggested in that two simultaneous angle measurements can be made for compound bevels by use of the two boxwood legs and the third, brass leg. Taken together, the scales and graduations of this versatile member of the measuring instrument family suggest that it was probably a product of the latter part of the 17th century or the early decades of the 18th century.

The author wishes to thank Carl Bopp of Audubon, New Jersey, and Philip Walker of Bungay, Suffolk, England, for helpful information. Particular thanks are due to Richard Knight of Birmingham, England. His explanation of the letters and figures of the sectors was essential for preparation of this article.

NOTES

1. Rålamb, Åke Classon. *Skeps Byggerij eller Adelig Ofnings.* Stockholm, 1691. Facsimile reprint, 1943. Plate K.

2. Anderson, R.C. *Seventeenth-Century Rigging; a Handbook for Model-Makers.* London: Percival Marshall, 1955, p.9.

3. Duhamel du Monceau, Henri Louis. *Elemens de l'Architecture Navale; Traite Pratique de la Construction des Vaisseaux.* Paris: C-A. Jombert, 1758. p.135.

4. Anderson. p.29

5. Underhill, Harold A. *Masting and Rigging the Clipper Ship & Ocean Carrier.* Glasgow: Brown, Son & Ferguson, 1949. p.258.

6. Story, Dana A. *The Building of a Wooden Ship.* Barre, Mass.: Barre Publishers, 1971. *(Two unnumbered pages describe and illustrate shipwright Arthur Gates making masts).*

7. Anderson. p.30.

8. The octagon, or eight square lines are commonly found on 19th and 20th century two foot, two fold carpenters' rules of both English and American manufacture. A middle line (M) and edge (E) scales are frequently provided.

The Country's Largest Tool Store at the Turn of the Century

By R. JAMES ABER

September 1971

One of the finest hardware stores at the turn of the century, Hammacher, Schlemmer & Co. spared no expense or effort to maintain the largest inventory of tools and hardware in the East. The selling floors of the store were handsomely appointed with wood-panelled, glass-topped counters and salesmen were impeccably dressed in dignified frock coats. The multi-drawer apothecary style cabinets along the walls held back-up supplies of hundreds of items, all meticulously organized for easy access. The store was roughly divided into two sections: tools were on the right of the entrance, hardware to the left; and in the rear were the widely-advertised tool cabinets and general supplies.

The store further distinguished itself with magnificent window displays in four solid mahogany panelled windows. They were changed frequently and the representative items of tools and miscellaneous supplies were planned to be instructive as well as interesting and provided an irresistible lure to the passerby. Understandably, they accounted for much of the success of the store.

Large stocks of items were also kept in separate warehouses enabling Hammacher, Schlemmer to make prompt shipments to factories and merchants who bought on a jobbing basis. Contrary to the general impression that the store was merely a retail outlet, a great bulk of its business was conducted on a wholesale scale. The start of World War I brought so much business that additional space was needed and the store was later honored with a government citation for its services as a prime supplier for the United States Army.

Hammacher, Schlemmer had become a giant in the tool and hardware field at the turn of the century. It had its beginnings in 1848 when William Tollner opened a small hardware business at 221 The Bowery, New York City. A few years later, in 1853, when a dingy packet ship made its way from Europe to New York, one of its passengers was eleven-year-old William Schlemmer, wide-eyed at his first view of houses on the land which was to become his home. The sight of land through the grey mist was welcome relief to the weary passengers who had been cooped up during the uncomfortable voyage of many weeks. The youngster arrived with few possessions and wore a tag tied around his neck with the address of his uncle, Tollner.

William went to work for his uncle selling tools and hardware from tables set up on the sidewalk in front of the store at a salary of $2 a week. In 1859 when Tollner wished to expand his business, Schlemmer, by then a promising young merchant, persuaded a wealthy friend, Alfred Hammacher to invest $5,000 in the enterprise. The store moved to larger quarters at 209 The Bowery, a commercial center with fine residential areas nearby. Business was good and the firm of Tollner and Hammacher prospered.

At the beginning of the Civil War when coins became scarce because people were hoarding them, the Government gave merchants permission to mint their own. Tollner and Hammacher were among the first to issue such coins. They were known as "copper heads" or "rebellion money." They bore the name and address of their store and were given as change for merchandise purchased. They were redeemable at the store from which they were issued. During the years 1862 and 1863 the city was flooded with millions of these tokens from the many stores. The resulting confusion and abuse caused the Government to put a stop to such coinage in 1864. Today, a coin bearing the inscription, "Tollner and Hammacher, 209 Bowery, New York" would be a choice collector's item.

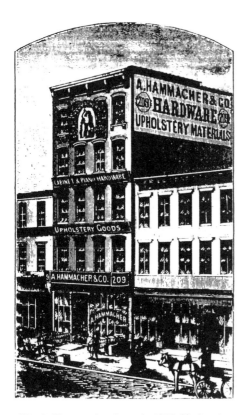

The A. Hammacher Store in 1881. Notice that the big emphasis is on upholstering materials and cabinet hardware.

In 1867, William Schlemmer was appointed a partner in the company. He gradually bought all his uncle's shares and after the founder's death, changed the name to Hammacher, Schlemmer & Company.

Alfred Hammacher never did take an active part in the company except to attend directors' meetings. In 1885 he sold most of his share to Schlemmer. Schlemmer, on the other hand, was completely dedicated to the store. For 64 years he worked to build it into the largest, most complete tool and hardware store in the country. In the early years he was at work by five in the morning, seldom leaving before midnight. In 1904 the expanded business was required to move to larger quarters again and settled at Thirteenth Street and Fourth Avenue.

William F. Schlemmer, the founder's son, joined the firm in 1893 at the age of fifteen. He decided he'd rather follow in his father's footsteps than go to college. He started learning the business right from the bottom as a stock clerk at $5 a week. His father was a stern man and promotions were hard-earned by the young apprentice. Eventually he was elected Vice-President in 1898 when the company incorporated. He held the office until 1914 when he became President and took on the added responsibilities of Treasurer upon the death of his father in 1916.

In 1899, young Schlemmer added automotive parts and tools to the store's vast stock of tools, tool chests, work benches, hardware, and an unusually large inventory of piano parts. A brisk wholesale and retail business continued to increase the store's profits and prestige.

In 1914, Hammacher Schlemmer issued their Catalogue No. 500. It took three men four years to assemble the huge 1,100 page volume which was a masterpiece of information. Profusely illustrated with minutely detailed line engravings, it contained merchandise from over a thousand different manufacturers. The customer had every assistance in making his selection, from the seemingly minute data of a standard nut and bolt or the economical use and directions for connection of a dry cell battery, to complicated, detailed mathematical formulas included for determining dimensions of gears. It was a beautiful catalogue, bound with a hard cover, filled with a staggering variety of tools and hardware. Catalogue No. 600, issued in the mid-1920's, was equally impressive in size and content. Unique and unequalled in its field, it is still used as a reference book by Naval Supply Depots.

A notable feature of the store's advertising was a campaign using bearded gnomes as an eye-catching device. The little men who became a

The famous HS gnomes from a 1923 advertisement. Redrawn by the author.

well-known trade mark of Hammacher Schlemmer, were illustrated in ads, buying and using the tools. They added a touch of whimsy to the advertising, but were dropped after the change-over of 1926 because New Yorkers began to associate them only with the hardware portion of the business.

In 1926 the firm opened its latest 12-story building on East Fifty-Seventh Street, its present location. It was used in addition to the Thirteenth Street store. The neighborhood was different and the location did not produce the demand for tools and hardware that had been anticipated. After a few months of poor business, William F. Schlemmer had to make a bold decision. He moved the tools and hardware into the basement and stocked the rest of the store with luxury items. This proved to be a wise business decision and has kept Hammacher, Schlemmer's name at the top of the field of high quality retailing.

By 1935 the Thirteenth Street store and downtown warehouses had been closed and in 1938 the wholesale-hardware and mill-supply departments were reduced in size. That year the store sold enough piano pins to one customer alone to build twelve thousand pianos.

With the advent of modern means of transportation and new marketing techniques, the tool and hardware manufacturers began to sell directly to the consumer. This necessitated further reductions in inventory. The hardware department was discontinued in 1955 and the store now deals in luxury and unusual gift and household items.

Today, with Dominic Tampone, a former star salesman, as its President, Hammacher, Schlemmer has built a reputation for the unusual. Tampone says that 98% of the store's items are functional with a small selection of purely decorative accessories. Functional items include such wares as an Electronic Oven the size of a portable TV, a small operating train which can be set up to transport guests around an estate, a Hot Dog Cart for the patio at a cost of $1,295 or a small Merry-Go-Round to amuse the children at $2,750.

Basically, the only tools available now are in the housewares line, and include commodities such as a stainless steel lobster cracker and shears, as well as other household necessities. As for hardware, the bath department offers a selection of gold towel bars and fixtures for its customers.

Truth in Advertising 1770-1776

Compiled by Rita S. Gottesman

Submitted by Paul B. Kebabian

December 1984

—*The New York Gazette and the Weekly Mercury*, April 16, 1771.

Isaac Heron. — At the Sign of the Arch'd Dial, by the upper End of the Coffee-House Bridge; Isaac Heron Watch-Maker, Has an Assortment of Watches, and the best, second, third, fourth, fifth, bad, and worse Sorts; some very neat, some very ugly, and others—so, so; most of them in plain, and a few in engrav'd, gold, silver, gilt, and shagreen'd, double and single Cases; some he warrants for a long Time, some for a shorter Time, and others for no Time, at all . . .

—*The New York Journal or General Advertiser*, December 24, 1772.

Isaac Heron, Watch-Maker, at the corner by the Coffee-House, sign an arched Dial; has for sale, a Few very fashionable warranted watches, in gold, silver, gilt and green cases, which he will sell as low as they retailed in London. Also, Ladies elegant steel watch-chains; mens do. seals, trinkets, glasses, springs, strings, and keys by the dozen; enamelled-dials; elegant do. for the Craft, Ancient, and Modern. Elegant broach-jewels for their honest breasts; sword knots, sundries, &c. with Sprigs, which more beauteous makes the fair; and lockets, various, for the hair.

These watches he repairs, he does as well and charges as low for, as his neighbours, i.e. as near the London prices as possible. To say more, would neither be prudent nor honest. As usual, he warrants their performance — not for ever, but one year, and if they be not very bad, for ten; after the first, at 4s. per ann. for each, cleaning; accidents and mismanagement of them excepted; but,

Should the all-sustaining hand him drop,
His movements all springs, wheels hands must stop!
Then, like the the tale of "a bear and fiddle,"
This bargain — "breaks off in the middle."

He is extremely grateful to his friends, and the indulgent public for their smiles and favours, and really means to merit a continuance of them . . .

News item in *The New-York Journal or the General Advertiser*, January 7, 1773:

The inhabitants are desired to be careful of their Doors, Windows, &c. as there are a set of House breakers now in Town. Some of them broke off one of the Shutters, and a Pane of Isaac Heron's Shop Window, on Sunday the 27th ult. so early as 9 o'clock in the Night. The Noise alarmed the Family and the Villains made off.

Isaac Heron presents his Compliments to those Gentry, and congratulates them on their getting so clean off, on a Night so very dirty and wet. If they choose to return during the cold Season, he will take care to provide them with a warm Reception. And will endeavour to prevail upon them to leave behind e're they go, some Proof of their identity; such as an amputated Limb, or, even a Snuff box full of Brains. The latter may be of Use to our very vigilant City Watch, though, it is thought they sleep pretty sound with the few they have got.

The New-York Gazette and the Weekly Mercury, April 18, 1774.

Isaac Heron. — Choice and Cheap, or Union of the Good and Beautiful, exemplified in an assemblage of warranted Watches, Isaac Heron, near the Coffee-House, has received by the *Rosamond.* Also a few for jockies, of that quality which a Frenchman would style patraques, to which a Dutchman would give the epithet Schlegt; a Scotchman that of littleguidwarth; an Irishman, neenshagamagh; an Englishman, poor-thing, runner, bauble, d—d bad, &c. Such to be sure he warrants—not, Patent Keys, trinkets, &c.

The EARLY AMERICAN INDUSTRIES ASSOCIATION (EAIA) hopes that you've enjoyed these selections from the last 58 years of our quarterly journal. If you would like to receive future issues of *The Chronicle*, with its fine articles on early tools, trades, and industries, we invite you to

JOIN THE EAIA.

Besides *The Chronicle,* the EAIA offers many other benefits:

■ A NEWSLETTER, published six times a year, packed with information on tool events, auctions, new publications, and much more.

■ BOOKS on tools and trades, at reduced prices.

■ AN ANNUAL MEETING featuring demonstrations, museum and restoration tours, dealer tool sales and an auction, etc.

■ A LIBRARY that offers a wide range of books and catalogs for loan by mail.

■ EUROPEAN TOURS - adventures tailored for all who are interested in exploring early tools and trades abroad, in a comfortable and congenial atmosphere.

TO JOIN:

Annual Dues are
$25.00 for an individual
37.50 for a husband and wife
Canadian and overseas add $9.00 to cover extra postage on mailings.
Dues subject to change after 1992.

Send your check or money order (U.S. funds only, please) to:
John Watson, Treasurer
P.O. Box 2128, E.S.P. Station
Albany, NY 12220

A brochure giving more details is also available from John Watson.

Index

A

B

G

H

J

K

L

M

N